The Biology of Physical Activity

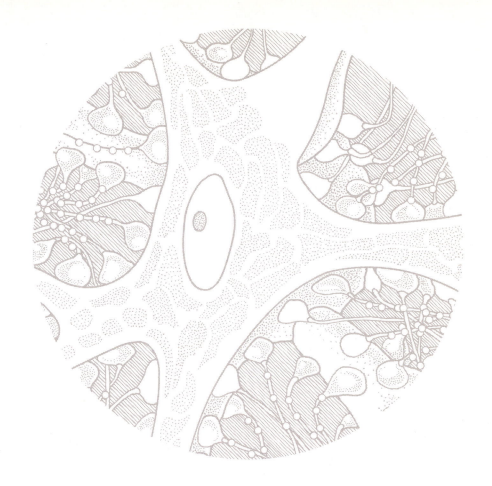

The Biology
of Physical Activity

D.W. EDINGTON

UNIVERSITY OF MASSACHUSETTS, AMHERST

V.R. EDGERTON

UNIVERSITY OF CALIFORNIA, LOS ANGELES

HOUGHTON MIFFLIN COMPANY   BOSTON

Atlanta   Dallas   Geneva, Illinois   Hopewell, New Jersey   Palo Alto   London

To those who came before us,
to those who are with us, and
to those who will follow us.

Printed in the U.S.A.

Library of Congress Catalog Card Number: 75-26095

ISBN: 0-395-18579-3

Contents

Unit IV Energy Support Systems of Movement and Exercise   111

Preface

Designed for use as a basic text in courses on the physiology of exercise, this book incorporates recent advances in biological research into the study of exercise.

Anticipating major advances in molecular biology, pharmacology, and other life sciences, we have presented basic biological concepts necessary for students to understand the mechanisms of exercise and training.

The book is unique in its treatment of the following topics: (1) general morphological, biomechanical, and physiological cellular responses to acute and chronic exercise, (2) mechanisms of metabolic control, (3) movement control treated as the functional motor unit, (4) detailed coverage of hormonal, renal, and digestive factors in exercise, (5) exercise effects on tissue growth, (6) nutrition in relation to exercise, (7) drugs and performance, (8) exercise and sex differences, and (9) coaching techniques.

The book is organized into five units. The first and second units describe concepts basic to our understanding of exercise science, the third unit covers the means through which exercise is initiated and controlled, the fourth unit describes physiological systems which support all movement, and the fifth unit applies concepts of activity to topics of everyday concern.

Key concepts are given for each chapter, there are examples and applications throughout the text, and end-of-chapter study questions which can be used for further student reinforcement. A glossary is provided at the end of the book.

The preparation of the manuscript required seven years, during which time we received encouragement from sources too numerous to completely recognize. We would like to acknowledge, however, the contribution of Walter P. Kroll (University of Massachusetts) for his critical reviews of the manuscript. A special thanks also goes to Wayne Van Huss

(Michigan State University), and to Ray Martinez and Bob Haubrick (East Carolina University) for igniting our undergraduate excitement on the topic; to Wayne Van Huss and Bill Heusner (Michigan State University) for intensifying our interests and providing a learning atmosphere; and to William Deal and Rex Carrow (Michigan State University) for augmenting our research skills.

A very special thanks goes to Donald Simpson (University of California, Los Angeles), a most talented and humanely gifted man who had an integral role in every phase of the book's development. The preparation of the manuscript could not have been completed without the help of two special people—Jackie Hartford and Charlene Solomon.

Thanks also to Paul Ribisl (Wake Forest University), Jimmie Grimsley, Robert Gantt, and Ray Martinez (East Carolina University), Roger Soule (Boston University), Jack Daniels (University of Texas), Wayne Osness (University of Kansas), and David Lamb (University of Toledo) for their critical review of the manuscript.

Finally, we want to take the opportunity to thank nearly two thousand students who have provided us guidance in content selection and in-depth discussion. One of our major regrets is that we have had to be selective in content and have therefore not been able to include all topics of interest to physical educators and exercise biologists. Throughout the book we have attempted to follow the concept: "If our students are not better educated than we were at that point, we have not done our job effectively."

D.W. Edington
V.R. Edgerton

Orientation to exercise

Key Concepts

• Exercises can be classified as a function of the nature of the demands upon the body.
• Specific exercises elicit specific biological effects.
• The energy source for an exercise depends upon the type of exercise.
• Training is the net summation of the adaptations induced by regular exercise.
• Fatigue and exhaustion are the result of the body's inability to meet the exercise demands.

Introduction

Biology of exercise is the study of how the human body responds to the demands of physical activity (*exercise*). Exercises can be categorized according to the specific responses created in the body that result from the demands of the exercise (*specificity of exercise*). Regular exercise (*training*) enables the body to make adaptations to the exercise stress and thus increases the performance capability. Fatigue and exhaustion result from the body's inability to meet specific demands.

Exercise

To one person the word "exercise" may mean calisthenics; to a second person exercise may mean a mile run; to a third person exercise may mean a long walk; and to a fourth it may mean a game of basketball. The term *exercise* can be applied to all of the above examples; however, we must realize that the effects within the body will be specific to the type of exercise performed. Our intention is to propose a relatively elaborate scheme for the classification of exercise. We have done this to emphasize the fact that, when we consider the biological effects of exercise, we must first describe what type of exercise is being used.

CHAPTER ONE

Classifications and definitions

If we examine the effects of activity on the human body, we find that the *specific* physical requirements of *specific* exercises regulate *specific* biological responses. An identical exercise stress will elicit varied responses in different people, and frequently the identical exercise stress will elicit varied responses in the same person at different times. We call this *individual specificity of exercise.*

To understand fully the capability of a given exercise, or to design an exercise regimen to meet a specific goal, we must understand the specific effects of that exercise—from the molecular level to the effects on the total body. For the most successful prescription of an exercise regimen for a given person, the psychological and sociological needs of the individual should also be considered.

Classification of Exercises

When the body is at rest, there exists a relative steady-state condition, which we will term the *resting state.* Any type of physical activity that requires the body to move from the resting state then becomes an *exercise stress.* If we use this broad definition, all activities can be classified as a form of exercise. To differentiate between the different types of exercise, we need a classification scheme. The method by which we choose to classify exercises is not as important as the realization that different exercises have different biological requirements and that the effect of the exercise can be predicted from a knowledge of the type of exercise. We have chosen to classify exercises according to the *speed* of movement, the *resistance* to that movement, and the duration or the *time* over which that movement has to be repeated. Initially we will discuss these considerations two at a time; later we will consider all three simultaneously.

Speed and Duration Considerations

Under normal conditions, the relationship between speed of movement and duration is shown in Figure 1-1. The faster the speed of movement, the shorter the time that that exercise can be maintained; conversely the slower the speed of movement, the longer the time the exercise can be performed. For example maximum running speed can only be maintained for a few seconds; however, a much slower speed can be maintained for longer periods of time—such as in a marathon competition.

Any point on the line in Figure 1-1 represents the maximum performance limit for a given individual, at a given speed and duration. Any point below the line represents a submaximal exercise, with no fatigue resulting from the performance of these exercises. We should emphasize that this graph will be different for each individual and for every type of activity (arm exercise, running, swimming).

With the proper training, it is possible to elevate the curve so that at a given speed of movement the trained person can exercise for an increased length of time. Obviously one of the goals of training for athletes is to perform at a faster rate for a longer duration.

Speed and Resistance Considerations

If we look at the general relationship between speed of movement and the resistance to that movement, we would see the relationship as drawn in Figure 1-2. In general, as the resistance to movement increases, the maximum speed of that movement decreases. With low resistances the speed of movement is maximal; with high resistances the speed at which movement can take place is reduced. For example it is possible to move the arm faster when throwing a baseball (low resistance) than when

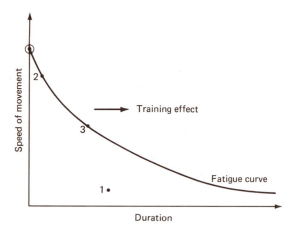

Figure 1-1. Relationship between speed of movement and time to fatigue (duration). The points on the graph represent: (1) jogging for 10 minutes, (2) running a fast 200 meters, and (3) running at half maximal speed until exhausted.

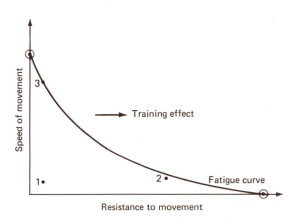

Figure 1-2. Relationship between speed of movement and resistance to movement. The points on the graph represent: (1) swimming a mile for relaxation, (2) weightlifting workout, and (3) running a 50-yard dash.

throwing a shot put (higher resistance). It is possible to increase the resistance to such a magnitude that no movement can be accomplished (*isometric* or *static exercise*).

If an athlete is given the proper training, a given resistance can be moved faster and probably easier than prior to the training period—in which case the limiting curve can be elevated or shifted to the right. Any point below the limiting line represents submaximal levels of exercise. Movements such as casually throwing a ball, doing a situp at a moderate speed, or pedaling a bicycle fall into this category.

Resistance and Duration Considerations

The general relationship between resistance to movement and total exercise duration is shown in Figure 1-3. The greater the resistance to movement, the less time that the movement can be endured. However, if the resistance is

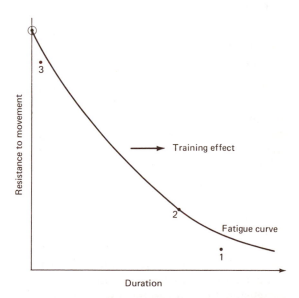

Figure 1-3. Relationship between resistance to movement and duration of the activity. The points on the graph represent: (1) jogging for 20 minutes, (2) swimming at half maximal speed until exhausted, and (3) doing three arm curls with near maximal weight.

decreased, the time that the movement can be performed increases. Proper training programs allow for a shift in the maximal-performance line so that for any specified resistance the performance can be maintained for longer periods of time.

Weightlifting activities involving heavy weights can only be performed for short periods of time—although these same activities can be maintained for an extended period of time if the weight or resistance is reduced. If the weight is reduced enough and if the time is short enough, the activity becomes submaximal; and the point representing this exercise falls below the limiting line. Pedaling a bicycle for 100 yards and walking up a flight of stairs represent activities that would be in the submaximal classification in terms of resistance and duration.

Speed, Duration, and Resistance Considerations

A more complete classification of exercises can be made if we simultaneously examine speed of movement, resistance to the movement, and the time over which the movement is to be continued. The necessity of the three-dimensional classification is obvious if we consider the biological effects of a specific exercise lasting for 10 seconds compared to the effects of the exact exercise (with the same resistance and speed) continued for 10 minutes.

The simultaneous consideration of the three classification components of exercise is shown in Figure 1–4. This figure is based on the two-dimensional classifications discussed previously. Exercises located near the origin in this figure represent the least strenuous exercises. As the point, representing an activity, moves in any direction from the origin, we progressively encounter more difficult exercise tasks. The boundaries of this three-dimensional figure represent the maximal performance level. This

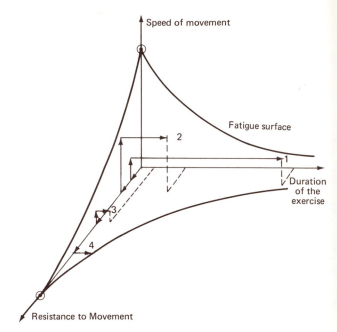

Figure 1-4. **Relationship between duration of exercise, speed of movement, and resistance to movement. The points on the graph represent: (1) jogging, (2) running 400 meters, (3) high resistance-low repetition weightlifting, and (4) isometric exercise until exhausted.**

curve naturally varies for each individual and can be altered through training (expanding specific areas of the curve) or through sedentary living (shrinking specific areas of the curve). We suggest that you locate on this curve several relatively uncomplicated exercises.

Given this scheme for the classification of exercises, or any other scheme, we can see that the term *exercise* represents a wide range of activities and related biological stress responses. In our consideration of the *biology of exercise*, we must be constantly aware of the differences between specific exercises when we discuss any specific effect of exercise on the body.

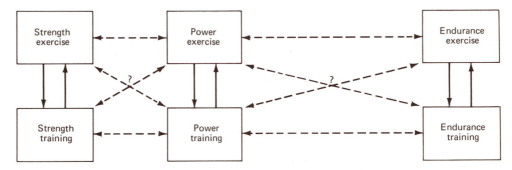

Figure 1-5. The specificity of exercise (training) theory is represented across the horizontal strength, power, and endurance exercise level. The specificity of training theory is represented by the vertical relationships in the diagram.

Unfortunately exercise classification is not as simple as we have just described it. A true classification scheme would, in all likelihood, take into consideration additional variables affecting the activity, including environmental factors. When an opponent and/or team sports are introduced into the consideration, the complexity of the activity and its classification are further increased. Other considerations—such as mental pressures, social pressures, and dietary considerations—add to the total consideration for exercise classifications.

Specificity of Exercise

"A specific exercise will elicit a specific response in a specific individual at a specific point in time." This statement implies that the biological end result of an exercise is directly determined by the specific exercise. Therefore a knowledge of the biology of exercise is essential in order to prescribe the correct set of physical activities to meet a given biological need of an individual.

Intuitively most of us understand that if we want to develop strength characteristics, we must stress the body with those exercises that demand strength. Similarly, if we want to develop endurance traits, we must perform exercises that improve our endurance. Figure 1-5 illustrates these relationships and demonstrates that the concept of *specificity of exercise* allows for a minimum amount of carry-over from one activity to another. For example, if we participate daily in one specific activity, we soon find that we are "trained" in that activity. However, upon switching to a new activity, we find ourselves unable to compete with others who have already been trained in the new activity. One telltale sign of the specificity of exercise is the amount of soreness that we experience in our muscles following the first few attempts at a new activity. The sore muscles are the ones more specific to the new activity: they were not used as extensively during the previous training periods.

The concept of specificity of exercise is reinforced by the fact that there are *specific energy sources* within each muscle that respond to specific types of exercise (Figure 1–6); that is, there is a difference between the types of energy production required for the various types of physical activity. For example, when we perform endurance exercises, we stress the oxidative-energy sources in our muscles; but when we perform strength and power activities, we involve the immediate sources of energy and the nonoxidative utilization of stored carbohydrate. (These energy considerations will be discussed more fully in Chapter 3.)

As an extension of our exercise classification schemes, we can estimate the specific energy sources available for specific types of activities. Figure 1–7 is a composite of earlier graphs, but we have added our estimates of the energy requirements. The energy for these specific types of exercises is derived from specific energy sources within the muscle. If training is to take place in one specific type of performance, then those specific energy sources for that exercise must be trained for increased energy production.

We also find that specific exercises have dif-

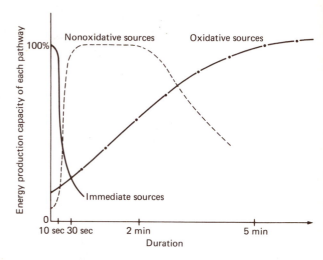

Figure 1–6. Sequence of the activation of energy sources within an active muscle.

ferent nervous system requirements; some of these include neuronal recruitment frequencies, nerve firing patterns, and neural factors that influence the control of metabolic and dynamic properties of skeletal muscles. These factors will also be eventually considered in our exercise classification scheme.

In summary our main emphasis will be that *the effect of exercise is specific to the type of exercise.* By understanding the biological mechanisms, we can begin to make an assessment of the biological effects of a particular exercise and of a training program.

Specificity of Training

Exercise is the act of performing a physical activity, *training* is "the result of biological adaptations achieved after repeated exercise bouts over a period of several days, weeks, or months of exercise." We have made the case for the specificity of exercise; by the same line

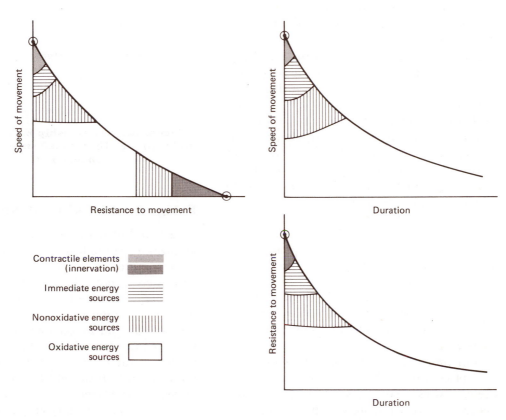

Figure 1-7. Estimated energy sources for specific types of exercises.

of reasoning we claim that *training is specific to the exercise being performed.*

Proper training results in an increase in the exercising capacity of an individual; that is, a given level of exercise intensity can be maintained for an extended length of time by an individual in the trained as compared to the nontrained state.

We know that there are physiological differences between a person in a nontrained state and that same person in a trained state. We assume that these differences account for most of the increased capacity for exercise. If we assume that the only way to achieve the trained state is through the performance of daily exercises, then those characteristics of the trained state must be the result of adaptations to the exercising condition.

The model presented in Figure 1–8 represents the method by which we feel that the trained state may be achieved. It is likely that the daily exercise session "sets up" the body so that subcellular mechanisms are stimulated to bring about those adaptative changes that characterize the "trained state." In other words something specific about the act of the exercise stimulates the cell to adapt so as to be more prepared to protect against this same exercise stress. It is highly likely that the exercise "sets up" the cell, while the actual adaptation occurs during the recovery phase. This sequence would appear probable since the exercising cell would at first be occupied with the exercise stress, but presumably during recovery the energies of the cell could be more efficiently pointed towards making these adaptative changes.

Both *short-term* and *long-term adaptations* take place within the cell in response to an exercise stress. Both types of adaptation primarily involve alterations in the active muscle. Short-term adaptations, which occur during the actual exercise, mainly involve the conversion of inactive to active chemicals. Long-term adaptations, which account for the primary training adaptations, are mainly concerned with increased amounts of chemicals, primarily proteins.

We know that strength exercises induce strength adaptations and that endurance exercises elicit endurance-training adaptations. These specific adaptations to specific exercises imply that there exists a *specificity of training*. The cellular mechanism by which specific exercises elicit specific training adaptations remains unclear at the present time.

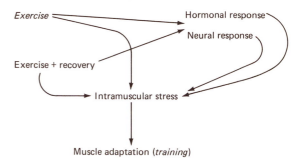

Figure 1–8. Relationship between exercise and training. The trained state represents a change in the active muscle.

Fatigue and Exhaustion

Fatigue and exhaustion may be thought of as an inability of the body to meet daily tasks or, more specifically, as an inability to continue with a given task. This effect can arise in a short period of time (i.e., after a strenuous effort) or over a longer period of time (i.e., after a lesser effort of longer duration). In general we apply the term *fatigue* to indicate a level of decrement of performance; we apply the term *exhaustion* to indicate the absolute end point of performance.

Does the actual biological site of fatigue vary in reference to the type of exercise that leads to fatigue? It would be unlikely that the site of fatigue resulting from an exercise requiring a slow speed of movement, encountering a medium resistance, and lasting until fatigue would be the same as that arising from an exercise requiring a high speed of movement, encountering a low resistance, and lasting until fatigue. We can hypothesize that there exists a *specificity of fatigue*: specific types of exercises lead to specific types of fatigue. Regardless of the type or site of fatigue a cessation of the activity must occur whenever the performance is such that the corresponding speed, resistance, and time are at a maximum.

The generally accepted theory that training increases our resistance to fatigue is explained by the body's ability to strengthen the weak links that are vulnerable to specific exercises. Whenever we exercise or train with the purpose of improving our performance capabilities, we are attempting to alter our *fatigue points*.

Summary

Exercise is a term applied to a wide variety of physical activities. To study exercise requires a recognition of the *specificity* of the response of the human body to specific forms of exercise. *Specific exercises elicit specific adaptations within the body to create specific training effects.*

Exercise can be classified according to the *speed* of the muscle movements, the *resistance* to the movement, and the *time* over which the exercise must be continued. Different combinations of speed, resistance, and duration lead to the utilization of different energy sources.

Fatigue and/or *exhaustion* may occur when the relative maximum values for the speed of movement, resistance, and performance time are reached simultaneously.

Study Questions

1. Why do we need to have a system to classify exercises?
2. How are speed of movement, resistance to movement, and duration of movement related?
3. What is the "specificity of exercise" theory?
4. How are exercise and training related?
5. Describe the importance of recovery from exercise.
6. Describe our "training adaptation model."
7. Under what conditions do fatigue and exhaustion occur?
8. Describe a "specificity of fatigue" model.

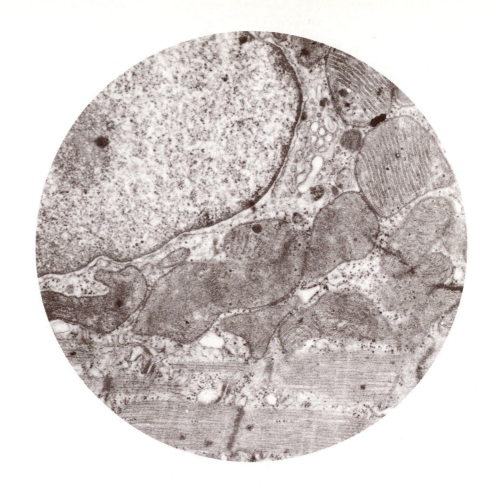

Muscle
and energy

Key Concepts

• The muscle cell is a highly organized and specialized cell with connections to the nervous and cardiovascular systems.
• The cell membrane regulates the transport of materials into and out of the cell.
• Transport systems within the cell augment simple diffusion processes.
• The genetic material in the nucleus controls the information processing in the cell.
• The mitochondrion is the site of oxygen utilization and energy production.
• The functional contractile unit of skeletal muscle is the sarcomere.

General Considerations

In general, cells within our bodies are the site of oxygen utilization, energy production, protein synthesis, carbon dioxide formation, and a host of other functions. In addition to these general functions, each type of cell within the body has specialized functions; the specific structural organization of each cell is varied in a way that allows for the performance of its specialized function. In our study of exercise, it is especially important to understand the nature of the *muscle cell*: which is the basic functional unit of movement.

Skeletal muscle is organized as shown in Figure 2–1. In general muscle fibers (cells) are arranged in a parallel fashion, and the contraction force is along the long axis of the fiber. The fibers are not only arranged in a parallel fashion but also the contractile elements (*myofibrils*) within the fiber are arranged parallel to the fiber axis. A high degree of organization exists within the skeletal muscle cell; the schematic diagram in Figure 2–2 illustrates the various subcellular structures we will discuss in this chapter. The electron micrographs in

The muscle cell

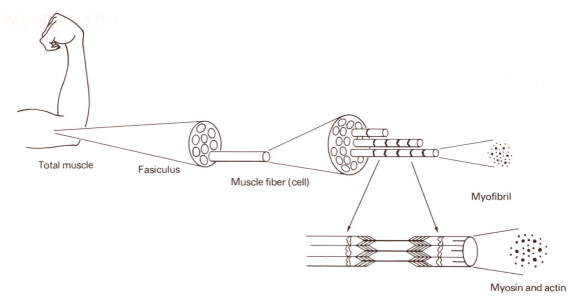

Total muscle

Fasiculus

Muscle fiber (cell)

Myofibril

Myosin and actin

Figure 2–1. Levels of organization within skeletal muscle. The protein organization represents an approximate magnification of 200,000 x.

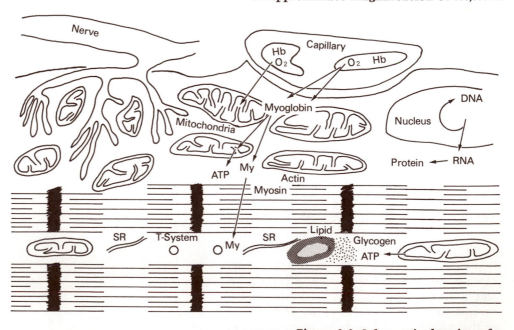

Figure 2–2. Schematic drawing of a longitudinal section of a muscle fiber, showing the relationship between the neuromyojunction, the vascular myojunction, and the subcellular components of the muscle fiber.

Figure 2-3. Electron micrographs of skeletal muscle. Approximate magnification 6,000 x.

Left photo courtesy of Brenda Eisenberg, UCLA.

Figure 2–3 show actual pictures of skeletal muscle; Figures 2–4 and 2–5 show the connections between the muscle cell and the neural system (*neuromuscular junction*) and the muscle cell with the vascular system (*capillary-fiber junction*).

Specific Components of the Cell

Membrane Characteristics

The cell membrane is composed of two parts. The inner membrane, or the *plasma membrane*, is intimately associated with the cells' cytoplasm and is necessary for the life of the cell. The outer membrane performs more or less a minor role and is less critical for the maintenance of cell life.

The plasma membrane is frequently described as a bimolecular leaflet with an inner lipoid layer (largely cholesterol and phospholipids) between two layers of protein (Figure 2–6). This bimolecular leaflet is the so-called *unit membrane*, characteristically 70–100 angstroms in thickness.

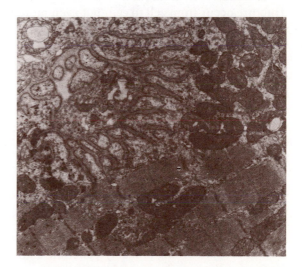

Figure 2-4. Neuromuscular junction. The terminal axon of the nerve divides into motor endplate feet (fingerlike projections) which invaginate into the muscle fiber.

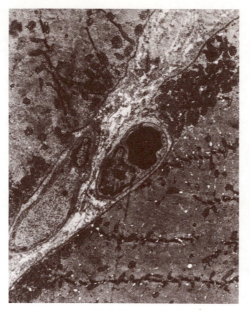 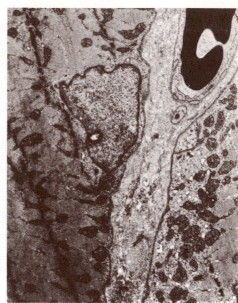

Figure 2-5. Muscle-vascular junction. The large dark cells within the capillary are the red blood cells.

Variations in membrane chemistry and structure occur from cell to cell; but even more important the membranes of specific subcellular structures (such as mitochondria, sarcoplasmic reticulum, and T-tubules) exhibit more variability. This variability is particularly obvious in light of the difference in specific enzyme activities associated with the various membranes found within a cell. The differences in membranes help to determine the specificity of structures within the cell.

Identification of *pores* or *plugs* of nonlipid substances (proteins perhaps), extending from one side of the membrane to the other in some areas, also adds to the complexity of the membrane structure. It is believed that membrane protein conformations probably penetrate into the core of the membrane and serve as a membrane stabilizer, thereby allowing ions and electrically polar substances to diffuse through water-filled electrostatically charged pores.

The understanding of many of the current problems of exercise are undoubtedly dependent upon a more thorough knowledge of biological membranes. It is unlikely that we are overemphasizing the importance of the membrane. Membranous components play significant roles as hormonal receptor sites in addition to their role as metabolic regulators in governing the flux of all materials into and out of the cell. For example, the regulation of ionic flux in neurons and muscle fibers for the control of impulse propagation and of muscular contraction is a necessity for the maintenance of life.

MEMBRANE TRANSPORT

Membrane *permeability* has long been recognized as a critical factor in the understanding of cell physiology. Although it is known that the membrane imposes limitations on the compounds that can enter or leave a cell or subcellular compartment, a fundamental understanding of the *mechanisms* that regulate the passage of water, electrolytes, and other compounds through membranes has essentially eluded the scientist.

We will develop throughout this chapter the concept of *subcellular compartments*. These compartments are relatively independent of each other, due to membrane selectivity. However, in the course of cellular metabolism, it is necessary to transport certain molecules from one location to another—since in many instances the product of one compartment is the substrate for or forms an integral part of the metabolism of a second compartment.

There are essential membrane-transport mechanisms that are critical to an understanding of energy metabolism. Three examples of the more critical transport mechanisms are: (1) the translocation of sodium and potassium ions (Na^+ and K^+), (2) the transport of oxygen from hemoglobin to mitochondrion and, (3) the transport of high-energy phosphate bonds (*adenosine triphosphate* or ATP) from the site of formation (*mitochondrion*) to the site of utilization (*contractile mechanism*).

Sodium-Potassium Pump. The membrane is permeated with electrostatically charged pores, which can be thought of as tunnels whose walls are covered with positively charged particles. These pores create a barrier where positive ions on the outside cannot pass through the membrane. The term *active transport* is used to denote that energy is required to move molecules against a concentration gradient or through an electrostatic barrier such as the one just discussed. Little is known about this mechanism other than that it requires energy. The sodium (Na^+) pump and the potassium (K^+) pump refer to the active transport of these ions such that sodium is exchanged for potassium. This ionic exchange is necessary to maintain

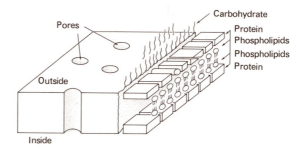

Figure 2-6. Structural model of a membrane.

the sodium (Na+) potassium (K+) gradient necessary for nerve impulse propagation and for muscle impulse conduction (see Chapter 4). This electrochemical gradient is a form of potential energy since the natural flow of ions would equalize the concentrations on each side of the membrane. Negative ions (e.g., Cl⁻) are not greatly affected by the positive pores so that the concentration of these ions exists in near equal amounts on each side of the membrane.

Oxygen Transport. For many years exercise physiologists assumed that oxygen transport was a simple matter of diffusion from the hemoglobin molecule to its arrival at the mitochondrion. It is now more generally accepted that oxygen transport is facilitated by a *myoglobin* (hemoglobin-like molecule in muscle) *shuttle system.* Just as individuals in a bucket brigade pass pails of water from one person to another, myoglobin passes oxygen molecules from a location near the capillary to the mitochondrion. This shuttle mechanism reacts to oxygen concentration (*tension*) and passes the oxygen from areas of high oxygen tension (near capillaries) to areas of low tension (near active mitochondria). The low tension is always on the side of the utilization site: as the mitochondrion uses oxygen, the oxygen tension falls in that immediate area. Under extreme exercise conditions, it is possible to imagine the demand for oxygen becoming so great that eventually the demand would exceed the capability of the capillaries to supply it. Under these conditions the myoglobin would become desaturated, and the cell could only produce ATP through nonoxidative sources.

ATP Transport. Although this mechanism is still under active investigation, we would like to propose that, for muscle, it operates in a manner similar to that of the myoglobin-facilitated transport of oxygen. The transport direction is from the mitochondrion (the site of ATP production) toward the contractile machinery (the site of ATP utilization). The ATP formed at the site of the electron transport chain is used to form another high-energy phosphate, *creatine phosphate.* This reaction requires the presence of an enzyme (*creatine phosphokinase*), which is equally distributed throughout the muscle cell. The distribution pattern makes it a very likely candidate for a shuttle enzyme. As the muscle contracts, the ATP concentration becomes low; the "flow" of high-energy phosphate bonds is then initiated from the high-concentration area (mitochondria) to the low-concentration area (contractile machinery).

Nucleus

The nucleus is the "brain" of the cell since all of the cellular processes are controlled by the action of the nucleus—for example the adaptations to physical training. Although the nucleus of most cells is spherical, muscle nuclei are more nearly cylindrical in shape (Figure 2–7). The nucleus is surrounded by a double nuclear membrane, which at some points may be continuous with the endoplasmic reticulum. Nuclear pores exist at intervals along the membrane; the number and position of the pores in the nuclear membrane are probably related to the biosynthetic activity of the cell.

The nucleus consists of a nuclear membrane, nuclear fluid, and chromatin—the strands of nucleic acids which make up the genetic material. The nucleus directs protein synthesis through deoxyribonucleic acid (DNA) via ribonucleic acid (RNA) which is concentrated in the nucleolus.

NUCLEOLUS

The *nucleolus* is a dense intranuclear area that contains much of the cell's RNA. The number of nucleoli per nucleus varies (from one to seven in the liver) being most abundant in the early stages of growth when protein synthesis is probably near maximum. Little is known about the effect of exercise and training on the size and number of nucleoli.

PROTEIN SYNTHESIS

Protein synthesis, along with DNA and RNA synthesis, will be mentioned repeatedly in this book; the importance of these processes cannot be overemphasized. The nucleic acids are essential to life, and the ability of the cell to maintain this essential process regulates the functional capability of the cell.

Whether protein synthesis is stimulated during the exercise response is questionable, but protein synthesis has been proven most definitely to be essential in the adaptation to training. The synthesis of structural proteins, contractile proteins, and enzymatic protein has been shown to occur in response to a wide variety of training programs. The type of protein adaptation varies from one type of training program to another. It is uncertain which of the specific combinations of exercise-induced stimuli initiates the protein-synthesis machinery necessary for the training adaptations.

The ultimate control of protein synthesis lies in the DNA molecule located in the nucleus of the cell. DNA (also referred to as *genes*, *chromosomes*, and *genetic material*) contains the information necessary to code for all of the proteins of the body. The DNA contained within the muscle cell is the same as that contained within the cells of the brain, liver, skin, nerves, kidney, and so forth. The mechanisms that allow cells to become specialized (muscle cell rather than liver cell) are unknown. The specific portion (*genome*) of the DNA that codes for the contractile proteins is turned on in the muscle cell but turned off in the brain cell. Furthermore this genome in muscle cells

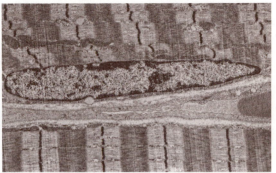

a.

b.

Figure 2-7. Muscle cell nucleus. (a) Photomicrograph of a cross section of a nucleus and its location near the cell membrane. (b) Photomicrograph showing the relationship between the nucleus and the surrounding cytoplasm.

Top photo courtesy of Brenda Eisenberg, UCLA.

can be activated through the process of strength training but not through endurance training. An ultimate understanding of adaptations to exercise is dependent on our developing the knowledge of how DNA and RNA are controlled.

There are at least two protein-synthesizing mechanisms available to the cell. The first and most important is in the nucleus; the second is in the mitochondrion. For our present purposes, we will primarily discuss protein synthesis associated with the nucleus.

In Figure 2–8 we have presented, in abbreviated form, a summary of protein synthesis. The details of the total mechanism are beyond the scope of this book. The original DNA is acquired through the process of *mitosis*. The DNA is surrounded by protein molecules (*histone* and *nonhistone proteins*), which appear to act as regulatory proteins in that they surround the DNA to protect it from being read until the proper stimulus releases the histone inhibition.

The DNA molecule contains many segments or *operons* (*genes*), which code for specific traits, enzyme systems, or other related proteins. Given the correct stimulus, the DNA molecule is capable of being transcribed into RNA. The specific RNA is thus determined by the portion of the DNA that is exposed. The DNA molecule is specific for at least four types of RNA (*nuclear, transfer, ribosomal,* and *messenger*). There is a specific *transfer RNA* (tRNA) for each of the 26 different amino acids. *Ribosomal RNA* consists of two parts that eventually form the structure that binds the mRNA, which sequences the tRNA-labelled amino acids. The specific sequence of the amino acid chain is determined by the sequence of nucleotides contained in the messenger RNA. The sequence of the amino acid chain determines the specificity of the protein.

The mechanisms involved in protein synthesis present many possible points of control, several of which are of interest to the exercise biologist. It is important to determine what type of subcellular environment will "turn on" protein synthesis; e.g., the selective availability of amino acids for the tRNA could act as a limiting factor in protein synthesis. The ultimate control of genetic expression lies in the control of DNA transcription to RNA. It is usually accepted that the synthesis of each protein is not controlled separately, but that proteins serving different aspects of the same function or enzymes of the same pathway are transcribed in response to the identical or similar control mechanisms; that is, there appears to be a regulator gene that can be controlled either to activate or to repress the main operon. Activation or derepression of the regulator gene excites a genetic operator that allows the transcription of the genes within the given operon. Only the genes within the given operon are transcribed; similar processes must be performed to transcribe additional operons.

In summary we should consider that the keystone to cellular biochemistry is the specificity of the DNA. Within its chain of nucleotides is encoded the information necessary to make every protein in the body. The fact that each cell has the same DNA molecule, whether it is skin, liver, or muscle, emphasizes the importance of *unknown regulators* that allow cellular specificity: certain proteins are made in one organ while a different set is made in another. Once the regulation of the DNA molecule is understood, a revolution in medical and genetic science is at hand. Cells could be made unresponsive to viral RNA infections, genetic diseases could be eliminated, and improvements in physical and mental capabilities could possibly be made.

Satellite Cells

Satellite cells are cells located on the periphery of muscle fibers—between the plasma and basement membrane of the fiber (Figure 2–9). This structure consists of primarily a nucleus—with very little cytoplasm.

The role of the satellite cell in normal muscle function is not clearly defined; we think that these cells are dormant *myoblasts* and, as such, are capable of being stimulated and incorporated into the actual muscle fiber. During times of muscle regeneration, the satellite cell is clearly stimulated and assumes a very active role. The role of the satellite cell during muscular exercise or training is unclear, but the possibilities exist for its involvement in the adaptation process.

Cytoplasm

A strict definition of the cytoplasm would include all cell constituents outside of the nucleus that are bounded by the plasma membrane. In muscle, we will consider the contractile mechanism, the cytotubule systems, and the mitochondrion as separate compartments. The cytoplasm then contains the "soluble" enzymes, or those enzymes that are not bound to cytomembranes. It is generally considered that myoglobin and at least one form of *creatine phosphokinase* are in this soluble cytoplasm. Traditionally the enzymes of nonoxidative glycogen degradation (*glycolysis*) have been thought to be "soluble" in the cytoplasm, but more recent evidence suggests that these enzymes are bound to portions of the ER or to the contractile elements.

CYTOTUBULE SYSTEMS

The major ultrastructural features of skeletal muscle fibers are illustrated in Figure 2–10. The *sarcolemma*, or muscle cell membrane, has tubular invaginations called *transverse tubules* (T-tubules) at regular intervals along the fiber. These T-tubules conduct depolarizing waves from the sarcolemma to the deep regions

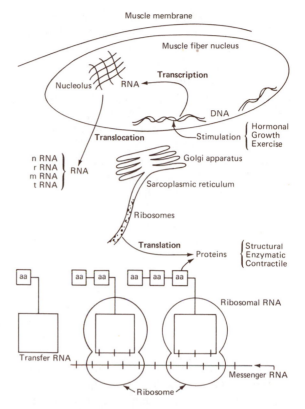

Figure 2-8. DNA—RNA—protein synthesis. Genetic information contained in the DNA molecule is translated to the formation of protein.

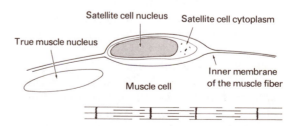

Figure 2-9. Schematic representation of a satellite cell consisting almost entirely of the nucleus. The heavy shading indicates an abundance of chromatin material.

of the fiber. Adjacent to the T-tubules but not structurally continuous, are numerous longitudinally oriented membranous sacs called the *sarcoplasmic reticulum* (SR). The SR is the storehouse or reservoir for the calcium ions necessary to stimulate muscle contraction.

In some cases at least, the SR appears to be continuous with the plasma membrane and the nuclear membrane. Some of these membranes appear granulated because of the adherence of ribosomes (involved in protein synthesis) and are called "rough" SR; the cytomembrane without these particles is referred to as "smooth" SR.

In addition to being associated with protein synthesis, the SR appears to bind enzymes related to glycogen metabolism. The enzymes *glycogen synthetase* (which adds glucose units to existing glycogen) and *phosphorylase* (which cleaves glucose units from glycogen) as well as *glycogen* seem to be in some way bound to the SR.

The response of the SR to acute bouts of exercise is largely unknown except for a dilation of the cisternae immediately after exhaustive exercise. The dilation-induced structural alterations are rather quickly returned to that state typical of resting conditions. Lipid soluble drugs, such as the anesthetic *phenobarbital*, induce marked temporary SR deformations in the liver. Thus the SR not only acts as a matrix for binding enzymes and as a site for protein synthesis but also seems to play some adaptive role to acute chemical and mechanical stress.

The junction of a single T-tubule and the terminal cisterna of the SR on either side of the T-tubule is called the *triad*. This is the point where the signal of the T-tubule is probably transferred to the SR—although this signal may not be in the form of the typical depolarizing wave as seen in the sarcolemma depolarization. This signal induces the SR to release the bound Ca^{2+} into the cytoplasm where it dif-

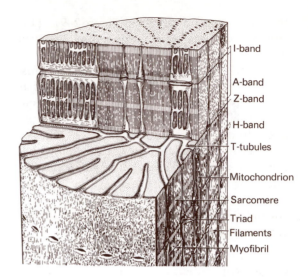

I-band

A-band

Z-band

H-band

T-tubules

Mitochondrion

Sarcomere

Triad

Filaments

Myofibril

Figure 2-10. Schematic representation of a muscle fiber.

fuses into the regions of the contractile filaments to initiate muscular contractions. The energy-dependent uptake of Ca^{2+} by the SR initiates muscular relaxation. The sites for Ca^{2+} release and accumulation in the SR differ. The terminal cisternae of the SR seem to be the sites of Ca^{2+} release; the middle tubular segments of the SR are the major Ca^{2+} accumulating sites and are therefore related to rates of relaxation.

FUNCTIONAL CONTRACTILE UNIT

A *myofibril* is the contractile subunit of the muscle fiber. It is very long but only about a micron in diameter; and hundreds are packed together to form a single muscle fiber. When observed longitudinally, myofibrils have a characteristic transverse banding that is

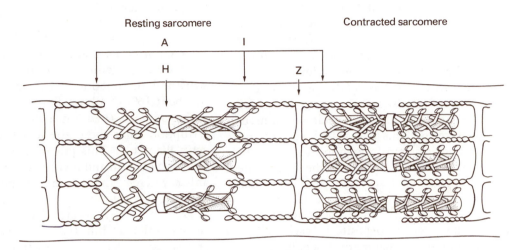

Resting sarcomere Contracted sarcomere

A I

H Z

Figure 2-11. Diagram of a resting and contracted sarcomere. The A-band contains the myosin, the H-band contains the M-substance, and the I-band contains the Z-disk and actin.

repeated every 2.5 microns in relaxed muscle (Figure 2–3). The repeating units are called *sarcomeres* and are demarcated on each end by a Z-line (thin dark lines in Figure 2–3). At low magnification two major transverse bands are prominent. They are called the *light I-band*, which basically contains the Z-line and the contractile protein actin, and the *dark A-band*, part of which contains myosin and part contains both actin and myosin contractile proteins (Figure 2–11). Another band within the sarcomere is the H-band, which is contained in the midportion of the A-band, and contains myosin but not actin filaments. The width of the H-band increases during relaxation and decreases during contraction. The M-line marks the center of each sarcomere and reflects the point at which the myosin molecules of each filament are reversed in their direction of orientation. The M-line consists of transversely and longitudinally oriented proteins, which apparently provide structural support for the myosin filaments (Figure 2–12).

Molecular Organization of Myofilaments. The thick myosin filaments consist of several overlapping myosin molecules braided together with projections extending at right angles to the filaments. These projections (heads of the myosin molecules) interact with the thinner myofilament, actin. This projection is the "active end" in that it has *ATPase activity;* that is, it can catalytically free the energy from ATP to be used during the initial stages of muscular contraction. This released energy initiates a morphological transformation in the myosin molecule, which interacts with the appropriate actin molecule in a way not yet understood.

The orientation of the actin and myosin molecules associated with the generation of contractile forces is opposite in each half of the sarcomere. Consequently, if a pulling force between actin and myosin is developed, the elements of force are additive within one sarcomere. The sum of the individual forces produce greater forces, which slide actin filaments stepwise over the myosin and pull the Z-discs toward the center of each sarcomere. These actions shorten the length of the sarcomeres and thus account for muscle shortening. It appears that the limit to muscle shortening is the point at which the myosin filaments touch the area of the Z-disc.

CONTRACTION OF SKELETAL MUSCLE FIBERS

Development of Force. The conformational change in myosin can theoretically account for muscle shortening. This actomyosin interaction must occur repeatedly during a normal contraction in a hand-over-hand pulling fashion. The rate of actin-myosin interaction, and consequently the speed of contraction, is determined in a large part by the ability of myosin to hydrolyze ATP (*actomyosin ATPase activity*). Some muscle cells have a higher proportion of myofibrillar ATPase activity; these fibers are referred to as *fast-twitch fibers* since they have an increased contraction speed. (Further discussion of this concept will be found in Chapter 4.)

Regulatory Proteins of Muscle. The successful interaction of actin and myosin is dependent upon other proteins, which are regulatory or structural in function. These proteins make up about 16% of all myofibrillar protein:

Tropomyosin (5%) is distributed without interruption along the entire length of the thin actin filaments. This protein inhibits the interaction of actin and myosin, creating a mechanism whereby the two myofilaments remain disengaged until the inhibition is released.

Troponin (3%), located in the actin filaments, is the calcium (Ca^{2+}) receptive protein and is responsible along with tropomyosin for the sensitivity of actomyosin to Ca^{2+}. Since troponin has high calcium affinity, it may undergo a transformational change after binding with calcium—thereby altering the tropomyosin in such a way that it can no longer inhibit actin and myosin from interacting.

Alpha-actinin (7%) is a protein found in the region of the Z-band. It can be divided into several fragments, the heavier of which is the active principle and is thought to cement actin filaments to the Z-band structure and to one another. Its direct involvement in actin-myosin interaction may be less important.

Beta-actinin (<1%) is found in the actin filament. Its function is unknown, but like alpha-actinin its major role appears to be related to actin filament structure and the Z-line.

M protein (<1%) has been isolated from the M-line. The M-line consists of three to five parallel striations perpendicular to the myosin filaments (Figure 2–12). These striations consist of M-bridges connecting six adjacent thick filaments in the form of a regular hexagon. This M protein serves to keep the thick filaments in proper orientation within each sarcomere.

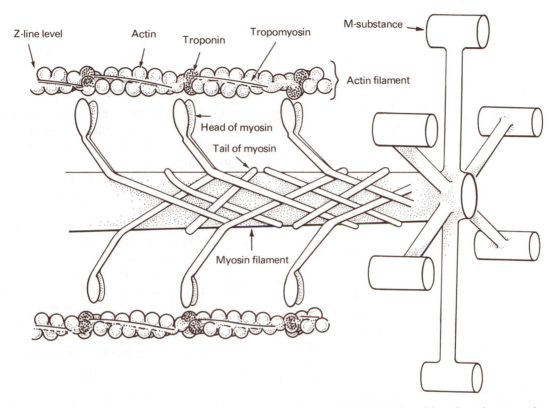

Figure 2–12. Relationship of actin, myosin, and M-substance. The controlling proteins, troponin and tropomyosin, are located on the actin chain. The myosin ATPase is located on the head of the myosin.

Summary of Events During the Contraction Process

1. Depolarization of sarcolemma by impulse from nerve
2. Impulse conduction down length of sarcolemma and through T-tubules
3. Release of Ca^{2+} from SR
4. Binding of Ca^{2+} with troponin causing a release of tropomyosin inhibition
5. Actomyosin interaction
6. Activation of ATPase
7. Breakdown of ATP and energy release
8. Conformational changes at the head of the myosin molecule

9. Actomyosin binding
10. Conformational changes at the actin-myosin linkage
11. Muscle shortening

The steps listed above allow each sarcomere along the full length of the myofibril to contract. Somehow the sequence of events from steps 6 through 11 is recycled so that additional stepwise, or hand-over-hand, shortening of the sarcomere continues until the calcium is taken up by the SR, exposing the inhibitory tropomyosin, which prevents bonding between myosin and actin; when this occurs, the sarcomere relaxes.

The exact microsequence of molecular movements, bond formation, and bond breaking is unknown; but the points we reviewed are important for an understanding of the types of adaptations that can occur with different training regimens. For example, the number and force of the actin-myosin bonds formed at a given instant can influence the maximal contractile force. Training programs that can successfully and specifically modify the events of contraction (e.g., actomyosin ATPase activity) are the types of programs that the "strength" and "power" athlete will want to consider.

Ribosomes

Cytoplasmic particles consisting of a complex of ribonucleic acids (RNA) and proteins are called *ribosomes*. They may be free ribosomes or bound to the sarcoplasmic reticulum. The size and structural composition of the ribosome is quite sensitive to magnesium (Mg^{2+}) concentration; this is one property that is utilized to control ribosomal dissociation. Ribosomal RNA comprises about 85% of the total cellular mass of RNA and is transcribed from DNA in the nucleus (Figure 2–8). The purpose of the ribosome is to provide the structure upon which proteins are synthesized. We know very little about the effect of exercise and/or training on ribosomal distribution, composition, or activity.

Golgi Apparatus

A compact network of double membranes with flattened cisternae is characteristic of the *Golgi apparatus*, a specialized area of the ER (Figure 2–8). It is composed of proteins and lipids (*lipoproteins*) and is primarily associated with secretory activity, as evidenced by its prominence when substances are being secreted by the cell. Glycoproteins are in abundance in the Golgi apparatus; mucus, proteins, carbohydrates, lipids, and hormones also have been found and identified in the Golgi apparatus. Some ribosomes associated with the ER are later transported to the Golgi region for aggregation and perhaps cellular extrusion. The Golgi apparatus may also condense certain materials absorbed by the cell prior to distribution. Like several other cellular membranes, the Golgi apparatus is largely unexplored in terms of its response to exercise and could prove to be an exciting new area for investigation.

Lysosomes

The cytoplasm of the cell contains electron-dense bodies (*lysosomes*) that contains *hydrolases* or *hydrolytic enzymes*, which aid in the digestion of materials ingested. Lysosomes are logically quite numerous in *phagocytic* (eating cell) *cells* such as *macrophages* and white blood cells (*leukocytes*). They sometimes demonstrate *autophagic capacities*; that is, they utilize part of their own cell cytoplasm during stages of starvation and in some diseased states. In relation to the number of mitochondria in a cell, we are likely to find only about one lysosome per 100 mitochondria in the liver; whereas in the kidney this ratio would be closer to one per 10 in kidney tubules. In normal muscle the lysosome is quite rare.

Probable functions of lysosomes include intracellular digestion of certain cellular components (acting, for example, in defense against microorganisms), cellular differentiation, and self-clearance of dead cells. The response of this organelle in muscles to chronic or acute exercise is not known, but their number and hydrolytic-enzyme activity are augmented in some atrophic muscular conditions.

Mitochondria

Probably the most specialized organelle of the cell is the *mitochondrion* (Figure 2–13). It has deservedly received much of the exercise biologists' attention since it is the *major* site of ATP production and the site of oxygen and food utilization. As the "energy unit," it is rightfully called the powerhouse of the cell. Its size and shape vary depending on the type of cell and the activity of the cell in which it is found.

The mitochondrion consists of a double membrane, the inner one consisting of inward folds called *cristae* (Figure 2–14). Arising from these folds are fingerlike projections on which submitochondrial particles have been identified. These subunits are chemically and morphologically distinct from the rest of the inner membrane.

Specific enzyme complexes have been identified on the inner and outer mitochondrial membranes, the most important of which are the citric acid cycle enzymes, enzymes of fatty acid oxidation, and enzymes associated with the electron transport chain.

One of the primary subcellular adaptations to endurance training is the increase in the number and/or size of mitochondria in chronically active muscle. This adaptation apparently increases the ability of the muscle to supply high-energy phosphate bonds for the performance of chemical and mechanical work.

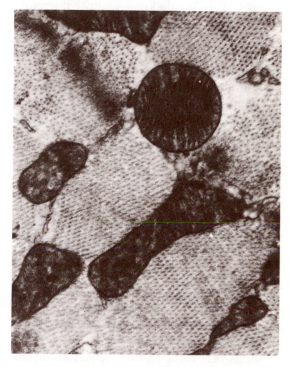

Figure 2-13. Mitochondria location surrounding a myofibril.

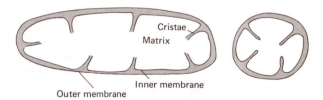

Figure 2-14. Schematic representation of a mitochondrion showing the inner and outer membranes, the cristae, and the inner matrix.

The mechanism by which mitochondria replicate themselves eludes investigators at the present time. We do know that the mitochondria possess the ability to increase in size, to change shape, to divide, and to join together. Mitochondria possess a nucleic acid-protein mechanism that is capable of synthesizing several proteins. The mitochondrially synthesized proteins are essential for the complete assembly of a functional mitochondrion, but the mitochondrial DNA is not capable of coding for the total structure. In concert with the nuclear mechanism, the mitochondria's protein-synthesizing machinery can replicate a new mitochondrion. From the available evidence, it appears that the nuclear nucleic acids code for the proteins of the outer membrane and for the enzymatic proteins of the inner membrane. The mitochondrial nucleic acids code for the membrane components of the inner membrane, which are necessary to bind the enzymatic proteins into a completed structural-functional relationship for *oxidative phosphorylation*. A knowledge of the subcellular stimulus that turns on this replication mechanism would be a key to understanding endurance adaptations. This subcellular stimulus, whatever it may be, arises only in response to acute exercise and recovery combinations.

Glycogen Particles

Glycogen particles can be clearly demonstrated with the electron microscope (Figure 2–15). Most particles are of a uniform size and exhibit a random population distribution. Enlargement of glycogen particles can be seen microscopically when glucose units are added to existing glycogen particles; they appear as small projections attached to the large glycogen particle. Glycogen particles are thought to be bound in some manner to the SR, at least in skeletal muscle—and perhaps even bound as a

Figure 2–15. Glycogen particles in skeletal muscle.

larger complex with the glycogen-degradative enzyme *phosphorylase* as well as with the glycogen-synthesizing enzyme *glycogen synthetase*. Some evidence exists that this intimate interrelationship of the substrate and enzyme complex is disrupted by physical exhaustion. The glycogen content of an endurance-trained muscle is elevated over its pretraining level.

Lipid Granules

Lipid droplets are easily visible microscopically and often appear as though they are fused with mitochondria ready for oxidation (Figure 2-16). Loss of endogenous lipid granules after exercise, as is the case with glycogen particles, has not been demonstrated convincingly—even though lipids are metabolized particularly during long-lasting exercise bouts.

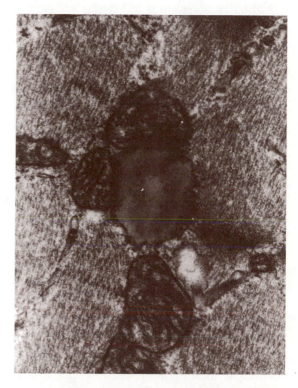

Figure 2-16. Lipid storage in muscle. Mitochondria are usually found adjacent to lipid droplets.

Training Adaptations

It is well accepted that to achieve the trained state it is necessary to "practice", "work out", "exercise", and so forth. What we would like to propose is that, when we exercise, we are "setting up" the muscle or body to those conditions that are necessary to stimulate the adaptive processes. In the muscle these adaptive processes are the DNA-RNA-protein mechanisms. Actual protein synthesis probably takes place during the recovery process, but it can only take place in response to being "set up" by the previous exercising condition.

As we have pointed out, the response to different exercises is specific to the exercise. A strength activity will create a different subcellular environment than that created by an endurance exercise. Therefore those conditions that stimulate strength adaptations will arise from strength exercises; and those conditions that stimulate endurance adaptations will arise from endurance exercises.

Muscular adaptations to exercise are dependent upon the type of exercise stimulus. The strength adaptations are primarily limited to adaptations in the contractile mechanisms, that is, the *myofibrillar proteins*. Strength training results in increased myofibrillar protein and a hypertrophy of the muscle fibers. This hypertrophy has been shown to be the result of an increased size and number of myofibrils per fiber.

Endurance adaptations are observed in the enzymatic capabilities of the metabolic pathways. In response to endurance training, the adaptative process is now known to include an increased concentration of at least 20 enzymes. Other factors, such as muscle-fiber types, capillarization, capacity for fatty acid oxidation, and mitochondrial alterations, also are involved in the training response.

Summary

An understanding of the functional structure of cells is an important basis for an understanding of exercise. The performance of an exercise depends upon the functional capacity of the individual cells, whether they be muscle cells, brain cells, endocrine cells, blood cells, or other cells. Furthermore the interaction of the components of the cell determines the functional capacity of each cell.

Membranes are essential to cell function since they provide cellular and subcellular specificity. The *nucleus* controls the information-processing system of the cell. The *mitochondrion* contributes to the energy source for the cell. The *contractile elements* provide for the contractile property of muscle cells. Each of these cellular components is necessary for the optimal function of the muscle cell.

Muscle contraction is the result of an intricate interaction between neural innervation, calcium concentration, contractile proteins, and energy production. Muscle shortening occurs through the energy-dependent interaction of actin and myosin muscle proteins, which results in a shortening of the thousands of sarcomeres. Relaxation of the muscle occurs when the neural stimulation is withdrawn, and calcium is removed via the sarcoplasmic reticulum.

Functional subcellular compartments exist within the cell; these are usually separated by membranes. Efficient communication links between compartments are essential, especially in times of extensive exercise demands. Specific exercises selectively alter specific subcellular compartments, which causes varying degrees of exercise capability and selectively altering the training adaptations. Therefore the *training adaptations are specific to the exercising conditions*.

Study Questions

1. Draw the subcellular organization of a muscle.
2. Why are membrane transport and transport mechanisms so important to a working muscle?
3. What role does the nucleus play in resting and working muscle?
4. Sketch the relationship of the contractile proteins and how they interact with the addition of calcium.
5. Describe the role of the sarcoplasmic reticulum.
6. What is so special about the mitochondrion?
7. Describe the specificity of training adaptations?

Review References

Ashhurst, D.W. The fine structure of pigeon breast muscle. *Tissue and Cell* 1:485–496, 1969.

Carlson, F.D. and D.R. Wilkie. *Muscle Physiology*. Englewood Cliffs, N.J.: Prentice Hall, 1974.

Cassens, R.G., ed. *Muscle Biology*. New York: Marcel Debber, 1972.

Fawcett, D.W. *The Cell*. Philadelphia: W.B. Saunders, 1966.

Fbe, T. and S. Kobayashi. *Fine Structure of Human Cells and Tissues.* New York: John Wiley, 1972.

Gauthier, G.F. The ultrastructure of three fiber types in mammalian skeletal muscle. In *The Physiology and Biochemistry of Muscle as a Food*, vol. 2, edited by E.J. Brisky, R.G. Casseus, and B.B. Marsh. Madison: The University of Wisconsin Press, 1970.

Murray, John M. and Annemarie Weber. The cooperative action of muscle proteins. *Sci. Amer.* 230:58–71, 1974.

Porter, K.R. and C. Franzini-Armstrong. The sarcoplasmic reticulum. *Sci. Amer.* 212:72–78, 1965.

The Mechanism of Muscle Contraction. Cold Spring Harbor Symposia on Quantitative Biology. Cold Spring Harbor Laboratory, New York, 1973.

Zachar, Jozef. *Electrogenesis and Contractility in Skeletal Muscle Cells.* Baltimore: University Park Press, 1971.

CHAPTER THREE

Key Concepts

• ATP (adenosine triphosphate) is the most immediate chemical source of energy for a cell.
• The formation of ATP within muscle involves the conversion of foodstuffs to carbon dioxide.
• Specific types of exercises activate specific energy-production pathways.
• The active muscle has available immediate, short-term, and long-term sources of energy.
• The control of energy metabolism during exercise is a function of the type of exercise and the degree of previous training of the individual.
• The mitochondrion has a very special role in energy production.

Introduction

There are specific energy sources available to the muscle that are activated in response to specific exercises. All exercise stresses are not identical in their demands on the muscle; our concept of specificity of exercise will be strongly supported as we analyze the energy-production potential.

Energy production in a tissue, such as muscle, can be schematically described (Figure 3–1). The circulatory system provides the *substrates* (fuel) needed for the function of the tissue. During times of low-energy requirements, the substrates from the circulatory system are added to tissue storage depots in the form of *glycogen* (a polymer of glucose) or *lipid* (fatty acids or triglycerides). During times of high-energy demands, the active tissue makes use of these stored intracellular materials while the vascular system provides additionel fuel and oxygen. The cell utilizes the fuel and oxygen in the metabolic reactions that generate ATP. The ATP is then used in the energy-requiring reactions (such as muscular contrac-

Energy production within muscle

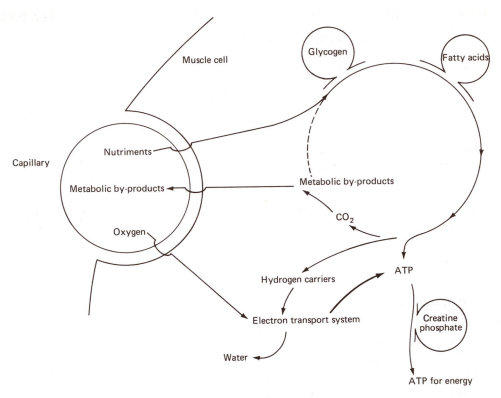

Figure 3-1. General view of energy metabolism. Nutriments (primarily glucose, fatty acids, and amino acids) are brought into the cell.

tions). The resulting metabolic by-products are returned to the circulatory system, where they are excreted from the body or reconverted by the liver into useful materials.

The muscle derives energy from the oxidation of food materials via various biochemical pathways (Figure 3–2). The oxidation of food materials involves the breaking of the chemical bonds to remove hydrogen atoms and to release carbon dioxide (CO_2). The hydrogen atoms are transported by *hydrogen-carrying molecules* to the mitochondria where the hydrogen is released to combine with oxygen to form water and, in the process, to provide the energy to form ATP.

When a chemical bond is broken, most of the energy that went into its formation is released, while some is lost as heat. During the enzymatic degradation of glucose or fatty acids, some of the energy (approximately 50%) is captured to be either immediately used or stored by the cell. The intracellular energy-capturing molecule is ATP, which is formed by energetically linking *adenosine diphosphate* (ADP) with a third *phosphate* (Pi). The resultant ATP can then be transported to various parts of the cell to meet the energy demands of the cell. ATP gives up its newly captured energy when it is enzymatically broken down to its previous state (ADP + Pi).

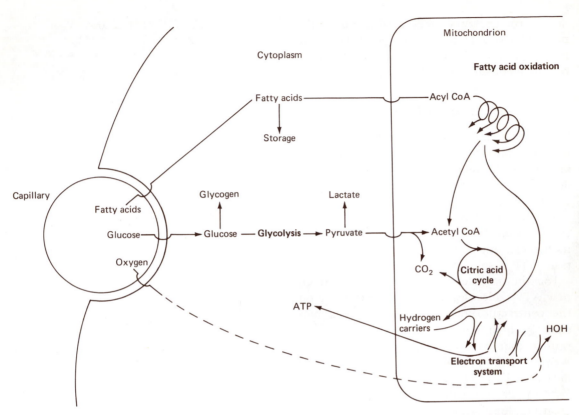

Figure 3-2. Relationships of biochemical pathways. The major biochemical pathways are glycolysis, fatty acid oxidation, citric acid cycle, and electron transport system.

Sources of Energy Within the Muscle

The energy for muscular contraction is provided, in general, by the *immediate, short-term*, and *long-term sources*. As we saw from Figure 1–6, the relative importance of each of these energy sources is time dependent. For the active muscles, the immediate sources of energy are the almost exclusive suppliers of energy for exercises lasting ten seconds or less; however, oxidative metabolism provides the bulk of the energy for the longer (more than five minutes) exercises.

Immediate sources of energy come from ATP and creatine phosphate stores in the muscle.

Short-term energy sources correspond to the nonoxidative ATP production potential of the muscle; long-term sources are linked to oxidative metabolism.

Immediate Sources of Energy

High-power and power-endurance activities have a high-energy demand over a short period of time (see Chapter 15). These activities must depend upon ATP that can be generated by rapid one-enzyme reactions.

ATP is stored in limited quantities within the muscle. At rest ATP concentration is at its highest; but with the initiation of contraction, ATP is split to form ADP and Pi.

For the muscle to continue to contract, new ATP molecules must move into place on the contractile cross bridges. Two enzymes exist in high concentration within skeletal muscle that have the capacity to regenerate ATP from the breakdown products, ADP and Pi (Figure 3–3). The enzyme *creatine phosphokinase* (CPK) is capable of producing ATP from creatine phosphate and ADP.

$$\text{creatine-P} + \text{ADP} \xleftrightarrow{\text{CPK}} \text{ATP} + \text{creatine}$$

This enzyme is known to exist throughout the muscle and more specifically near the contractile filaments. With the formation of ADP, resulting from muscular work, creatine phosphokinase acts to transfer Pi from creatine phosphate to ADP. This enzymatic action will continue to form ATP and creatine until the muscle is exhausted of its creatine phosphate supply.

The second enzyme assisting in the rapid resynthesis of ATP via a one-enzyme reaction is *adenylate kinase* (AK). This enzyme utilizes two ADP molecules and converts them into ATP and AMP (*adenosine monophosphate*).

$$\text{ADP} + \text{ADP} \xleftrightarrow{\text{AK}} \text{ATP} + \text{AMP}$$

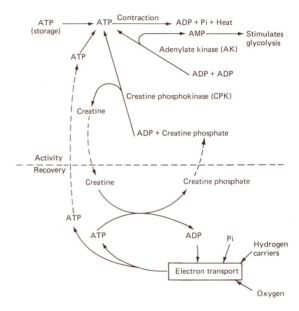

Figure 3–3. Immediate sources of ATP during short-term activity and the regeneration of ATP during recovery.

Although adenylate kinase provides ATP for use in contraction, the enzyme's primary responsibility appears to be to provide a source of AMP, which serves as a stimulus for the initiation of *glycogenolysis* and *glycolysis*. Increased levels of AMP activate the enzymes *phosphorylase* (the enzyme which breaks down glycogen into *glucose-1-phosphate*) and *phosphofructokinase* (the first enzyme of the glycolytic pathway). The activation of these enzymes leads to the mobilization of the next most immediate source of energy, the short-term or nonoxidative sources.

The immediate sources of ATP provide the energy for muscular contraction during the first few seconds of activity. Exercises involving strength and activities requiring high power rely primarily on these immediate energy sources; therefore training techniques for these

activities should emphasize the development of an increased energy-production potential involving ATP stores, creatine phosphokinase activity, and adenylate kinase activity.

Short-Term Sources of Energy

For activities lasting longer than a few seconds, a source of energy (*carbohydrate*) is needed besides the immediate sources. During intense activity, stored intramuscular glycogen is broken down into *glucose-6-phosphate* (G-6-P) units. The enzymes of glycolysis utilize G-6-P molecules and in the process of converting G-6-P to lactate, form ATP molecules (schematically depicted in Figure 3–4). This nonoxidative formation of ATP is the primary energy source, within muscle, for activities lasting between 30 seconds and two minutes.

The overall effect of strenuous power-type exercise on carbohydrate metabolism is to increase the rate of flow of molecules through glycolysis (up to a 100-fold increase). As a result of this increased metabolism, the concentrations of alpha-glycerol phosphate and lactate increase dramatically in muscle. These molecules could be called "dead end" molecules because they cannot be readily metabolized in the muscle during times of high-energy demand.

At the cessation of the exercise, the alpha-glycerol phosphate molecule can either be reconverted to glycogen or be further metabolized to lactate or carbon dioxide and water. The lactate molecule can either be utilized by the muscle to form carbon dioxide and water or be transported to the liver where two lactate molecules can be converted to a glucose molecule. The resulting glucose is either stored as glycogen in the liver or released into the blood stream and transported to the muscle for utilization. This lactate-glucose cycle has been named the *Cori cycle* in honor of the scientist who first described it.

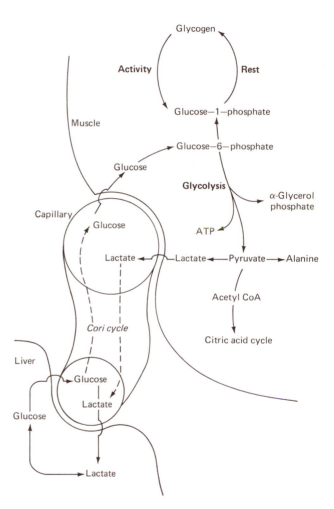

Figure 3–4. Glycolysis, Cori cycle, and the central role of glucose-6-phosphate and pyruvate.

We know that during exercise the intramuscular concentrations of G-6-P, α-GP, pyruvate, lactate, and alanine increase. It appears that these intermediates are critical to the operation of the glycolytic pathway.

GLUCOSE-6-PHOSPHATE CONSIDERATIONS

When muscle utilizes glucose from the circulation, it immediately must convert the glucose to glucose-6-phosphate (G-6-P), utilizing one ATP as the source of the phosphate. G-6-P is a key metabolic molecule: it is a branch point between glycogen synthesis (during times of rest) and glycolysis (during times of activity). At rest the muscle cell contains low levels of AMP, ADP, Pi, and G-6-P—and high levels of ATP. This intracellular environment is optimal for providing stimulation to glycogen synthetase for the utilization of glucose (taken up from the blood in the presence of insulin) and for the ultimate formation of glycogen.

As muscular activity begins, ATP is split into ADP and Pi; the result is the activation of the enzymes *creatine phosphokinase* and *adenylate kinase*. The latter enzyme also forms AMP, which serves to activate the enzyme *phosphorylase*, normally found in a relatively inactive state in resting muscle. In the presence of AMP, phosphorylase is more active, and the degradation of glycogen is increased, resulting in the release of G-6-P molecules. This G-6-P is now available for utilization in glycolysis.

The control of G-6-P metabolism is central to the control of carbohydrate metabolism: during times of *rest* (low AMP and low G-6-P), the available G-6-P is converted to glycogen; during times of *activity* (high AMP and high G-6-P), the G-6-P is utilized in the energy-producing pathway *glycolysis*.

ALPHA-GLYCEROL PHOSPHATE

We know that the concentration of *alpha-glycerol phosphate* increases during exercise.

There exist at least three possible explanations for this increase. First, the alpha-glycerol phosphate (α-GP) may serve as a reservoir for excess molecules that cannot be metabolized by the enzymatic reactions further down the glycolytic chain. After exercise, the accumulated α-GP may be reintroduced into glycolysis or used for the resynthesis of glycogen. Second, the formation of α-GP may serve as a method to carry hydrogen atoms into the mitochondria during times of oxidative metabolism, since α-GP is permeable to the mitochondrial membrane. A third possibility is to provide for the formation of the oxidized form of *nicotinamide dinucleotide* (NAD), a molecule formed from the vitamin *niacin*, which is needed for glycolysis.

PYRUVATE

Assuming that α-GP is not formed, the overall action of glycolysis is to convert one G-6-P molecule into two pyruvate molecules with the formation of three ATP. The ATP is obviously necessary for those energy-requiring reactions, such as muscle contraction.

$$\text{Glucose-6-phosphate} \xleftrightarrow{\text{glycolysis}} 2\text{ pyruvate}+3\text{ATP}$$

Pyruvate, an extremely important molecule, is a branch point for the formation of three other important molecules: lactate, mentioned previously in terms of the Cori cycle; alanine, an amino acid; and acetyl Co-A, which enters the citric acid cycle (Figure 3–4).

LACTATE

The formation of lactate in the muscle is very similar, in purpose, to the formation of α-GP, since lactate can serve as a fuel reservoir. The formation of lactate is tied to the formation of NAD, which can be utilized to maintain glycolysis (an earlier step in glycolysis is dependent upon a constant supply of NAD).

This system of nonoxidative glucose metabolism (G-6-P to lactate) is referred to as the *nonoxidative* or *anaerobic* phase of glycolysis. This method of producing ATP is adequate for the short-term operations required in power-endurance type activities.

ALANINE

A second alternative from pyruvate is the formation of *alanine*. Alanine has not received the degree of recognition given lactate, but may serve as an alternative to lactate formation, thus decreasing the intensity of the possible fatigue effects of lactate. Alanine may be reconverted into pyruvate at some later time.

Although alanine is a key amino acid in protein synthesis, it is highly unlikely that it is used for protein synthesis during times of exercise. During recovery from exercise, this pyruvate-alanine interconversion may be a very significant factor in protein metabolism.

It is known that the muscles of the body are a key supply depot for alanine; it has also been shown that, during times of starvation or long-term activity, alanine is released from the muscles. The released alanine is carried by the blood stream to the liver for conversion to glucose (requires two alanine molecules) via the process of *gluconeogenesis* (the new formation of glucose). It is obvious that alanine plays a role in energy metabolism, but the exact role has not yet been defined.

ACETYL CO-A

The third possibility for pyruvate metabolism is for pyruvate to be taken up by the mitochondria. An enzyme, *pyruvate dehydrogenase*, splits off one carbon from the three-carbon pyruvate molecule to form CO_2 and combines the remaining two-carbon part with *Coenzyme A* (a molecule formed from the vitamin *panththenic acid*) to form *acetyl Co-A*. The formation of acetyl Co-A represents the link between nonoxidative (short-term sources of energy) and the oxidative (long-term) sources of energy.

Long-Term Sources of Energy

The metabolic processes that utilize oxygen to generate ATP are the last of the ATP-generating mechanisms to be activated during exercise. These oxidative-related sources generate ATP by coupling the utilization of *oxygen* with the oxidation of *hydrogen carriers* in the electron transport system of the mitochondrion.

Oxygen, stored and transported in chemical combination with myoglobin, is stored in the muscle in a very limited amount. Therefore a continuous vascular-dependent supply of oxygen is necessary for any exercise situation lasting more than approximately two minutes.

The continuing supply of *hydrogen carriers* (NADH) arises primarily from two possible sources: fatty acid oxidation and the oxidation of acetyl Co-A in the citric acid cycle.

FATTY ACID OXIDATION

As with G-6-P, the sources of fatty acids are intracellular (tissue stores) as well as extracellular (supplied via the vascular system). The fatty acids are stored in the cell and must be mobilized and transported across the mitochondrial membrane. Once inside the mitochondrial membrane, the fatty acid is available to the enzymes of fatty acid oxidation, which are located on the mitochondrial membranes.

The biochemical pathway for the oxidation of fatty acids is depicted in Figure 3–5. The fatty acid oxidation cycle consists of a series of reactions that result in a sequential splitting off of acetyl Co-A groups. Each completed cycle results in two hydrogen carriers (one NADH and one FADH), one acetyl Co-A, and a fatty acid that is two carbons shorter. Subse-

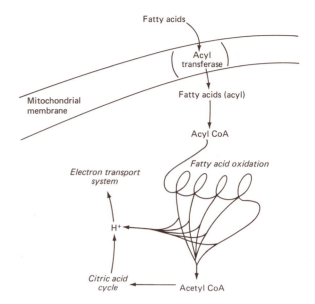

Figure 3-5. Oxidation of fatty acids (acyl groups) to form acetyl Co-A and reduced carriers.

quent oxidations of the modified, shortened fatty acid chain split off additional hydrogen carriers and acetyl Co-A molecules.

During times of high rates of fatty acid oxidation in the *liver*, the production of acetyl Co-A exceeds the utilization capacity of the citric acid cycle. When this happens, the liver stops the complete breakdown of the fatty acids at the 4-carbon molecule, *β-hydroxybutyrate*. This molecule is sometimes converted into *acetoacetate*, and both—along with *acetone*—are classified as *ketone bodies*. The excessive production of these ketone bodies can occur during times of extensive fatty acid mobilization, such as may occur during exercise, or during pathological conditions, such as diabetes.

Ketone bodies can be utilized by skeletal or cardiac muscle for energy utilization; they are metabolized by the fatty acid oxidation cycle.

CITRIC ACID CYCLE

The acetyl Co-A formed from pyruvate or fatty acid oxidation is the substrate for the *citric acid cycle*. The acetyl Co-A molecule, as it enters the cycle, combines with a four-carbon molecule, *oxaloacetate*, to form citrate and a free Co-A molecule. Through the series of reactions illustrated in Figure 3–6, one ATP molecule, four hydrogen carriers (three NADH and one FADH), and two CO_2 molecules are formed during the citric acid cycle. (You should now be able to account for all of the CO_2 formed during energy production.)

In response to endurance training, there is an increase in the concentration of many of the enzymes associated with fatty acid oxidation and the citric acid cycle.

OXYGEN AND HYDROGEN CARRIER MOLECULES

So far in our discussion of energy production for muscular contraction, we have not demonstrated the role of oxygen. The pulmonary system extracts the oxygen from the air; the cardiovascular system transports it to the cell; there the oxygen is used by the mitochondria. This utilization is by a series of enzymes and *cytochromes (electron transport chain)* that have the capability of performing a coupled series of oxidation-reduction reactions. The electron transport chain initially utilizes one of the previously formed hydrogen carriers and ends with the utilization of oxygen to form water. For every hydrogen carrier (NADH) that enters into the electron transport system, *three ATP* molecules and one water (H_2O) molecule are formed.

ATP PRODUCTION

In essence the hydrogen carriers (NADH and FADH) have a high *potential energy* value; oxygen has a low potential energy value. Throughout the various steps of the electron

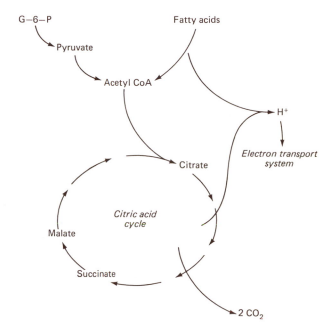

Figure 3-6. Citric acid cycle.

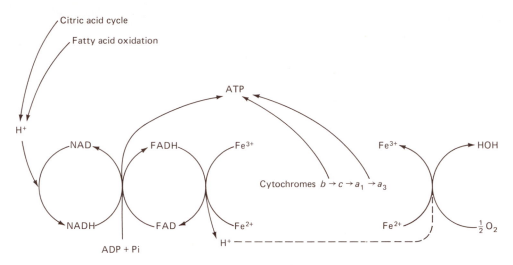

Figure 3-7. Electron transport system to couple oxygen utilization to ATP production (oxidative phosphorylation).

transport system, the high-energy molecules are transformed to lower-energy molecules with the concomitant release of energy at each step (Figure 3-7). The topology of the mitochondrion is such that the release of energy at key steps can be controlled and used by the mitochondrion thereby resulting in the coupling of phosphate (Pi) and adenosine diphosphate (ADP) to form an adenosine triphosphate (ATP) molecule.

$$ADP + Pi \xrightarrow{\text{energy}} ATP$$

The components of the electron transport system are very critical to energy production; many of these electron transport components are increased in the muscular adaptation to endurance training.

INHIBITORS

The electron transport system couples the utilization of *hydrogen carriers, ADP, Pi,* and *oxygen* to form *ATP, heat,* and *water.* Obviously

this system is essential to life as we know it in our oxygen environment. The electron transport system is the site of action of many of the more commonly known poisons. *Cyanide* inhibits the action of the terminal cytochrome (a_3); *arsenate* inhibits the coupling of the energy released from the electron transport system to the formation of ATP; and *amytal* and *antimycin A* inhibit other aspects of the electron transport system.

CONTROL OF RESPIRATION DURING EXERCISE

The activity of the electron transport system can be altered by the lack of hydrogen carriers, the lack of oxygen, the lack of ADP or Pi—or the presence of hormones or inhibitors such as cyanide.

During times of rest, oxygen is readily available; but ADP and Pi are present only in limited quantities. Not unexpectedly the cell would have all the ADP and Pi tied up in the form of ATP during rest. Obviously there is a basal metabolism (*basal metabolic rate*, BMR) that maintains minimal muscle activity and other basic functions necessary for life; however, the ATP requirement for these processes is very minimal.

At the initiation of exercise, the ADP and Pi become available because of the splitting of the ATP molecule at the site of muscular contraction. As the result of a shuttle mechanism, such as using *creatine phosphokinase*, the ADP soon becomes available to the mitochondrion. Whether exercise ceases or continues for a time, eventually the electron transport system will be called upon to replace the lost energy: the storage materials within the muscle (or in the liver or adipose tissue) will have to be activated and delivered to the specific metabolic pathways. When this happens, the hydrogen carriers formed from the metabolism of these storage materials (glycogen and fatty acids) will be utilized by the electron transport system to form ATP.

If muscle contraction is such that oxygen is not available to the muscle cell (because of, for example, an impaired blood flow), the electron transport system will shut down—even though the cell is loaded with the other necessary components, NADH, ADP, and Pi. When this happens, the muscle can still produce a limited amount of ATP through the nonoxidative system of glycolysis. The total capacity of the nonoxidative system is at an energy level equal to less than 5% of the amount available through the oxidative processes of the mitochondrion.

Control of Energy Metabolism

The *control of energy metabolism* is a highly complicated subject—and beyond the scope of this book. To understand the control of energy metabolism completely, it would be necessary to interrelate all the energy systems discussed in this chapter. Furthermore the type of control exerted depends upon the type of exercise (specificity of exercise), the degree of previous stress, and many other factors. With these complexities in mind, we will present only the basic chemical control mechanism that the body (muscle, in this case) has at its disposal to regulate the flow of energy.

Endocrine Control

It is recognized that hormones can and do exert significant control on the rate of energy production within the muscle. *Epinephrine, thyroxine, growth hormone,* and *cortisol* are influential during exercise. *Insulin, testosterone,* and *antidiuretic hormone* are influential during rest and recovery from exercise.

Epinephrine, thyroxine, and growth hormone are responsible for alterations in the rate of energy production and partially for the selection of the substrate of preference during the various types of exercise. Epinephrine increases the rate of the breakdown of glycogen within the muscle and increases the rate of the release of fatty acids from adipose tissue. Thyroxine aids in the release of fatty acids from adipose tissue; at the level of the muscle, this hormone increases the rate of oxygen uptake—due to its ability to uncouple oxidative phosphorylation. Growth hormone is known to increase the blood levels of fatty acids and to increase the uptake of amino acids by the muscle. The latter effect seems to be of less significance during exercise since, in fact, there is a net release of amino acids (mainly in the form of alanine) from muscle during exercise.

Activation of the Energy Pathways

The first contraction during exercise necessarily causes a splitting of the ATP molecule to form ADP + Pi. The presence of ADP in the vicinity of stored creatine phosphate induces a phosphorylation of the ADP molecule to form ATP + creatine. At the same time, some

of the ADP is in the vicinity of the enzyme *adenylate kinase,* which immediately catalyzes the combination of two ADP molecules to form ATP + AMP. The initial muscle contraction thus lowers the ATP-to-ADP ratio and at the same time increases the concentration of AMP. These two factors are known to effect the activity of key enzymes of glycolysis, increase glycogen breakdown, and increase the activity of glycolysis.

The exercise-induced decrease in the ATP-to-ADP ratio in the muscle could be hypothesized to stimulate the action of the electron transport system. At *rest,* as we have pointed out, ATP, hydrogen carriers, and oxygen levels are high in the mitochondrion, and the limiting factor to respiration is the availability of ADP and Pi. With the initiation of *exercise,* the ATP molecule is split, thus increasing the ADP and Pi levels within the muscle. As soon as this information is passed to the mitochondrion, the electron transport system increases its ATP production rate, depending upon the availability of hydrogen carriers and oxygen. It is possible to hypothesize that the limitation in the ability of the muscle to produce ATP during exercise lies in the ability of the muscle to supply any of the previous factors.

The Special Role of the Mitochondrion

If there is any one structure within the muscle carrying the greatest responsibility for energy production, it is the mitochondrion. Within this structure are the enzymes of the *citric acid cycle,* the enzymes of *fatty acid oxidation,* and the *electron transport system.* The mitochondrion is the site of the intracellular utilization of oxygen and the production of most of the ATP generated within the muscle. The mitochondrion has been most appropriately called the "powerhouse of the cell."

During exercise the size of the mitochondria within the muscle is altered. The exact mechanism for this phenomenon remains obscure, but the size increases are not as great as those increases observed with injections of thyroxine. In the case of thyroid toxicosis, the swelling uncouples oxidative phosphorylation (normally oxygen utilization is closely linked to ADP phosphorylation). It is unlikely that exercise brings about the same uncoupling effect.

Mitochondria take up calcium and other divalent ions from the cytoplasm during exercise. Part of the increased oxygen demand during recovery may be needed to pump these ions from the mitochondria. The active membrane pump could account for a sizable portion of the extra oxygen utilized after exercise. It should be kept in mind that the resynthesis of creatine phosphate is energy-dependent and requires a large portion of the oxygen used during recovery from exercise. The postexercise oxygen uptake also can be related partially to the elevated muscle temperatures.

The biological oxidations within the mitochondrion have been compared to the action of a campfire, in which the wood provides the fuel and the oxygen is the oxidizing agent. To stop a fire, one can remove the wood or smother the fire by eliminating the source of oxygen. As with the fire, when the body runs out of food or oxygen, the process of ATP production ceases.

Summary

There must be a readily available source of energy for a cell to function and especially for a muscle cell to contract during exercise. The specific source of energy is dependent upon the type of exercise being performed (Figure 3–8).

Immediate sources of energy for activities lasting less than 10 seconds come from stored ATP and creatine phosphate. For activities lasting 30 seconds to two minutes, the primary

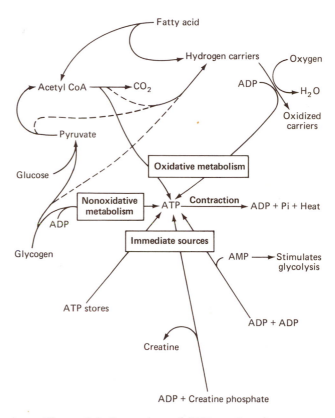

Figure 3–8. Summary of ATP production, showing the three primary methods of resynthesizing ATP: immediate sources, nonoxidative and oxidative energy production.

energy source is the nonoxidative use of carbo-hydrates (glycogen). Oxidative metabolism (utilizing glycogen and fatty acids) provides the energy source for activities longer than five minutes. The time periods, 10 to 30 seconds and 2 to 5 minutes, represent transition periods from immediate sources to nonoxidative and from nonoxidative to oxidative metabolism respectively.

Immediate sources of energy involve the action of no more than a one-enzyme reaction to supply ATP for muscular contraction. *Short-term, nonoxidative* energy production generates ATP through the breakdown of glycogen via the enzymatic pathway of glycolysis, with the end formation of lactic acid or of alpha-glycerol phosphate. *Long-term, oxidative* energy production utilizes oxygen to combine with the hydrogen ions gathered from carbohydrates and fat metabolism. Key enzymatic systems for oxidative metabolism involve glycolysis, fatty acid oxidation, citric acid cycle (*Krebs cycle*), and the electron transport system.

During times of rest and recovery from exercise, the body replaces the energy sources used during an exercise bout. Eventually all the replacement energy must come from oxidative metabolism. This oxidative energy production takes place within the mitochondrion; thus the mitochondrion has a special function within the muscle as the "powerhouse" of the cell.

Study Questions

1. What are the possible fuels for energy production?
2. What role does ATP play in energy-requiring reactions?
3. What is the sequence of energy activation?
4. Under what conditions are the immediate sources of energy most important?
5. How and when is glycogen important as a source of energy production?
6. What does lactate production signify?
7. When are long-term sources of energy most important?
8. What roles do fatty acids and glycogen play in long-term exercises?
9. How are oxygen utilization and ATP production related?
10. How is energy metabolism controlled?
11. What role do hormones play in energy production?
12. Describe the special role of the mitochondrion.

Review References

Gollnick, P.D. and L. Hermansen. Biochemical adaptations to exercise: Anaerobic metabolism. In *Exercise and Sport Sciences Reviews*, vol 1, J.H. Wilmore, ed. New York: Academic Press, 1973.

Holloszy, J.O. Biochemical adaptations to exercise: Aerobic metabolism. In *Exercise and Sport Sciences Reviews*, vol 1, J.H. Wilmore, ed. New York: Academic Press, 1973.

Hoyle, G. "How is muscle turned on and off?" *Sci. Amer.* 222:84–93, 1970.

Kuel, J.; E. Doll; and D. Keppler. *Energy Metabolism of Human Muscle.* Baltimore: University Park Press, 1972.

Lehninger, A.L. *The Mitochondrion.* New York: W.A. Benjamin, 1965.

Margaria, R. The sources of muscular energy. *Sci. Amer.* 226:84–91, 1972.

Miller, A.T. *Energy Metabolism.* Philadelphia: F.A. Davis, 1968.

Pernow, B. and B. Saltin, eds. *Muscle Metabolism During Exercise.* New York: Plenum Press, 1971.

Simonson, E., ed. *Physiology of Work Capacity and Fatigue.* Springfield, Ill.: Charles C Thomas, 1971.

Wilkie, D.R. *Muscle.* New York: St. Martin's Press, 1968.

Neurological aspects of movement

CHAPTER FOUR

Key Concepts

• A motor unit consists of a motor nerve and all the muscle fibers innervated by that alpha-motoneuron and is the functional unit of voluntary skeletal muscle.
• Motor units can be divided into two populations on the basis of muscle contractile speed and into three categories when metabolic properties of the muscle are combined with muscle speed.
• Endurance training will cause changes in the metabolic properties of some motor units, but changes in contractile speed have not been shown to occur as a result of a physiological training program.
• The site of neuromuscular fatigue with prolonged work depends on the nature of the work demands.

Dynamics of the motor unit

Introduction

The functional aspect of muscle control is the *motor unit*. The motor unit consists of an *alpha-motoneuron* and all of the skeletal muscle fibers that are functionally connected (*innervated*) by that neuron. The cell body of the alpha-motoneuron, which contains the cell's nucleus and protein-synthesizing machinery, is located in the spinal cord (Figures 4–1 and 4–2). There are thousands of direct and indirect neural connections to the cell body of the motoneuron, but once an impulse is generated this motoneuron determines the final neural output through which movements are controlled. A long single *axon* extends from the cell body within the spinal cord to the muscle where it finally branches to innervate from 10 to 1000 individual muscle fibers. All muscle fibers are innervated by only one motoneuron; all muscle fibers of any one motor unit have similar physiological, biochemical, and morphological properties.

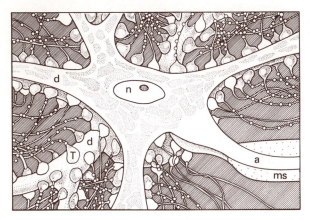

Figure 4–1. A motoneuron cell body showing the terminal axons (T) of other nerves on the dendrites (d) and cell body. The nucleus (n) is easily seen in the center of the cell body. Note the myelin sheath (ms) on the axon (a).

From J.P. Schadé and D.H. Ford. *Basic Neurology*, 2d ed. Elsevier, Amsterdam, 1973.

The Motor Unit

Types of Motor Units

All motoneurons transmit impulses from the spinal cord to the muscle fibers they innervate, and all of the muscle fibers of the specific motor unit contract (referred to as a *twitch*) when a neural impulse arrives. However, motor units differ in regard to (1) speed, (2) force, and (3) endurance or the length of time a muscle fiber can contract without loss of tension (Figure 4–2).

SPEED

It is clear that two different populations of motor units account for the speed with which the muscle can react to a stimulus. Some motor units can reach a peak twitch tension (response to a single impulse) as much as twice as fast (*fast-twitch*) as other units (*slow-twitch*). The neural impulse for these fast-twitch units is generally transmitted down the axon more rapidly than the impulse for the slow-twitch units.

FORCE

Another difference in motor units is muscle-fiber force. The greatest force can be exerted by the larger muscle fibers, which are innervated by the larger motor neurons (Figure 4–2). These high-force fibers are also fast-twitch. The motor units that produce the lowest force are innervated by small motoneurons, which have slow transmitting axons and slow contracting muscle fibers. Some motor units can exert only moderate force—even though they are fast-twitch units. The force-yielding capacity of a motor unit depends on a number of characteristics: muscle-fiber number per motor unit, muscle-fiber size, and perhaps the type of myofibrils in the particular motor unit.

ENDURANCE

A third very important property that differentiates motor-unit types is the ability to resist fatigue. The slow-twitch units are more resistant to fatigue than most fast-twitch units (Figure 4–2). However, certain fast-twitch fibers are almost as fatigue resistant as the slow-twitch units. As a matter of fact, the fatigue resistance of fast motor units depends almost entirely on the state of physical-endurance training of the individual: a fast-twitch, fatigable unit can be converted to a fast-twitch, fatigue-resistant unit in response to endurance-type training.

To summarize, the physiological properties of different types of motor units are: (1) fast-twitch, high tension, and fatigable muscle fibers; (2) fast-twitch, moderate tension, and fatigue-resistant muscle fibers; and (3) slow-

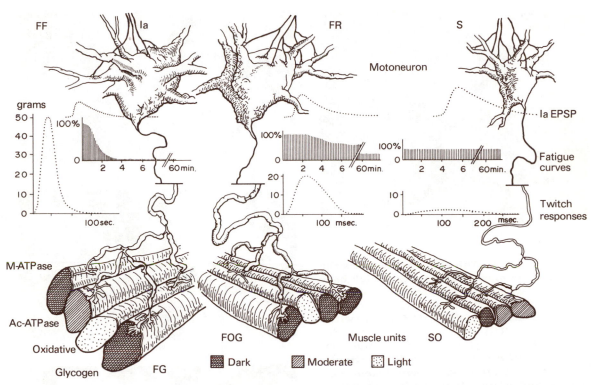

Figure 4-2. Motoneurons may be *phasic* (fire rapidly but with short bursts) or *tonic* (slow and continuous). Axon diameter size is directly related to conduction velocity. The muscle fibers have been stained for a different biochemical: myosin ATPase, acid ATPase, succinate dehydrogenase (an oxidative enzyme), and glycogen. Note the differences in twitch tension for each motor unit: FF, fast, fatigable; FR, fast, fatigue-resistant; S, slow; Ia EPSP, sensory excitatory, post-synaptic potentiation; FG, fast-twitch glycolytic; FOG, fast-twitch high-oxidative glycolytic; and SO, slow-twitch oxidative muscle fibers.

From V. Reggie Edgerton and the
Neuromuscular Research Laboratory, UCLA.

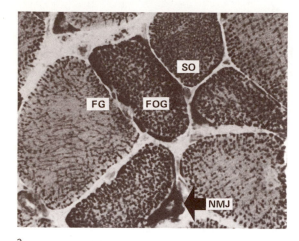

a.

Figure 4–3. (a) The photograph shows the three basic muscle fiber types (which have been stained for malate dehydrogenase) found in small mammals: fast-twitch oxidative glycolytic (FOG), fast-twitch glycolytic (FG), and slow-twitch oxidative (SO). The arrow points to a neuromuscular junction (NMJ) innervating that particular fiber. Photos b–e are serial transverse sections of muscle in the human biceps.

The fibers are stained for a different biochemical: (b) NADH-diaphorase, (c) alpha glycerophosphate dehydrogenase (α-GPDH), (d) myosin ATPase, and (e) myosin ATPase reversed.

twitch, low tension, and fatigue-resistant muscle fibers (Figure 4–3). Other motoneuronal properties and characterization of the metabolism of these motor-unit types are consistent with these physiological properties.

Dynamics of Muscle Contraction

The Motor Unit and the Physiological Properties of Muscle

In Figure 4–2 note that the fast-twitch fatigable units have a large motoneuronal cell body but

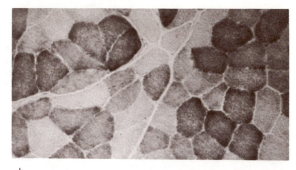

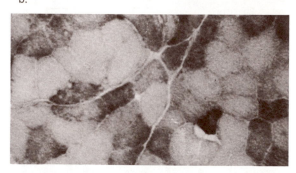

b.

c.

d.

e.

a little sensory input spindle (Ia). Both the large cell-body size and the low sensory input are responsible, in part, for the infrequent activation of fast-twitch muscle fibers. The large cell body requires a higher neural electrical threshold to initiate an impulse, yet the fewer sensory inputs provide fewer stimulations. The higher-threshold neurons innervate muscle fibers with the least oxidative capacity, but with moderate to high glycolytic capacity. The terminal axons of the neuromuscular junctions of the units with the larger cell bodies are greater in length and number than in the other unit types. This unit innervates fast-twitch muscle fibers that are relatively fatigable; the muscle fiber of this unit is generally largest.

The muscle-fiber property that is largely a determinant of muscle speed is reflected in the amount of a certain enzyme in muscle that breaks down ATP. This enzyme is called *myofibrillar ATPase* or *adenosine triphosphatase* (M-ATPase). The fibers with the highest activity of the M-ATPase enzyme contract the fastest.

We have seen that the biochemical properties of muscle fibers reflect the physiological properties discussed previously; that is, M-ATPase activity is directly related to contraction speed. Mitochondrial content or the oxidative enzyme activity of a muscle fiber is directly related to fatigue resistance. Glycolytic capacity is also related to muscle speed—for a reason not yet ascertained. A precise biochemical measure for muscle tension is not known; but, in a general way, the fibers with the higher M-ATPase activity exert the greatest amount of tension. By combining the biochemical properties with the physiological characteristics of muscle fibers, the nomenclature *fast-twitch oxidative-glycolytic* (FOG), *fast-twitch glycolytic* (FG), and *slow-twitch oxidative* (SO) can be used.

Muscle Fiber Types in Exercise, Training, and Performance

It is probable that in all movements no one type of motor unit is used exclusively. It appears that we preferentially recruit slow-twitch units in low tension, slow-moving, or enduring exercises, while we involve fast-twitch units in fast, powerful movements. The *soleus muscle*, which consists of a relatively high proportion of slow-twitch fibers, fatigues at a slower rate than the *gastrocnemius muscle*, which is essentially fast-twitch; thus the soleus is more naturally used in the longer enduring types of exercises. This physiological use is consistent with the dynamic and neurological properties of these fibers (Figure 4–4).

Several kinds of data lend support to the idea that specificity of training should be considered in regard to fiber types. This information suggests that there is selection of specific fiber types in specific types of exercise. For example it has been demonstrated that 83% of the fibers in the leg of a weightlifter and 78% of the fibers in the leg of a sprinter were of the fast-twitch types. In contrast, in the leg of a distance runner, and in the arm of a swimmer and a canoeist, the number of fast-twitch fibers ranged from 26% to 16%. It seems that the proportion of slow- and fast-twitch muscle fibers is different in muscles of the untrained, the endurance-trained, and those trained for power or strength. However, we should be hesitant in concluding that specific kinds of training will induce the fiber changes suggested above. Another possibility is that the demonstrated differences are genetic and that these particular athletes were successful because of the muscle-fiber composition of the muscles involved in the performance. The fact that no one has yet demonstrated a before-and-after training difference in fiber-type population may be due to an insufficient length of traing period.

Endurance-related properties of fiber types

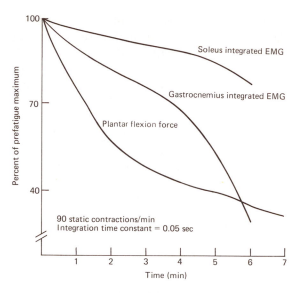

Figure 4–4. Fatigue of human gastrocnemius and soleus muscles.

Data collected by Robert Ochs and the Neuromuscular Research Laboratory, UCLA.

are responsive to training. For example, with endurance training, there is a decrease in fast-twitch glycolytic fibers. This means that endurance training augments the oxidative capacity of fast-twitch glycolytic fibers by increasing the mitochondria content of the fibers to the point where the fiber has a high oxidative capacity after training (Figure 4–5). This is not to be interpreted exclusively as a selective training of fast-twitch glycolytic fibers because, as we stated earlier, all muscle-fiber types are involved in most movements; and the amount of a training effect that occurs in a particular muscle-fiber type is basically a reflection of the frequency of use of the alpha-motoneuron. For example, if fast-twitch oxidative-glycolytic motor units are recruited twice as much as fast-twitch glycolytic units, the training effect will be about twice as great in the oxidative as in the glycolytic fast-twitch muscle fibers.

The question of how much voluntary control we have over the order of recruitment of

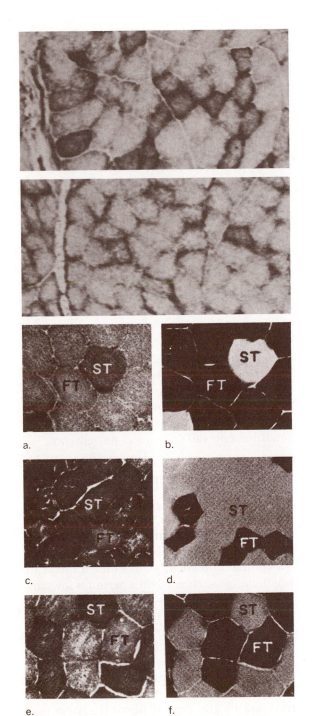

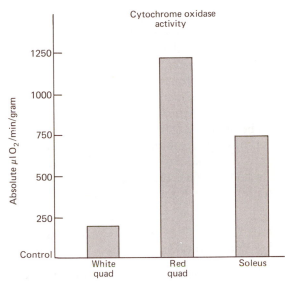

Figure 4–5. The top photos are cross sections of an untrained (left) and endurance-trained (right) rat muscle. The same is shown for humans in photos a–f: photos c and d show endurance-trained subjects; photos a and b were controls. Muscle sections (for rats, and photos c and e for humans) are stained for NADH-diaphorase, an indicator of mitochondrial oxidative ability. Photos b, d, and f are stained for myosin ATPase. The darkly stained fibers are fast twitch; the lightly stained fibers are slow twitch. The graph shows the relative effect of endurance training on each fiber type: white quad, FG; red quad, FOG; and soleus, SO fibers.

Photos, top left: Courtesy of V.R. Edgerton; L. Gerchman; and R. Carrow. Histochemical changes in rat skeletal muscle after exercise. *Experimental Neurology* 24:110–124, 1969. Photos, bottom left: Courtesy of P.D. Gollnick; R.B. Armstrong; C.W. Saubert; K. Piehl; and B. Saltin. Enzyme activity and fiber composition in skeletal muscle of untrained and trained men. *Journal of Applied Physiology* 33:312–319, 1972. Line art from K.M. Baldwin; R.L. Klinderfuss; R.L. Terjung; P.A. Mole; and J.V. Hollszy. Respiratory capacity of white, red and intermediate muscle: Adaptation response to exercise. *American Journal of Physiology* 222:373–378, 1972.

specific types of motor units is under investigation—with two divergent opinions prevailing. One opinion is that the order of recruitment is based on the size of the cell body of the alpha-motoneuron, with the smaller motoneuron (slow oxidative and fast oxidative-glycolytic) always activated before any larger one can be recruited (fast glycolytic). Others think that this fundamental recruitment sequence can be modulated voluntarily. The authors of this book believe that evidence for the latter opinion is much more convincing.

Does Chronic "Exercise" Cause Muscle to Hypertrophy?

Although a more thorough treatment of the effect of chronic exercise is discussed in Chapter 13, an overall view will be presented here. Exercise does not always induce hypertrophy of skeletal muscle fibers. It appears that endurance training causes minimal changes in muscle size; and it has been shown repeatedly that endurance-type training of a muscle causes little change in maximal strength. There is good evidence that selective but moderate hypertrophy of the more frequently used fibers occurs, while overall muscle mass does not change. This selective hypertrophy has been demonstrated in endurance-trained guinea pigs, in nonhuman primates, and in humans.

In essence, some hypertrophy of muscle fibers can occur with endurance training, but it is selective for the high-endurance fibers that are worked the most. Total muscle strength usually will not change—perhaps because the less frequently recruited fibers decrease in size or perhaps because the overall slight muscle enlargement is not due to the contractile but to the endurance-related components. The opposite holds true for strength-related training: there is an enlargement in the contractile but not the endurance components. This comparison again demonstrates the *specificity of the effect of exercise*.

Static and Dynamic Contractions

Muscles can contract either in a static (*isometric*) or dynamic (*isotonic*) manner. A static contraction is one in which there is no overall change in muscle length—such as when a person holds a weight in a steady position. Dynamic contractions are accompanied by muscle shortening and by limb movement—such as takes place when a person is moving a weight. Both types of contractions occur even in the most simple movements, as well as in more complex athletic events. Although a muscle may perform a static contraction during one phase of a movement and a dynamic contraction during another phase, we should realize that it is unlikely that a muscle is contracting wholly in either a static or a dynamic manner. Rather we should think of muscle contractions as falling on a continuum between the two extremes: zero shortening to maximal shortening. It is seldom, if ever, that the extreme conditions prevail in normal movements.

Concentric and Eccentric Contractions

Muscle contractions have also been described as being *concentric* or *eccentric*. Concentric contractions are those in which the muscle actually shortens as tension is developed. Eccentric contractions are those in which muscles are lengthening—such as when a person is lowering a weight against gravity. This type of contraction has also been referred to as *negative* work. The necessary strength for a concentric contraction—to lift a weight—is greater than the strength required for a comparable eccentric contraction—to subsequently lower the same weight. Consequently fewer motor units are recruited by the central nervous system, and less energy expenditure is required in eccentric contractions. This fact has been demonstrated by recording the electrical characteristics of muscles (*electromyography* or *EMG*) (Figure 4–6).

Muscle Speed and Tension

When a single electrical impulse of sufficient voltage is applied to skeletal muscle, either via the nerve (indirect stimulation) or via the muscle (direct stimulation), it responds with a characteristic contractile response called a *twitch*. A twitch is a single contraction of one or more motor units in a muscle. From the isometric *twitch tension curve*, the "time-to-peak tension" speed of the muscle can be determined (Figure 4–7) by the amount of time it takes the muscle to reach peak tension.

Another indication of muscle speed is the amount of time it takes to relax. Since the second phase of relaxation is slow, the rate of relaxation is usually measured as 1/2 *relaxation time* (1/2 RT). This term means the time required for the muscle to relax to one-half of its maximal twitch tension. The maximal tension produced by the muscle is the peak tension exerted by that muscle.

Muscle speed, as measured in *isotonic* contractions, is represented by the maximum velocity at which the muscle shortens. Theoretically this maximal speed occurs when there is no external resistance to the shortening (Figure 4–8). As loads are added to the muscle, the velocity decreases to the point where the load on the muscle equals the maximal force that the muscle is capable of exerting; this load is equal to the maximal isometric tension.

Muscle Length

The amount of tension a muscle can exert is related to the muscle length at the time of con-

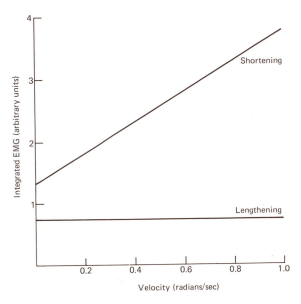

Figure 4–6. Integrated electromyography (EMG) at various velocities of movement in *concentric* (upper line) and *eccentric* (lower line) contractions.

From S. Bouisset, EMG and muscle force in normal motor activity. In *New Developments in Electromyography and Clinical Neurophysiology*, vol. 1 (J.E. Desmedt, ed.). Karger, Basel, 1973, pp. 547–583.

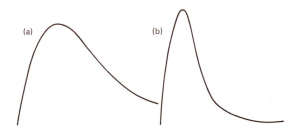

Figure 4–7. Twitch curves of slow (curve *a*) and fast (curve *b*) muscles. Note the greater time to peak tension and half relaxation time in curve *a* than in curve *b*. See Figure 4–2 for comparison of relative tension from slow and fast motor units.

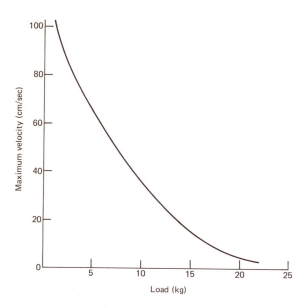

Figure 4-8. Hypothetical relationship between contraction velocity and load.

From S. Bouisset, EMG and muscle force in normal motor activity. In *New Developments in Electromyography and Clinical Neurophysiology*, vol. 1 (J.E. Desmedt, ed.). Karger, Basel, 1973, pp. 547–583.

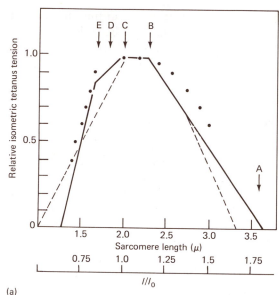

(a)

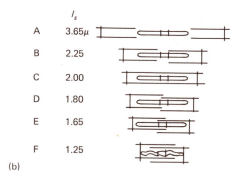

(b)

Figure 4-9. (a) Effect of sarcomere length on the tension developed by single muscle fibers (solid line and dots), and whole muscle (dashed line). (b) Note the overlap of myosin (thick) and actin (thin) filaments at various sarcomere lengths. Compare with diagram (a).

traction (Figure 4–9). It has been shown that, when a muscle is shortened to a point where the sarcomeres are only 1.9 microns long, the myofibrils in the center of the fiber are inactivated; the result is a decrease in tension production. There are several possible explanations for this loss of tension at very short sarcomere lengths. It appears either that the cross links of actomyosin filaments (contractile proteins) are disrupted or that the T-tubules and the sarcoplasmic reticulum are distorted to the point where the signal to release calcium is incapacitated. On the other hand muscle fibers can also be elongated to such a point that the actin and myosin overlap is reduced;

the result is lower tension. This length-tension relationship holds true for whole intact muscles as well as for isolated muscle fibers.

Motoneuronal Firing Patterns

The nature of the stimuli transmitted to a muscle determines the type of muscle response. If an initial stimulus is followed closely by another, before the muscle can fully relax, the two twitches summate. Furthermore the shorter the interval between two stimuli, the greater the summation. Whether several motor units are synchronously or asynchronously activated is an important determinant of the tension produced. The smoothness of the natural movement resulting from the muscular contraction is also dependent upon the patterned firing of the individual motor units (Figure 4–10).

The tension produced by a single twitch can be more than doubled in fast-twitch muscles and slightly less than doubled in slow-twitch muscles by applying successive stimuli at higher frequencies. When a train of impulses—with a frequency high enough to prevent the characteristic relaxation phase of the twitch—is sent to a muscle, the muscle is *tetanized* completely; that is, the twitch tension from one stimulus is unable to relax before the next contraction. Fast-twitch muscles must be stimulated at a greater frequency than slow-twitch muscles in order to be tetanized. The frequency at which *tetany* is reached is called *fusion frequency*. It corresponds to the stimulus frequency that prevents any relaxation at all and is a measure of the maximum force that can be generated by a given muscle (Figure 4–11).

Tetany rapidly induces fatigue in a motor unit. Tetanized muscle not only utilizes ATP at a maximal rate but also seriously curtails blood flow in the capillary beds. Fatigue is less evident when the repeated twitches are at a rate just below the frequency necessary to

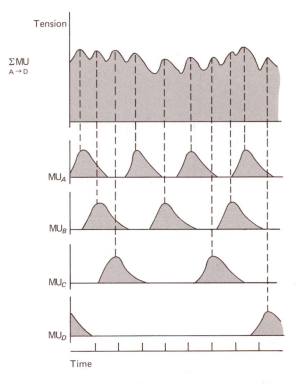

Figure 4-10. Note that single motor units (MU) firing at different times (MU_A-MU_D) can produce a relatively smooth and summated contraction, as shown at the top of the diagram. The total tension is greater than that for each separate motor unit.

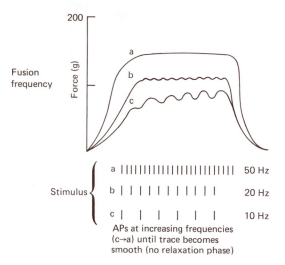

Fusion
frequency

Stimulus

a ||||||||||||||||||||||||| 50 Hz

b | | | | | | | | | | | 20 Hz

c | | | | | | | 10 Hz

APs at increasing frequencies
(c→a) until trace becomes
smooth (no relaxation phase)

**Figure 4-11. Effect of frequency of
stimulation on muscular force and the
attainment of tetany (curve *a*).**

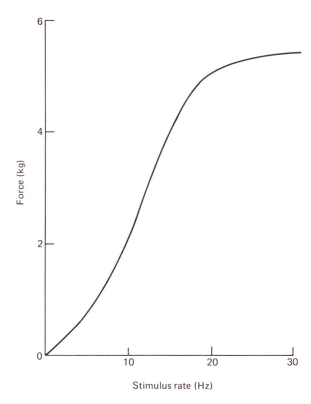

**Figure 4-12. Relationship between frequency
of stimulation of the ulnar nerve and the
resulting force generated (presumably by
the first dorsal interosseous muscle).**

From H.S. Milner-Brown, R.B. Stein, and R. Yemm.
Changes in firing rate of human motor units during
voluntary isometric contractions. *Journal of Physiology*
230:359–370, 1973.

cause fusion of individual contractions. In fact, in repetitive contractions, muscle fibers with a high oxidative metabolic capacity can maintain almost normal twitch-tension levels for several hours when an adequate blood flow is maintained.

Muscle Temperature

The mechanical response of a muscle is affected by the temperature of the muscle. When cooled below normal, the speed and maximum tension are decreased. When a muscle contracts, heat is produced from the breakdown of ATP to ADP + Pi + heat. Muscular temperature as high as 43°C has been recorded in heavily exercised animals. The first few contractions of a muscle seem to be enough to elevate the temperature, resulting in maximum tension increases. However, this relationship of muscle temperature and tension may be incidental; other factors might be the real explanation for

the rise in tension with successive twitch contractions.

All-or-None Principle of Muscular Contraction

Muscular contraction generally responds in an *all-or-none* fashion to muscular action potentials: when a muscle fiber contracts, it contracts fully—or not at all. If a nerve impulse travels down the axon from the cell body, all of the muscle fibers of that motor unit contract synchronously.

There are some interesting exceptions to the all-or-none principle concerning the contraction of skeletal muscle fibers. Individual twitches can be summated, as pointed out earlier; and rapid rates of stimuli augment the tension that a motor unit or single muscle fiber can produce (Figure 4–12). The reason for this summated tension production may be the incomplete return of Ca^{2+} to the SR between stimuli, resulting in more Ca^{2+} being available to the actomyosin ATPase and, consequently, greater tension.

Metabolic and Tension Responses to Training

Response to Endurance Training

A muscle of a trained individual can maintain a relatively greater tension over a longer period of time than a muscle of a nontrained individual. Over 20 enzymes in skeletal muscle have been shown to respond to endurance training. Most of these enzymes are associated with the metabolic reactions that are related in some way to the ability of the muscle to utilize oxygen. For example the enzymatic capacity of the citric acid cycle is augmented. This metabolic cycle provides hydrogen carrier molecules for the electron transport chain in the mitochondria, essential for the production of high-energy phosphate (ATP) in the presence of oxygen. The capacity of muscle to metabolize lipids is also responsive to chronic endurance training.

The greater enzymatic capacity of the endurance-trained muscle is reflected in the enhanced rate of oxygen uptake. This increase in total oxygen uptake can be explained partially on the basis of a greater mitochondrial mass per gram of tissue in the trained muscle.

A greater amount of mitochondria means that more mitochondrial enzymes are available for the total muscle to produce energy. Since the membranes of these mitochondria house these enzymes in well-organized aggregates, the ratio of individual enzyme concentrations tends to remain constant in the mitochondrial population of a given type of muscle.

Response to Training for Strength

Overloading skeletal muscle with high-resistant, low-repetitive exercises supplies a stimulus sufficient to induce muscle hypertrophy.

Although protein synthesis is augmented in both endurance and strength training, the difference is in the specificity of the response. During strength training, the enhanced synthesis is primarily specific for those proteins that are directly responsible for muscle shortening or for tension development. It is likely that myosin and actin filaments are added to myofibrils until the myofibril enlarges to such an extent that it divides into two more normally sized myofibrils. Therefore, a muscle hypertrophied due to weight training has essentially the same number of muscle fibers; but the individual fibers have enlarged by increasing the size of the myofibrils and/or by adding additional myofibrils within its membrane boundary.

There is some evidence from the study of trained rat muscle that muscle fibers can split and thereby give rise to a larger number of muscle fibers (Figure 4–13). This evidence must not be interpreted to mean that splitting occurs to a significant degree in a normal type of training program in humans. The evidence is convincing that increased cell number cannot totally account for the tension-related adaptation usually associated with strength training. No evidence is available on the effect of training on the proteins that play a regulatory role in muscular contraction, such as *tropomyosin*, *troponin*, and *actinin*.

The effect of a strictly strength-related training program on the endurance capability of a muscle is negligible. Although metabolic adaptations to this type of training have not been studied extensively, it is unlikely that significant energy-producing elements of the fiber are altered. For example activities of enzymes related to oxidative metabolism are not altered in response to strength training. Generally the glycolytic pathway seems to be unresponsive to muscular training—to either, *high-repetitive*, *low-resistant* endurance training or to *high-resistant*, *low-repetitive* strength training. Specific enzymatic adaptations to these training regimens were discussed in more detail in Chapter 3.

Indirectly the strength of a muscle can affect the velocity of shortening. As a result of increasing the strength of a muscle, the velocity of shortening will not diminish as rapidly when the resistance is increased. In effect the force-velocity relationship for a trained and nontrained muscle can be distinguished as shown in Figure 4–14. It is well known that maximal tension capacity is enhanced by the muscle hypertrophy accompanying strength training.

Response to Training for Speed

The maximal speed of muscular contraction does not seem to be affected by chronic-overload strength training; however, the question remains open at this time. If any change in muscular speed does occur, it appears that muscular speed is decreased. No changes in the proportion of slow- and fast-twitch fibers of guinea pigs, rats, or humans have been found with endurance training. Observations of similar parameters after a training program for power events are scarce.

The mechanism to induce any change in the speed of movement that may be realized as a result of training probably lies within the

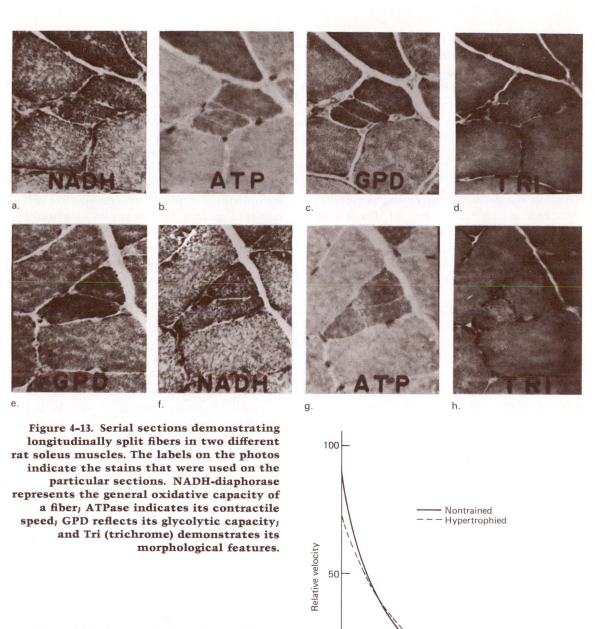

Figure 4-13. Serial sections demonstrating longitudinally split fibers in two different rat soleus muscles. The labels on the photos indicate the stains that were used on the particular sections. NADH-diaphorase represents the general oxidative capacity of a fiber; ATPase indicates its contractile speed; GPD reflects its glycolytic capacity; and Tri (trichrome) demonstrates its morphological features.

Figure 4-14. Force-velocity relationship in hypertrophied rat muscle compared to the control or nontrained muscle.

From R.A. Binkhorst and M.A. van't Hof. Force-velocity relationship and contraction time of the rat fast plantaris muscle due to compensatory hypertrophy. *Pflugers Arch.* 342:145–158, 1973.

nervous system. Such an effect, theoretically, could take place anywhere, from the supraspinal level of the central nervous system (CNS) to the neuromuscular junction.

Response of Muscular Tissue and Joints to Disuse and Use

MUSCULAR CHANGES WITH DISUSE

The rate of muscle atrophy during immobilization depends on: the degree of disuse compared with normal usage, the muscle-fiber type, duration of immobilization, and the fixed muscle length. Metabolic changes that result from long-term immobilization generally are minor. The glycolytic system, in large part, changes very little, if any. Glucose utilization is elevated in chronically immobilized muscles, as are oxygen uptake and blood flow. These changes probably reflect high catabolic rates of proteins, because amino acid uptake is depressed in immobilized muscles. After several months of immobilization, myoglobin and oxidative enzyme activities generally decrease or change in proportion to loss of muscle weight. Slow-twitch fibers seem to atrophy somewhat more rapidly than fast-twitch fibers.

The chronic (or resting) tension level and/or the resting length of a muscle determine, in part, the success of muscle-fiber maintenance during immobilization. Muscles immobilized in shorter than normal positions atrophy faster than muscles fixed in a stretched position. Athletic trainers should be aware of this potential differential atrophy in advising athletes during the rehabilitating process following immobilization. Clinically the immobilization should be such that the joints are fixed in a neutral position so that neither the agonists nor the antagonists atrophy severely.

Unusually high levels of circulating *glucocorticoids* can induce muscle atrophy. An analog of these hormones is used commonly in treatments of muscular and joint discomfort. *Cortisol*, a naturally occurring glucocorticoid, causes muscle atrophy by selectively lowering the availability of free amino acids to fast-twitch muscle fibers, thereby limiting the synthesizing capacity. However, a trained muscle is probably more resistant to the protein-degrading effect of glucocorticoids. The question of whether it would be advantageous for individuals undergoing rehabilitative glucocorticoid therapy also to engage in a training program has not been adequately studied.

Changes in the contractile machinery of muscles that are atrophied from disuse are to be expected. In slow-twitch muscle fibers, absolute maximum tension is lowered in the atrophied muscle because of loss of muscle mass—if not because of other complicating factors. The loss in muscle mass results from the loss of myofibrils within existing fibers. Tension losses in fast-twitch muscle fibers occur much more slowly than in slow-twitch fibers.

Another intriguing result of atrophy is the effect of immobilization on contractile velocity. In guinea pigs, disuse of slow-twitch muscle fibers results in an increased contractile speed, while the fast-twitch fibers tend to become slower with disuse. Figure 4–15 illustrates the effect of immobilization of the knee and ankle of a primate on the *vastus intermedius* muscle.

In an everyday situation, atrophied muscles are more susceptible to fatigue than they were prior to the muscle atrophy, since they tend to exert proportionally more than the usually required tension to perform a given task. Since the absolute tension it is capable of exerting is less, the atrophied muscle must perform at a level closer to its maximum—when it is called upon to carry its usual load. Consequently the rate of decline in tension (fatigue curve) will be greater. In cases of muscle atrophy due to immobilization for a few months, we usually

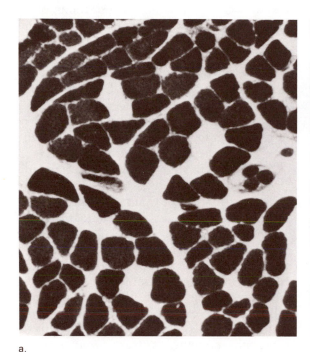

a.

b.

Figure 4–15. Effect of immobilization of a limb on muscle speed. The darkly stained fibers are fast twitch (high myosin ATPase activity); the negatively stained fibers are slow twitch (low myosin ATPase activity). Note the sparcity of slow-twitch fibers in the immobilized muscle (photo a) compared to the control (photo b). A muscle spindle can be seen in both muscle sections (middle right in a and lower right in b).

From V.R. Edgerton; R.J. Barnard; J.B. Peter; A. Maier; and D.R. Simpson. Properties of immobilized hind limb muscles of the *Galago senagalensis. Experimental Neurology 46:115–131, 1975.*

are not too concerned; the rate of recovery of normal motor activity of formerly atrophied muscle is rapid once mobilization is achieved.

Muscles of the limbs that are immobilized adapt quite rapidly to their fixed length so that an optimal amount of overlap of the myofilaments exists in the sarcomeres. If a muscle is fixed at a shortened length, it loses sarcomeres; if it is fixed in a lengthened position, sarcomeres are added. When a limb is immobilized in such a position that the muscle is at a shortened length, there are also concomitant changes in the surrounding connective tissues of the muscle. This new connective tissue serves to help prevent extreme stretching once the muscle is permitted its normal range of motion.

JOINT CHANGES WITH DISUSE

It is a common experience that long-term immobilization of a joint leads to a loss in range of motion. Extensive micro and macro structural changes are induced within a joint, such as the knee, after chronic immobilization. The range of motion limitation is due primarily to joint capsular and precapsular contractures. The joint becomes filled with fatty and fibrous tissues during lengthy periods (several months) of immobilization; in fact, the cartilage may be replaced by fibrous connective tissue. The bones lose their *osteoblastic* (bond-synthesizing) activity, and the patella may actually adhere to the femoral condyles.

A joint can practically be destroyed by immobilization lasting for a year. The actual cause of these immobilization-induced changes is due to a loss of nutrition to the cartilage and the synovial membranes of the joint. Forced manipulation of immobilized joints reduces the magnitude of the adherence of the joint tissues.

Immobilization of the joints of rats, lasting for only a few weeks, also causes weaknesses in ligamentous strength. This fact clearly suggests that, following joint damage and repair, the limb should be *mobilized* as soon as possible in order to minimize the disuse changes and to maximize the rate of repair of the ligaments. Consistent with the above findings is the observation that trained rats have stronger ligament-to-bone attachments than control rats. Both testosterone and estrogen seem to effect the strength of ligament-bone attachments. Testosterone has an anabolic effect on collagen metabolism that is consistent with its apparent effect on the strength of ligament-to-bone attachments.

Neuromuscular Fatigue

The site of neuromuscular fatigue is a common topic of discussion among exercise biologists.

Muscle fatigue can be associated with lack of glycogen, creatine phosphate, ATP, or oxygen; disturbance of functional connections of T-tubules and the SR; interaction of actin and myosin; simple redistribution of Ca^{2+} within the SR; and many other conditions. There is also good evidence that fatigue occurs at the neuromuscular junction (see Chapter 6). It is highly unlikely that the same fatigue factor or site exists in all circumstances; the nature of fatigue is undoubtedly related to the specific type of exercise and the individual involved.

The Ca^{2+} activation process of the actin and myosin system is a site of fatigue under some finite conditions. For example a single muscle fiber of a frog can be fatigued to the point where it can produce only 5% of its resting twitch tension, even though the electrical impulse reaching the muscle is not significantly impaired (electromyographic activity). These experiments demonstrate that neuromuscular fatigue can be caused by the inability of the *action potentials* of the sarcolemma to activate contractile proteins. It is analogous to a person's turning on an electrical switch leading to a light bulb in a lamp: if the bulb is unscrewed, there is no electrical linkage and therefore no light. The electrical plug into the wall would be analogous to the neuromuscular junction.

Other experiments suggest that a lack of glycogen, oxygen, creatine phosphate and/or ATP is not always a limiting factor since fatigued muscle can be induced to contract to the maximal prefatigue level by chemical activation (0.1 m KCl). Thus the immediate energy sources are sufficient to support a limited number of contractions.

It now is clear that low concentrations of glycogen in skeletal muscle correlate very well with the point of muscular exhaustion in humans exercised at a vigorous intensity (about 75% of their maximum oxygen uptake). The same dependence on glycogen can be demonstrated when electrically stimulating a single motor unit.

The results above point out a few of the specific locations within a muscle fiber that are susceptible to fatigue. It is likely that the *specificity of exercise* is again an all-important factor: the site of fatigue resulting from a specific exercise is specific to the particular demands of that exercise in that individual.

In unfatigued muscle, a direct relationship exists between EMG activity and force. A loss of electromyographic activity (relative to the force exerted) during fatigue supports the concept of neural involvement in neuromuscular fatigue. For example, as the loss of tension occurs during *maximal* and prolonged neuromuscular function, the EMG falls proportionately during the first minute (Figure 4–4). After that point the electrical activity plateaus, while voluntary force continues to fall (EMG/force ratio increases).

Prolonged maintenance of *submaximal* contractions results in a rise in integrated EMG, probably due to a loss of efficiency of muscular function of the active fibers; consequently more muscle fibers (therefore more EMG activity) are needed to maintain the same submaximal tension. The point at which EMG begins to increase markedly is closely associated with local impairment or fatigue of the muscle. This general muscular relationship between functional contraction and fatigue is suggested by the fact that the EMG/force ratio is higher in fatigued muscles even at low tensions (Figure 4–16). The rise in EMG is also associated with an accelerated progression of pain levels.

The ratio of EMG activity to force generation varies with the strength of the individual. Figure 4–17 illustrates that EMG/force ratio in the weaker subjects essentially means that a greater stimulus is needed for a given increase in force than is needed for the stronger subjects. This strength-to-EMG relationship is one reason why a stronger muscle can perform a

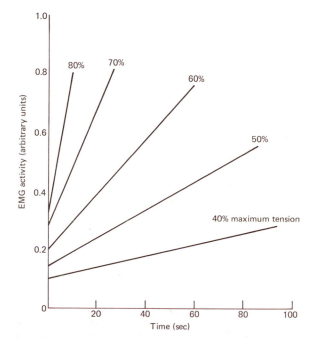

Figure 4-16. Note the rise in EMG level with maintenance of static contractions at percentages of maximal tension.

From J. Vredenbregt and G. Rau. Surface electromyography in relation to force, muscle length and endurance. In *New Developments in Electromyography and Clinical Neurophysiology* vol. 1 (J.E. Desmedt, ed.). Karger, Basel, 1973, pp. 607–622.

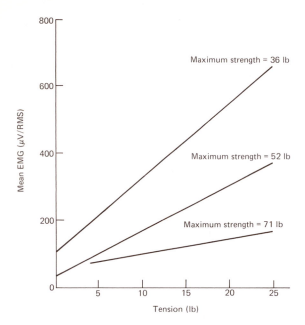

Figure 4–17. Mean EMG level of elbow flexors from human subjects who have different maximal strengths. Note the low rate of EMG increase with increasing tension in subjects with higher maximal strength.

daily chore for a greater duration: the stronger muscle is more likely to be working at a *lower* percent of maximum for that muscle group and therefore should fatigue less rapidly.

Summary

The contractile components of muscles are organized into functional entities called *motor units*. A motor unit consists of an *alpha-motoneuron* and of all of the muscle fibers innervated by that motoneuron. There are distinct physiological, biochemical, and morphological differences that collectively fit into a tripartite population distribution of motor units: (1) slow-twitch muscle fibers, which depend pre-

dominantly on oxidative metabolism; (2) fast-twitch muscle fibers, which rely predominantly on glycolytic metabolism as a method of providing adequate energy; and (3) fast-twitch muscle fibers, which have both a high oxidative and a high glycolytic metabolic capacity available to supply the energy needs. Slow-twitch oxidative and fast-twitch oxidative-glycolytic muscle fibers are relatively fatigue resistant; the fast-twitch glycolytic fibers are easily fatigued. One major advantage of fast-twitch glycolytic fibers is their ability to produce high forces rapidly (power). The fast-twitch fibers, which are fatigue resistant, can produce moderate forces. It appears that the slow-twitch motor units produce the least tension.

There is evidence of selective *atrophy* of slow-twitch muscle fibers resulting from *disuse*. Disused muscles and joints should be exercised as soon as possible after the disuse period for the quickest recovery. Selective, although slight, *hypertrophy* of slow-twitch fibers results from several months of *endurance training*.

When training for power events, fast-twitch fibers may selectively hypertrophy. Although a physiological endurance overload can induce fast-twitch glycolytic fibers to take on more of the metabolic characteristics of fast-twitch oxidative-glycolytic fibers, there is no evidence that a physiological overload, over a period of a few months, can cause muscle speed to decrease or increase. Muscle disuse of sufficient severity and duration causes slow-twitch muscle fibers to become fast-twitch; the mechanisms involved in these changes are unknown.

Muscular contractions may be classified into those that result in actual movement (*dynamic* or *isotonic*) and those that provide tension without marked movement (*static* or *isometric*). Isotonic means "same tension"; isometric means "same length"; practically all movements involve combinations of both of these contractions. Movement may also be classified in a third category, *isokinetic* ("same movement or velocity"). We can further categorize isotonic muscular contraction as *concentric* (muscular contraction during shortening) and as *eccentric* (muscular contraction during lengthening). Work done during these movements frequently is referred to as positive and negative, respectively. "Negative" work requires less muscular effort, consequently less of a metabolic demand, than "positive" work.

Muscular force is controlled by the varying of the number of motor units recruited, by the timing of recruitment of motor units, and by the frequency at which each motor unit fires.

The specific biochemical adaptations induced by training are dependent on the specific type of overload. Endurance-type exercise activates metabolic adaptations that enhance the oxygen-utilization capacity—or the synthesis of cellular components that enhance the ability of the muscle to provide energy. Power-type training activates adaptations related to the contractile components of the fiber, for example, myofibrils.

Neuromuscular fatigue may occur in the *contractile machinery* of the muscle fiber, in the *energy-production pathways*, at the *neuromuscular junction*, and also in the *central nervous system*. These possible fatigue sites are suggested by a number of experiments showing, for example, a drop in EMG activity with the development of fatigue while maximal tension is being exerted, and showing a rise in EMG with prolonged maintenance of submaximal tension.

Study Questions

1. What is a motor unit?
2. Describe the twitch characteristic of fast and slow motor units.

3. Describe the metabolic characteristics of fast and slow motor units.
4. How are the different motor units used during exercise?
5. What is the relationship between muscle-contraction force and motor-unit recruitment?
6. How does strength and endurance training influence motor units?
7. What factors determine muscle speed and tension characteristics?
8. How do synchronous and asynchronous firing patterns alter the muscle-contraction characteristics?
9. How can individual muscle fibers be summated?
10. How does endurance training effect the oxidative capacity of a motor unit?
11. What are some possible fatigue sites?
12. What can the EMG tell you about fatigue?

Review References

Buchthal, F., and Schmalbruch, H. Contraction times and fiber types in intact human muscle. *Acta Physiol. Scand.* 79:435–452, 1970.

Burke, R.E., and V.R. Edgerton. Motor unit properties and selective involvement in movement. In *Exercise and Sports Sciences Reviews* (J. Wilmore and J. Keogh, eds.) New York: Academic Press, 1975, pp. 31–83.

Close, R.I. Dynamic properties of mammalian skeletal muscles. *Physiol. Rev.* 52:129, 1972.

Edgerton, V.R. Exercise and the growth and development of muscle tissue. In *Physical Activity: Human Growth and Development* (G.L. Rarick, ed.). New York: Academic Press, 1973.

Edgerton, V.R., et al. Glycogen depletion in specific types of human skeletal muscle fibers after various work routines. In *Metabolic Adaptation to Prolonged Physical Exercise.* (H. Howald and J.R. Poortmans, eds.). Basel, Switzerland: Birkhäuser, 1975.

Gollnick, P.D., et al. Effect of training on enzyme activity and fiber composition of human skeletal muscle. *J. Appl. Physiol.* 34:107–111, 1973.

Hill, A.V. *First and Last Experiments in Muscle Mechanics.* Cambridge, England: Cambridge University Press, 1970.

Knuttgen, H.G. and K. Klausen. Oxygen debt in short-term exercise with concentric and eccentric muscle contractions. *J. Appl. Physiol.* 30:632, 1971.

Komi, P.V. Relationship between muscle tension, EMG and velocity of contraction under concentric and eccentric work. In *New Developments in Electromyography and Clinical Neurophysiology*, Vol. 1 (J.E. Desmedt, ed.), pp. 596–606. New York: S. Karger Press, 1973.

Peter, J.B., et al. Metabolic profiles of three fiber types of skeletal muscle in guinea pigs and rabbits. *Biochemistry* 11:2627–2633, 1972.

Saltin, B. Metabolic fundamentals in exercise. *Med. Sci. Sports* 5:137–146, 1973.

CHAPTER FIVE

Key Concepts

• The neuron is metabolically dependent on glial cells.

• Significant changes occur in neuronal morphology and biochemistry during a single bout of exercise and with training.

• The brain's internal environment has a special protective mechanism (blood-brain barrier) to prevent access of undesirable products that might alter normal neuronal function.

• Important supraspinal influences on alpha-motoneurons are the pyramidal tract neurons, red nucleus, vestibular nucleus, and the reticular formation.

• The cerebellum plays an essential part in the regulation of normal movements.

• The autonomic nervous system plays an essential role in the body's ability to adapt to a stress, such as exercise.

The central nervous system in movement control

Introduction

Neurons initiate and direct all movements including voluntary and involuntary movements as well as most supportive functions of movement. Consequently we need to gain some understanding of the fundamentals of *neuronal function* in order to gain insight into the mechanisms of human movement.

Gross and Microscopic Anatomy of the Nervous System

The nervous system can be viewed as consisting of a *central nervous system* (CNS) and a *peripheral nervous system* (PNS). Both the CNS and PNS are made up of *neurons*, which transmit messages to and from all tissues. It is also made up of *glial cells*, which serve a supportive role to the neurons. Structurally the CNS can be thought of as the cement (glial cells) between the rocks (neurons) of a walk-

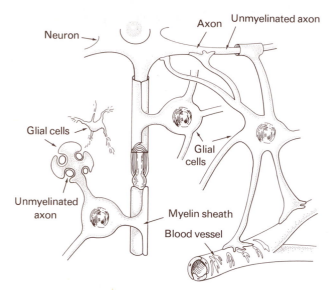

Figure 5-1. Schematic view of the relationship of glial cells to axons within the central nervous system.

way. Functionally the glial cells assist the neurons in assuring an adequate supply of certain critical chemical components essential for the neuron to maintain its functions. Examples of several types of glial cells are illustrated in Figure 5-1. In the PNS, glial cells also act as an electrical insulator for individual neurons, as does the rubber or plastic insulation around a copper wire in any electrical cord.

The Neuron

Structurally the neurons are unique—compared to other cells of the body—in that they are large and have long extensions (*axons* or *dentrites*) extending from the enlarged central portion (*cell body*) that contains the nucleus. The *dendrites* receive messages from other

neurons or from peripheral sensory receptors via special nerve endings or receptors. The sensory receptors may be sensitive to touch, smell, pressure, joint position, and so forth. The *axons* receive messages from the cell body and transmit them to other neurons or to special end organs, like muscle fibers.

A *motoneuron*, which innervates muscle fibers, has many dendrites extending from the cell body but only one axon. The axon is longer than its dendrites: for example, motoneuronal axons may reach from the cell body, located within the spinal cord, to the most distant muscles of the feet. The opposite is true of *sensory neurons*. For example, pain sensations from the feet are transmitted to the cell body, near the spinal cord, via dendrites. The axon is short since it connects to a neuron within the spinal cord, which then transmits the message to the brain. The volume of cytoplasm of the axons and dendrites is about 1000 times greater than that of the cell body, due to the great length and number of axons and/or dendrites.

The Impulse

In general terms a neural message is an *electrical event* that is self-propagating once it is initiated. The electrical event consists primarily of a movement of intracellular *sodium ions* to the extracellular compartments and of *potassium ions* from extracellular to intracellular positions. These transmembrane movements of ions cause a momentary decrease in the voltage difference between the inside and outside of the cell, a difference which exists at all times in neurons and muscle fibers. The electrical event is propagated from one position of the neuron to the next because the localized electrical activity excites the adjacent part of the membrane. This cell membrane consequently becomes more permeable to the movement of ions, and a reduction of the transmembrane voltage takes place in the membrane area.

The resting membrane is referred to as being *polarized* because it has a transmembrane voltage difference. The resting transmembrane voltage difference is only about 80 millivolts or 0.08 of a volt. At any given point along the neuron, the inside of the fiber is negatively charged (more negative ions than positive ions) with respect to the outside of the cell (more positive ions than negative ions). When the increase in permeability to ions occurs, the membrane becomes *depolarized*. The depolarization of the membrane reverses the electrical charge so that the inside of the cell becomes more positive than the outside.

Threshold

As stated above, the approximated voltage difference across neuronal membranes is 80 millivolts. If the membrane permeability to sodium and potassium ions is increased to the point where the voltage difference is reduced from −80 to −50 millivolts (threshold), the localized permeability change will cause the same electrical event in the adjacent portions of the membrane; the result is a self-propagating impulse that will be transmitted throughout the length of the neuron. This membrane voltage potential, recorded upon passage of the impulse, is called an *action potential* (Figure 5–2). Membrane-permeability changes of a lesser magnitude (subthreshold) will not result in a transmitted signal. The depolarization due to the passage of the action potential is localized and very brief, lasting only a few milliseconds.

Myelination

In the peripheral nervous system, there is a special type of glial cell, called a *Schwann cell*

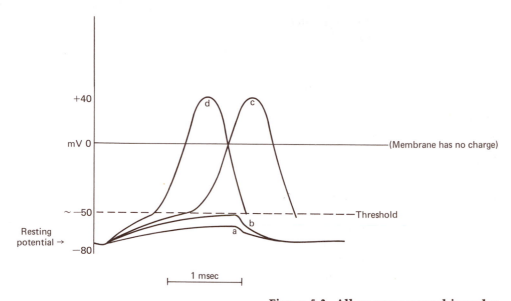

+40

mV 0 ————————————————————————(Membrane has no charge)

~ −50 ————————————————————————Threshold

Resting potential →

−80

|⊢—— 1 msec ——⊣|

Figure 5-2. All-or-none neural impulse conduction. Curves *a* and *b* are two different subthreshold stimuli. (There is not enough membrane depolarization, so there is no action potential.) Curves *c* and *d* are two different suprathreshold stimuli. (There is impulse propagation of the same magnitude.)

(Figure 5–3). The Schwann cell wraps itself around an axon, and the cytoplasm is forced into the outermost wrapping—much like a tube of toothpaste rolled from the bottom around a pencil. The interior wrappings, which contain little cytoplasm, are collectively called the *myelin sheath*. The multiple wrappings of a Schwann cell around a single axon electrically insulate the axon and enhance impulse-conduction velocity. Even axons that are not myelinated have a Schwann cell (but not multiply wrapped), which partially envelops the axon (Figure 5–3). The total peripheral length of the axon is insulated by a number of Schwann cells lying end to end. Each end-to-end juncture of two Schwann cells around a single axon is called the *node of Ranvier*. It is at these junctions that depolarization of the axon occurs, as impulses are transmitted down *myelinated fibers*. The jumping of the depolarizing wave from node to node proves to be one of the beneficial effects of myelinated fibers in augmenting conduction rate. The greater depolarizing and repolarizing activity at the nodal region

is reflected metabolically in a higher concentration of mitochondria in that region of an axon.

The myelinated neuron is specifically well adapted to conduction of impulses over great lengths. The rate at which the impulse can travel is related directly (1) to the diameter of the axon, (2) to the thickness of the myelin sheath and (3) to the distance between the nodes of Ranvier. *Internodal distance* is a factor since the impulse jumps from node to node (*saltatory*) rather than running smoothly down the cable, as an electrical current would run down a copper wire. Therefore axons with the myelinated sheath can conduct an impulse at a faster rate than equal-diameter axons without this insulation. The thickness of the myelin sheath is proportional to the axon diameter.

Environmental Influences on Neurons

The responsiveness of nerve-fiber diameter to use and disuse has been tested in animals. Some evidence suggests that nerve-fiber diameter decreases when the muscle atrophies and increases when the muscle hypertrophies or enlarges. However, more research must be done before this can be clarified. Direct measures of the conduction velocity of nerves leading to atrophied or hypertrophied muscle should be made in order to determine the potential physiological significance of changes in nerve-fiber size—assuming changes occur.

In adults, the number of neurons cannot be increased in the CNS; the number of functional *synapses* is more responsive to change. During the early developmental stages, a significant increase in functional synapses seems to occur in the CNS; conversely the number of synaptic contacts on neurons appears to decrease with advancing age. The effects of exercise, training, and inactivity, have yet to be investigated. However, a greater number of synapses were found in the iris of the eye of rats raised in environmentally enriched conditions; this evidence suggests that increased functional activity is directly related to neuronal growth.

When we realize the changes that can be induced in the brain by altering environmental conditions for only a few weeks, the plasticity of the brain is impressive. Rats that had spent four to ten weeks in either enriched or impoverished environments differed significantly. The enriched animals had (1) a greater cerebral cortex weight, (2) greater *acetylcholinesterase* activity (degrades acetylcholine, a synaptic neurotransmitter), (3) greater *cholinesterase* (mostly in glia and blood vessels), (4) more glial cells, (5) larger cell bodies and nuclei, (6) less DNA/mg of tissue, and (7) the same RNA concentration. The occipital region of the brain was the one most affected when enriched-

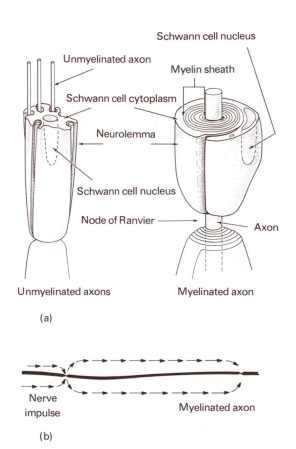

Figure 5-3. Diagram of (a) unmyelinated and myelinated nerves, and (b) saltatory nerve impulse between nodes on axon drawn in elongated position.

environment laboratory rats were compared to rats kept outdoors for one month.

Forced exercise apparently enhances amino acid uptake into proteins of the brain. Yet it seems certain that exercise, particularly exhaustive exercise, depresses amino acid incorporation into neuronal proteins; this may mean that exhaustive exercise will greatly prolong the usual time needed for metabolic recovery of the neurons. The known loss of RNA may be due either to a greater flow down the axons or to a decreased rate of synthesis. With exhaustive exercise, it may take more than two days before motoneuronal RNA is back to normal. *Training*, or daily sessions of moderate neuronal hyperactivity, elevates neuronal RNA, much of which is ribosomal.

The *pyramidal cells* (large neurons of the motor cortex of the brain involved in motor control) have been compared in rats of enriched and impoverished environments. These neurons normally have dendritic spines; however, these are more abundant in rats from enriched environments. The synaptic junctions are also about 50% larger in cross section, but there are fewer synapses.

Blood-Brain Barrier

The blood-brain barrier should be thought of as a physiological and morphological phenomenon that regulates the internal environment of the brain. The general lack of permeability between the vasculature of the brain and the extravascular tissue is demonstrated vividly by the lack of staining in the brain, unlike other tissues, after an intravascular systemic infusion of dyes. The barrier is very selective in the substances that are permitted to pass from the blood to the neuronal tissue. Our current understanding of the structural and physiological properties of the related structures is not sufficient to account totally for the effectiveness of the barrier.

The blood-brain barrier is readily permeable to *lactate*—an important consideration since the brain produces considerable lactate during *ischemia*. The lactate transport system of the membrane seems to be able to transport about three to four times as much lactate as that which is present in the resting subject. Although it is not known if the lactate carrier functions in both directions—blood to brain and brain to blood—this might be important since blood lactate levels can easily rise fivefold during strenuous exercise.

Other substances pass the blood-brain barrier readily. Glucose enters the brain rapidly; this transport is mediated by a membrane carrier. Some amino acids and long chain fatty acids are taken up rapidly by the brain. Generally, essential amino acids are more permeable to the blood brain barrier than nonessential amino acids: the six least permeable amino acids are nonessential amino acids.

Supraspinal Organization in Motor Control

The motor area of the outer layer (*cortex*) of the frontal lobe of the cerebrum consists of thousands of neurons, each of which may have up to 60,000 synapses to communicating neurons. Signals from these motor cells in the cortex can be radiated further by the interconnecting interneurons. Although full understanding of such a complicated communications system is beyond our current knowledge, some fundamental aspects of this network are known.

Descending Pathways of Motor Control

The nerve fibers that descend from the motor cortex collectively form a bundle of fibers called the *corticospinal tract* (Figure 5–4). As the name implies, most of these axons extend from the cell body of the motor cortex to the spinal cord level, at which point synaptic contact is made with an alpha-motoneuron. This important nerve tract is also known as *pyramidal tract neurons* (PTN). Other descending nerve tracts that are involved in motor control are: the *rubrospinal tract*, which originates in the red nucleus; the *vestibulospinal tract*, which originates in the vestibular nucleus; and the *reticulospinal tract*, which begins in the reticular formation. In general the majority of supraspinal neurons involved in motor control terminate indirectly on motoneurons. In other words, most neurons originating in the higher brain centers synapse, at the appropriate spinal cord segment, with an interneuron, which relays the impulse to the appropriate alpha-motoneurons. The interneurons further expand the effect of an impulse since this interneuron may synapse with several other neurons. This complex of neurons within the spinal cord has a remarkable ability to sustain normal locomotor functions even without supraspinal influences.

Physiological Motor Control Features of Supraspinal Neurons

PTNs differ in their firing patterns: some are *tonic*; others are *phasic*. The tonic PTNs tend to fire with a relatively low frequency but over longer periods than the phasic PTNs, which fire with short, rapid bursts of impulses. It is tempting to conclude that phasic PTNs converge on phasic alpha-motoneurons, which innervate *fast-twitch fibers* and that the tonic PTN synapse predominantly on alpha-motoneurons, which innervate *slow-twitch muscle fibers* (Figure 4–2); however, this has not been adequately demonstrated. The slow-twitch muscle fibers seem more efficient in maintaining certain postural positions (isometric contractions), whereas isotonic contractions are probably more efficiently performed by the fast-twitch muscle fibers.

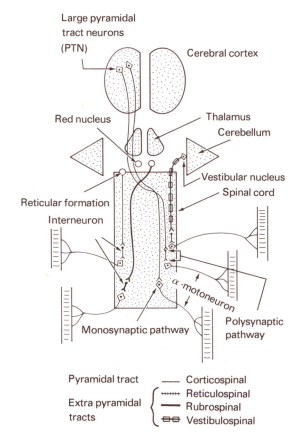

Figure 5–4. Supraspinal control of motoneurons.

Slow- and fast-twitch muscle fibers may be controlled differentially by the specific portions of the descending spinal tracts that are being excited or inhibited. For example, inhibiting the slower-responding antigravity postural locomotor systems, while simultaneously facilitating the more phasic motor units, could maximize one's potential speed of movement and power output. This inhibiting pattern is particularly prominent in the ankle and knee flexors; little inhibition is present in the forelimbs where antigravity muscles are of less concern.

Within flexor or extensor muscle groups, the smaller alpha-motoneurons generally fire before the larger ones. It seems that the action of supraspinal centers can override this general order of recruitment by inhibiting smaller and activating larger alpha-motoneurons when a quick, forceful movement is desired. In this situation the large PTNs, because of their higher stimulating thresholds, can fire prior to or without involvement of, the smaller PTNs, which have a lower stimulating threshold.

The PTNs receive numerous signals from other centers of the CNS, as well as from *proprioceptors* (environmental sensors: heat, pain, pressure, and so forth). The PTN "electrically evaluates" the total signal input and responds accordingly. In one sense the pyramidal tract neuron is to the supraspinal level of organization as the alpha-motoneuron is to the spinal level of organization.

Afferents (sensory nerves) from muscle spindles—one kind of proprioception—project to the cerebrum (motor cortex), providing one pathway through which information on the tension-length status of the muscle can be monitored by the cortex. This input is helpful for the regulation of voluntary neuronal firing patterns in a controlled movement and is necessary if maximal performance capacity is to be achieved.

Skill and Tension Control

Skill and dexterity in primates is related to the fundamental neurological organization. Intricate muscles of the hands and the more distal muscles of the forelimbs are under more direct cortical control than the larger limb muscles and the more proximal ones. PTNs are more likely to synapse directly (rather than indirectly through interneurons) on alpha-motoneurons of the more intricate muscles. Also there is less cortical inhibitory influence on the muscles that are involved in more intricate, skillful movements.

The graduated size of increments in tension output by different motoneurons is another factor in fine control of movement. For example the number of muscle fibers per motoneuron in some extra-ocular muscles, where precise muscular control is essential, is less than 10 compared to motor units consisting of almost 3000 muscle fibers in the *gastrocnemius*. Smaller motor units permit very fine increments in augmenting the tension produced by a muscle. This ability to exert a very fine tension control is more commonly found in muscles with a low total maximal-tension output capacity. Table 5–1 shows the number of motor units approximated in muscles of varying size and location in humans.

Cerebellum

Like other aspects of the CNS, the cerebellum's most fundamental operations defy our understanding. Hypotheses, however, have been presented that attempt to explain some aspects of movement control by the cerebellum. The cerebellum is intricately involved in fine muscular control and the coordination of movements. Malfunction of the cerebellum would make it difficult for a person to touch his nose with his finger. Excessive clumsiness is a com-

Table 5-1. Number of Motor Units in Muscles of Varying Sizes and Location

Muscle	Number of Motor Units	Number of Muscle Fibers	Number of Muscle Fibers/Motor Unit
First lumbrical (middle back)	100	10,000	110
First dorsal inter-osseus (finger)	120	41,000	340
Tibialis anterior (front of lower leg)	450	270,000	600
Medial gastrocnemius (calf)	580	1,030,000	1900
Brachioradialis (upper arm)	330	129,000	410

From B. Feinstein, B. Lindegord, E. Nyman, and G. Wohlfart, "Morphologic studies of motor units in normal human muscle," *Acta Anat.* 23:127–142, 1955. With permission of S. Karger AG, Basel.

mon symptom of cerebellar-lesioned subjects—as are movements involving simultaneous action in more than one joint.

Given the anatomical data currently available on the cerebellum and its interconnections, given the observations from electrophysiology of individual neurons, and given general animal behavior, it appears that: (1) once a movement is initiated, it will continue until the specific PTNs are inhibited directly or through a feedback loop via the medulla of the cerebellum; (2) the electrical patterns are such that the PTNs recognize patterns of input from the cerebellum feedback loops and determine the most appropriate subsequential firing patterns for themselves in order to execute a movement; (3) sensory and feedback loop corrections can occur within the time period required to be effective in the initial stages of a movement; and (4) the cerebellum is intricately involved in the learning of movement patterns.

The cerebellum receives impulses from a vast supply of muscle sensory receptors responsive to muscle stretch, tension, and pressure, as well as joint afferents, which provide information on joint positions. We should then ask just how important sensory (*afferent*) input is in motor control. It has been adequately demonstrated that motor performance is impaired when afferent input (sensory input) to the central nervous system is abolished. And yet we know that de-afferented monkeys learn and re-learn motor skills. Therefore we cannot conclude that most motor skills cannot be accomplished with limited afferent input, but it appears that the execution of these movements is completed with less ease and efficiency. We should imagine the sensory feedback channels as copiers of learning circuits—for the purpose of repeating various patterns of movement. Consequently, if these feedbacks are interrupted, we might expect some effect on the capacity to initiate, learn, or relearn skillful maneuvers.

Autonomic Nervous System

Many functions of the body are not normally controlled voluntarily, e.g., vasoregulation, glandular secretions, and heart rate. These functions are regulated by the checks and balances of the *autonomic nervous system* (ANS). The principle control of the ANS is in the *hypothalamus*; the ANS should be considered an integral part of both the CNS and the PNS. The autonomic system consists of *sympathetic* and *parasympathetic* components acting to balance one another. The sympathetic component is, in general, excitatory to its effectors (e.g., elevates heart rate); the parasympathetic is inhibitory (e.g., depresses heart rate). One may think of the sympathetic component of the ANS as concerned with energy expenditure and of the parasympathetic division as concerned with energy conservation. Interestingly enough, the parasympathetics are essential to life—the sympathetics are not. However, complete extirpation of the sympathetics would require marked metabolic readjustments to maintain homeostasis and would

necessitate supplementary protection against stresses.

Parasympathetic neurons regulate the lacrimal glands (secretion), the heart (cardiac deceleration), the eyes (pupillary constriction and near vision), the salivary glands (secretion), the lungs (bronchial constriction), the gut (contractions), the pelvic viscera (bladder and colon contraction, penile erection), and the skin (vasoconstriction, sweating). Sympathetic neurons assist in modulating vasoregulation, pupillary dilation and accommodation for far vision, secretion of salivary glands, cardiac acceleration and enhancement of stroke volume, bronchial dilation, inhibition of peristalsis of the gut, and secretion of catecholamines from the adrenal medulla.

Although the ANS is considered to be primarily an effector, it does respond to afferent feedback from autonomically regulated structures. For example, compensatory responses in the heart to elevated blood pressure are predominantly parasympathetic and mediated by a major parasympathetic nerve, *cranial nerve X* (the *vagus nerve*). The response to decreased blood pressure is mediated predominantly via the sympathetic part of the ANS. More specifically, the action of the ANS is through the action of a *reflex arc*. This reflex transmitted centrally induces an *effector* (*efferent*) *response*, which tells the appropriate organ to correct the temporary imbalance. High blood pressure detected by the *carotid sinus* (pressure-sensitive nerve endings at the carotid bifurcation) sends impulses to the cardiovascular center, which excites the parasympathetic vagus nerve. The vagus nerve, in turn, inhibits the heart rate and the strength of contraction, which reduces the amount of blood pumped by the heart—a sequence that tends to lower blood pressure to a more normal level.

The parasympathetic neurotransmitter released at the nerve endings of the structure being controlled is *acetylcholine*. However, the *catecholamines* (*epinephrine* and *norepinephrine*, predominantly the latter) are released at the sympathetic nerve endings. Some notable exceptions to this generalization are discussed in Chapter 9. The adrenal medulla receives stimulation via sympathetic impulses, which induce the release of catecholamines predominantly epinephrine) into the systemic circulation. The release of epinephrine (*adrenalin*) results in a mobilizing effect in the body (see Chapter 11). In most other cases, the direct effect of the ANS is very localized.

The Role of the ANS in Exercise

The autonomic nervous system not only serves a regulatory function but also, during exercise, assists the body metabolically to make energy sources available upon short notice. The autonomically-induced cardiovascular changes during exercise are consistent with the augmented needs of the muscles in terms of blood flow and metabolic rate. The ANS is regulatory in that it modifies skeletal as well as cardiac muscle tensions, and affects neuronal firing and digestive factors.

Physical activity of animals maintained in the laboratory is related to the circulating level of catecholamines. Depletion of catecholamines generally induces hypoactivity, while elevated catecholamine levels stimulate physical activity. The relationship of *specific dose response* catecholamine levels in the brain or blood to the level of physical activity has not been established. Amphetamines and cocaine elicit elevated dopamine (a biochemical precursor of the catecholamines) turnover in the CNS, which is thought to be a factor in elevated physical activity.

It has been suggested that the fluctuations in daily rhythm of brain norepinephrine are closely related to spontaneous physical activity and the sleep-wakefulness mechanisms. Exhaustive stress lowers norepinephrine in the hypothalmus, but its rate of resynthesis is higher during the recovery stage. In mice, chronic or acute stresses—such as daily exposure to fighting—lead to a more effective mechanism for utilizing norepinephrine.

The parasympathetic nervous system is involved in the training adaptation; it helps to induce *bradycardia* (slowing effect on heart rate) in response to training. This effect may be a reflection of elevated levels of acetycholine release, increased acetylcholine sensitivity, or more effective binding of acetylcholine to its receptors.

Summary

The *neurons* and *glial cells* of the nervous system are closely related to one another structurally, and the metabolism of the two types of cells is interdependent. A single bout of exhaustive exercise causes a number of marked motoneuronal changes, notably a loss of RNA, smaller cell body size, and enlarged nucleolus.

Principle movement effectors of the brain are the *pyramidal tract neurons* (PTNs) of the cortex. Other supraspinal motor control centers are the red nucleus, vestibular nucleus, and the reticular formation. The cerebellum plays an essential role in motor control and the learning of motor skills, utilizing proprioceptive feedback, integrating this information, and using it to program future movements.

The *autonomic nervous system* (ANS) controls many bodily functions that are supportive, but essential to movement. The ANS consists of a *sympathetic* and *parasympathetic system*, usually acting in opposition to one another to assure a balance in the functioning of heart rate, glandular secretions, and other "involuntary" functions.

Study Questions

1. What are the different type of cells found in the CNS?

2. Sketch the structural organization of the neuron.
3. What is the function of the "threshold level"?
4. What does the number of synapses tell you about the activity of a specific neuron?
5. How do motoneurons respond to training?
6. How is the supraspinal organization related to movement control?
7. Describe the role of phasic and tonic pyramidal tract neurons in movement control.
8. How is fine movement control different from the control of large muscle movements?
9. How is the cerebellum important in coordination control?
10. How does the autonomic nervous system influence movement?

Review References

Barondes, S.H. *Cellular Dynamics of the Neuron.* New York: Academic Press, 1969.

Castro, A.J. Motor performance in rats. The effects of pyramidal tract section. *Brain Res.* 44:313–323, 1972.

Eccles, J.C., *The Understanding of the Brain.* New York: McGraw-Hill, 1973.

Evarts, E.V. Relation of pyramidal tract activity to force exerted during voluntary movement. *J. Neurophysiol.* 31:14–27, 1968.

Evarts, E.V. Brain mechanism in movement. *Sci. Amer.* 229:96–103, 1973.

Granit, R. *Mechanisms Regulating the Discharge of Motoneurons.* Springfield, Ill.: Charles C Thomas, 1972.

Grillner, S. Locomotion in vertebrates: Central mechanisms and reflex interaction. *Physiol. Rev.* 55:247–295, 1975.

Orlovsky, G.N. Activity of rubrospinal neurons during locomotion. *Brain Res.* 46:99–112, 1972.

Rosenzweig, M.R., E.L. Bennett, and M.C. Diamond. Brain changes in response to experience. *Sci. Amer.* 226:22–29, 1972.

Tiplady, B. Brain protein metabolism and environmental stimulation. Effects of forced exercise. *Brain Res.* 43:215–225, 1972.

Key Concepts

• The synapse and neuromuscular junction play an essential role in movement control.
• Synapses and neuromuscular junctions adapt morphologically and physiologically to prolonged stimulation and to chronic disuse and use.
• There is evidence that what we classify as fatigue may occur in synaptic components of the cerebral cortex, at the spinal synapses, and/or at the neuromuscular junction.
• Depletion of vesicular sacs within the presynaptic nerve terminals may reach such a critical level that the synapse fails to respond to normal stimuli.
• The neuromuscular junction is an important element in the neurotrophic influence on skeletal muscle-fiber properties.

Introduction

Human movement is initiated by impulses originating in the central nervous system (CNS) and relayed through a chain of neurons to a muscle. A single neuron is arranged so that the impulse is received by the dendrites, passed through the cell body, and sent down the axon to some end organ (such as, muscle) or to the dendrites of the next neuron.

The transmitting junction between two neurons is a *synapse*. In mammals, synapses are characterized by a gap between two excitable cell membranes across which an impulse must be transmitted by a chemical substance called a *neurotransmitter*. The gap does not permit an impulse to be transmitted uninterrupted as, for example, an impulse is sent along an electrical cable.

Impulses are transmitted, unidirectionally, from an axon of one neuron to a dendrite of another. Actually the cell body and even the axon can receive input, forming what are

CHAPTER SIX

The synapse and neuromuscular junction in movement control

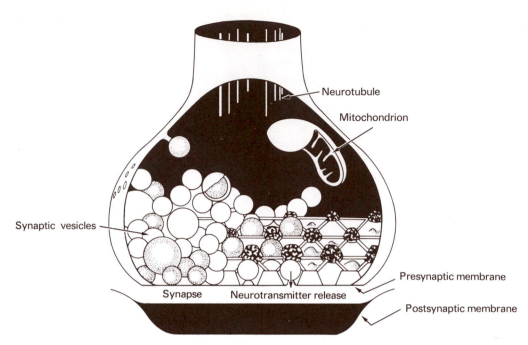

Labels on figure:
Neurotubule
Mitochondrion
Synaptic vesicles
Presynaptic membrane
Synapse
Neurotransmitter release
Postsynaptic membrane

Figure 6–1. Diagram of a nerve ending.

From K. Akert, K. Pfenninger, C. Sandri, and H. Moor. Freeze etching and cytochemistry of vesicles and membrane complexes in synapses of the central nervous system. In *Structures and Functions of Synapses* (G.D. Pappas and D.P. Purpua, eds.). New York: Raven Press, 1972, Fig. 15, p. 83.

called *axon-somatic* and *axo-axonal junctions.* The cell membrane of the transmitting axon forms the presynaptic membrane, and the membrane of the receiving neuron forms the postsynaptic membrane. The presynaptic terminal, or an axon, is characterized by the presence of many saclike structures called *vesicles* (Figure 6–1). Impulses traveling down the axon cause a release of the vesicular contents into the synaptic gap. This neurotransmitting chemical, released from the vesicles, crosses the synaptic gap, attaches to the postsynaptic membrane, and induces a permeability change to sodium and potassium ions in the receiving neuron. This altered permeability initiates the wave of depolarization from the postsynaptic membrane that will be propagated throughout the receiving neuron.

Synapse and Neuromuscular Junction

Neurotransmitters

The transmitter at the *neuromuscular junction* (NMJ) is *acetylcholine* (Ach). Some of the known potential transmitters in the CNS are epinephrine, norepinephrine, glycine, serotonin or *5-hydroxy-tryptamine* (5-HT), *gamma aminobutyric acid* (GABA), glutamate, and acetylcholine. The role of the catecholamines in the function of the autonomic nervous system (ANS) was discussed in Chapter 5. Acetylcholine is the transmitter released at all neuromuscular junctions in mammals.

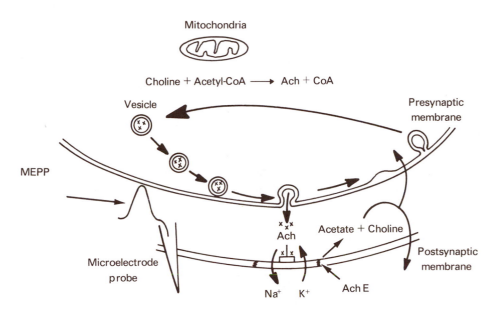

Mitochondria

Choline + Acetyl-CoA $\longrightarrow$ Ach + CoA

Vesicle

Presynaptic membrane

MEPP

Microelectrode probe

Ach

Acetate + Choline

Postsynaptic membrane

Na$^+$ K$^+$ Ach E

Figure 6-2. Summary of events of neuromuscular transmission. Vesicles make contact with the presynaptic membrane and the release of acetylcholine (Ach). Ach is synthesized from choline and acetyl CoA. It is hydrolyzed into acetate and choline which can be taken up by the nerve ending or muscle.

Adapted from J.I. Hubbard, Mechanism of transmitter release for nerve terminals. *Ann. N.Y. Acad. Sci.* 183:131–146, 1971.

Postsynaptic Membranes

The postsynaptic membrane is "specialized" at the synaptic area and is dense in relation to the nonsynaptic membrane areas. The high density of this specific region of the membrane may be caused by the presence of a special protein known as a *transmitter receptor.* It is this receptor that binds the transmitter, and it is this binding that initiates the change in membrane permeability to Na$^+$, K$^+$, and Cl$^-$ ions that, in turn, activates the postsynaptic propagated impulse. This transmitter receptor is very specific to the neurotransmitter released from the presynaptic ending (Figure 6-2). Without the receptor, the neurotransmitter is nonfunctional; in fact, it is this phenomenon that is exploited in the utilization of certain drugs for pain relief, muscle relaxants, cardiovascular control, psychotherapy, and so forth. Several reptilian venoms cause paralysis and subsequent death by permanently binding to postsynaptic receptors and preventing the normal effect of the neurotransmitter.

Functional Aspects of Synapses and the NMJ

Neurons are continuously being bombarded by excitatory impulses, resulting in the partial release of the neurotransmitter contained within the nerve. This low residual release causes a partial depolarization of the postsynaptic membrane. The release of the contents of a few vesicles at any given time is not usually sufficient to induce a depolarizing effect sufficient to initiate a propagated impulse or *action potential* in the postsynaptic membrane. As noted in Chapter 5, the magnitude of the localized or nonpropagated membrane depolarization is variable or graded and depends on the number of vesicles released at a given time and also on the spatial distribution of the vesicular release. For example the arrival of impulses from one or from even more than 800 synapses on a single large motor cortex cell may not be enough to induce it to fire (to propagate an *action potential*). However, if a specific threshold level is reached, there will be a propagation of the impulse (action potential) along the neuronal membrane. The subthreshold change is called an *excitatory postsynaptic potential* (EPSP). The amplitude of the EPSP is directly related to the quantity of the neurotransmitter release, to the number of vesicles that are stimulated to release their neurotransmitter, and to the size of the postsynaptic cell. EPSPs *summate* in an alpha-motoneuron only if they are generated within about 5 msec of one another. As might be expected, the summation may lead to an alteration in the membrane potential such that an action potential is more likely to be initiated. Thus the threshold EPSP level is reached as a result of facilita-

tion of varying magnitudes by several effector neurons and by indirect influences.

It should be recognized that neurons can be influenced by some neurotransmitters that are inhibitory: upon the release of the neurotransmitter by the inhibiting neuron, the postsynaptic membrane becomes *hyper*polarized rather than *hypo*polarized. An inhibitory effect on the neuron can also be achieved through a stabilizing effect on the membrane; that is, the membrane temporarily becomes less responsive to excitatory neurotransmitters. The hyperpolarizing neurotransmission effect is called *inhibitory postsynaptic potential* (IPSP). GABA (*gamma-aminobutyric acid*) is one neurotransmitter known to have an inhibitory effect.

In summary, interneuronal synapses are characterized by the following qualities: (1) they can transmit an impulse in only one direction; (2) a volley of impulses arriving at a presynaptic ending may or may not be sufficient to initiate a postsynaptic propagated impulse; (3) there is a finite time interval for the movement of the transmitter across the synaptic gap; and (4) presynaptic activity can hyperpolarize or hypopolarize the postsynaptic membrane.

Anatomy of the Neuromuscular Junction

The contact point of a motoneuron and the muscle fiber it innervates is called the *neuromuscular junction* (NMJ). As in the synapse, there is a gap between two cell membranes; however, the NMJ is a more specialized gap than the one characteristic of neuron-neuronal synapses. If we exclude some extraocular muscle fibers or muscle spindles, there is only one NMJ per skeletal muscle fiber. As the axon approaches the muscle fiber, the axon divides (where the final Schwann cell ends) into about three terminal branches or terminal axons (Figure 6–3). The terminal axons lie within depressions or synaptic grooves of the sarcolemma. These grooves are marked by numerous secondary furrows, called *subneural folds* (Figure 6–3). The secondary furrows are oriented perpendicularly to the synaptic grooves. These primary and secondary folds serve to increase the potential surface area of the postsynaptic membrane that can react to the neurotransmitter. The specific functional advantage for this morphologic feature is uncertain. The overall shape of the postsynaptic contact area is shown in Figure 6–4; its size in relation to muscle-fiber diameter and sarcomere length can also be appreciated from this photograph. The branching of a terminal axon (presynaptic membrane) and a nerve trunk is also shown in Figure 6–4. Numerous vesicles and mitochondria in each terminal axon can be seen in Figure 6–3.

The *motor endplate* (MEP) is the postsynaptic or sarcolemmal portion of the NMJ. The MEP contains an enzyme, *acetylcholinesterase*, which inactivates the depolarizing effect of acetylcholine by breaking it down into acetate and choline (Figure 6–2).

The NMJ structure is also specialized in regard to muscle-fiber type. In fast-twitch muscle fibers, the terminal axons are longer, but the muscle-fiber size is also somewhat larger. During muscle hypertrophy or atrophy, the size of the NMJ changes in proportion to muscle-fiber sizes. Another distinguishing feature between muscle-fiber types is the subneural postsynaptic folds; they are more complex and the vesicles are more concentrated in the NMJ of slow-twitch than of fast-twitch muscle fibers. Functionally there is more transmitter to be released for each action potential reaching the NMJ in the larger fast-twitch fiber. There also is about a 20% greater number of postsynaptic membrane receptor sites per postsynaptic membrane area in fast-twitch fibers.

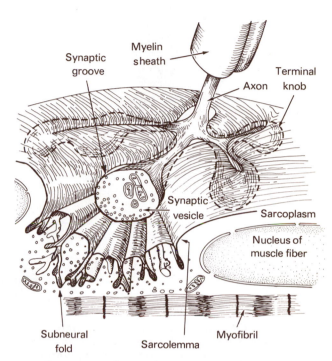

Figure 6–3. Schematic representation of a neuromuscular junction.

From H. Elias and J.E. Pauly, *Human Microanatomy*, 3d ed. Philadelphia: F.A. Davis Co., 1966.

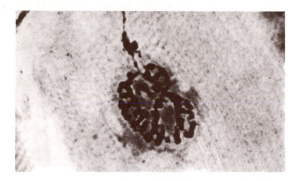

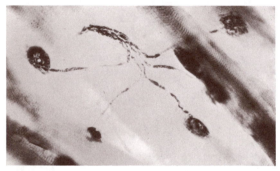

Figure 6–4. The top photo shows a single motor endplate with the axon leading to it. The sole plate or postsynaptic membrane is stained for gold chloride. In the lower photo, several motor endplates are demonstrated with the branching from a nerve trunk. (There is one motor endplate per muscle fiber.)

As in the case of the neuron-neuronal synapse discussed previously, small packets (*vesicles*) of neurotransmitter are released randomly at the NMJ; this release causes subthreshold alterations in the potential across the postsynaptic membrane. These events of the NMJ are called *miniature endplate potentials* (MEPP) and are similar to EPSP in that both are caused by the vesicular release of a neurotransmitter and both are facilitory and subthreshold. MEPPs occur randomly, even though the muscle fiber is at rest.

A postsynaptic membrane potential (either EPSP or MEPP) at or above the threshold level induces a propagated action potential in the nerve or along the muscle fiber. For example, if a sufficient quantity of Ach at the NMJ is suddenly released, the depolarization will induce a propagated depolarizing impulse down the sarcolemma. The depolarized wave will enter the muscle fibers centrally via the *transverse tubules* of the muscle fiber, initiating those events that lead to muscle contraction. Details of the process of muscle contraction were discussed and illustrated in Chapters 2 and 4.

Facilitation and Muscle Strength

From the earlier discussion in Chapter 4, you may recall that the tension produced by a muscle group depends upon: (1) the number and kind of motor units activated, (2) the frequency at which they are stimulated, and (3) the synchrony of recruitment. Normally we are unable to utilize all the motor units in a group of synergistic muscles at any one time. Therefore the maximum number of motor units that can be recruited at one time may be an important limiting factor in maximal strength or power performances. Perhaps the cortex can "learn" to increase strength by more effectively synchronizing motor units. The extent of cortical involvement in recruitment of motor units is evident by the intense concentration necessary by athletes preparing for maximal strength or for explosive events.

If a group of motoneurons is facilitated because of emotional excitement, greater muscular tensions are more easily attained. This is true because the emotional excitement produces a high level of electrical activity in the CNS, which means that more motoneurons are nearer their threshold for activation. Therefore a neural signal from the CNS is more likely to activate a larger number of motoneurons—and yield greater muscular tension. This same reasoning can be applied to the common observation that a tense person, when startled, jumps more vigorously than a relaxed person. Also "maximal" strength can be augmented if a sudden loud noise is heard when the muscular effort is being exerted.

The preciseness of motor control is similarly affected by general neuronal facilitation (*subthreshold hypopolarization*). Neuronal facilitation must be "accommodated" because most movements require a very narrow range of tension for optimal performance. In all performance activities, from those of a violinist to those of a shot-putter, the normal voluntary motoneuron recruitment pattern must be modified to compensate for the greater resting facilitation—that arises from the emotions of pre-competition stress. Without this modification, the force produced is likely to overshoot the requirement for that movement. Neural recruitment modifications are governed to an extent by the cortical network of the cerebrum, where repeated movement patterns are successively refined with the aid of feedback from the cerebellum. The cerebellum may be involved in selectively facilitating the appropriate motoneurons.

The Synapse and NMJ as a Factor in Fatigue

Neurons communicate with other neurons and muscle cells through synapses and NMJs, which are necessarily important in movement-control mechanisms during exercise. These junctions may act as filters or buffers to post-synaptic cells and help to prevent irreversible exhaustion of the effector cell. If synapses of the CNS or the NMJ become less effective with prolonged use, the effector cell is exposed to less stimulation. This would, in essence, be an example of neural fatigue, which protects the effector cell from extended use. Thus it would be possible to interpret neuromuscular fatigue, in this case, as an essential phenomenon in maintaining proper long-term functioning of the organism.

The basic question on this topic is, Does a cell (a muscle cell in particular) stop or reduce its level of performance because of the inherent limitations of that cell and its energy supply or because of inadequate neural messages to it? There is sufficient evidence available for us to state that transmitter depletion of terminal axons has to be considered one factor that limits the performance of skeletal muscle. This is not to imply that synaptic vesicles must be depleted in order for this effect to be possible—because only a small portion (1/500) of the total acetylcholine in the nerve terminal is immediately accessible for release. Consequently only a small loss of a *critical pool* of acetylcholine could limit NMJ function.

Electrical stimulation causes a marked vesicular depletion of presynaptic terminals in rat neurons. The area of the terminals nearest the presynaptic membrane shows the greatest loss of vesicles. In addition to these presynaptic effects of stimulation, there is also a decrease in sensitivity of the acetylcholine receptors in the postsynaptic membrane; however, recovery from this effect may be complete within one second of the termination of stimulation. Changes in the vesicular component of the rat cerebral cortex have been observed after stimulation. Although vesicle size and number appear to be stable, the number of large complex vesicles are decreased after stimulation. In effect, it seems very feasible that fatigue can occur at almost any level of the nervous system that is involved in motor control.

The neural element in neuromuscular fatigue is apparent when a person voluntarily works to such a point that the total muscular tension that can be generated decreases to practically zero. At this stage we can still elicit a vigorous stretch reflex (*patellar*). Following fatigue, induced by voluntary contractions, a near prefatigue tension can be elicited by direct electrical stimulation—which again suggests neural involvement.

The normal recruitment pattern of motoneurons during exercise is known to be altered at the onset of fatigue. For example, during near-maximal running, EMG activity decreases in the easily fatigable fast-twitch glycolytic fibers more than in the slow-twitch oxidative or fast-twitch oxidative-glycolytic fibers. The reduced EMG may result from decisions made in the CNS, inadequacy in synapses of the CNS, or "fatigue" at the NMJ.

A rise in EMG activity during neuromuscular fatigue has been reported by some investigators, and these experiments have been erroneously interpreted as evidence against NMJ fatigue. If, for example, we attempt to maintain a steady submaximal tension in a muscle, the EMG amplitude and intensity rise as the muscle tires (Figure 4–16). This rise can be explained by the reduction in the effectiveness of the contracting muscle fibers to produce tension. Consequently a neural compensation of recruitment of more *motor units* (increased EMG activity) is essential if the same muscle force is to be maintained. If, on the other hand, a maximal contraction is maintained—such that there are no new motor units to recruit—the exerted tension will decline, and the EMG activity will decrease. These experiments clearly suggest an element of neural fatigue.

Apparently both *neural* and *muscular* factors are involved in fatigue, and adjustments within the CNS are extremely sensitive to the performance capacity of a muscle. There exist mechanisms (barely known as of now) within the CNS for sensing when the muscle fibers of a unit are reaching their limits, and there appear to be alternative mechanisms that can be utilized to spare these muscle fibers.

Chronic Effects of Use and Disuse on the Synapse and NMJ

Larger NMJs are found in enduranced-trained than in nontrained animals. The terminal axons are longer, providing a greater synaptic area for the effective release of neurotransmitters. No one has yet determined whether or not the number, size, or concentration of vesicles is affected by training. The motor endplate area (sole plate, synaptic contact area of the muscle fiber) seems to expand in proportion to the increase in terminal axon length in hypertrophied muscle; this expansion implies that the acetylcholinesterase activity increases. After endurance training, during which muscle-fiber size changes are minimal, acetylcholinesterase activity increases significantly at the NMJ. The exact physiological consequence of these changes is not known, but these adaptations show that the NMJ is part of a dynamic and adaptable system and that the NMJ is in all probability stressed by chronic endurance exercise; otherwise it would not go through these adaptive changes.

Data on immobilized animals imply that the NMJ is functionally responsive to the amount of use of a muscle, since immobilization makes the muscle more susceptible to fatigue at the site of the NMJ. However, there are compensatory mechanisms that help to avert the greater disuse fatigue susceptibility of the NMJ: afferent responses are enhanced by more than 50% with prolonged muscle disuse. This enhancement makes any given afferent signal more effective on the motoneuron to which it may synapse. In this case a sensory adaptation helps to compensate for a decrement of the motor segment of the neuromuscular system.

"Trophic" Influences on Muscle Adaptation

We can no longer think of the NMJ as merely the site at which muscle fibers are depolarized by nerve fibers. Substances other than the neurotransmitter pass out of the terminal axons of the NMJ. The identity of these substances and how they induce their effects are questions of considerable importance to exercise biologists as well as to clinical neurologists.

It is a fact that the motoneuron has a great deal of control over the proteins synthesized in a muscle fiber. The characteristics of the motoneuron determine the speed of muscle contraction and the predominate form of metabolism (number of mitochondria, and so forth) of the muscle fibers innervated by that motoneuron. The motoneuron also determines how often the muscle will contract and, by some mechanism, influences how much enzymatic protein will be available to provide the supportive energy metabolism. Consequently growth and development of muscle and training of a muscle may be controlled largely via the final common neural pathway, the *alpha-motoneuron.* In other words, the rate and kind of muscle development in growing children and the nature of an exercise training effect (enhanced capacity of a trained muscle to utilize oxygen or augmented muscle strength) may be mediated and controlled by chemical substance or substances within the alpha-motoneuron.

The concept of neurotrophic influences is attractive because of the continuous process of bidirectional axoplasmic flow (distally and centrally) in motor axons. Proteins, amino acids, mitochondria, ribosomes, and practically all cellular organelles systematically flow along the axon at a rate generally specific to each organelle. In large part, neural substances fall into a category of slow (few mm/day) or fast (100 mm/day) axoplasmic transport.

The neural influence on muscle may be only a manifestation of the frequency (or *pattern*) of stimuli by the alpha-motoneuron and not some neurotrophic substance. At the present time, no one has been successful in suitably separating the neurotrophic "substance" hypothesis from the "impulse frequency" or "pattern" hypothesis. Some partially successful attempts are described in Chapter 4 in relation to the change in muscle-fiber types with training and in Chapter 13 in relation to factors that affect muscle growth.

If the controlling mechanisms (which seem to be neurally influenced) of muscle growth, development, and adaptation were known, more efficient programs could be developed to assure the proper growth and development of the neuromuscular system. Furthermore we could maximize our efficiency for rehabilitating muscular diseases, augment normal neuromuscular systems to perform maximally in routine daily life, and train, when desired, for ultimate physical performance.

Summary

A *synapse* is a specialized region of a neuron-to-neuron apposition where impulses are transmitted. A *neuromuscular junction* (NMJ) is the apposition of a motoneuronal axon to a muscle fiber over which the neuronal signal is transmitted. These specialized cellular junctions are characterized by a *gap* across which a neurotransmitter from the presynaptic cell diffuses and induces an alteration in the postsynaptic membrane potential by varying its permeability to specific ions. The neurotransmitter is stored in membranous sacs called *vesicles.* The transmitter may *depolarize* (facilitate) or *hyperpolarize* (inhibit) the postsynaptic cell. If the depolarization is large enough and reaches threshold, an *action potential* (impulse) will be initiated in the postsynaptic cell.

Precompetitive excitement is a state of general *neuronal facilitation;* consequently motoneurons are more apt than usual to be activated by a given stimulus. The synapses of the cortex, the synapses involved in tendon reflexes, and the neuromuscular junction can be a source of fatigue under a given circumstance. There is also evidence that the postsynaptic membrane of the NMJ becomes less responsive with repetitive stimulation.

Chronic use and disuse affect synapses. Even though the NMJ is more susceptible to fatigue after prolonged disuse, *monosynaptic* (*sensory*) responses on motoneurons are enhanced after prolonged disuse. In general the size of the NMJ is proportional to muscle-fiber size in normal, atrophied, and hypertrophied muscle fibers. However, there are anatomical and functional differences in the NMJ of the different muscle-fiber types.

The NMJ seems to be important in ways other than the transmission of impulses. The motoneuron has a definite influence on the amount of protein synthesized in skeletal muscle. This influence may be exerted in the form of neuronal substance or substances that are exuded from the nerve endings or may be related to the neural firing patterns transmitted to the skeletal muscles.

Study Questions

1. What is the function of the neurotransmitter in the central and peripheral nervous systems?
2. Discuss the postsynaptic membrane and the depolarization that occurs.
3. Draw and label a neuromuscular junction.
4. How are the neuromuscular junctions related to fiber type?
5. How is facilitation related to muscle strength?
6. How does EMG activity correspond to fatigue?
7. How does training affect the neuromuscular junction?
8. Discuss the possible "trophic" effects of nerve on muscle.

Review References

Bloom, Floyd E.; Leslie L. Iversen; and Francis O. Schmitt. Macromolecules in synaptic function. *Neurosciences Res. Program Bull.* 8:325–455, 1970.

Guth, Lloyd. "Trophic" influences of nerve on muscle. *Physiol. Rev.* 48:645–687, 1968.

Jones, D.G. and H.F. Bradford. Morphology of synaptosomes. *Brain Res.* 28:491–499, 1971.

Kaiserman-Abramof, Ita R. and A. Peters. Some aspects of the morphology of Betz cells in the cerebral cortex of the cat. *Brain Res.* 43:527–546, 1972.

Korneliussen, H.; J.A.B. Barstad; and G. Lilleheil. Vesicle hypothesis: Effect of nerve stimulation on the synaptic vesicles of motor endplates. *Experientia* 28:1055–1057, 1972.

Kowower, E.M. and R. Werman. New step in transmitter release at the myoneural junction. *Nature New Biol.* 233:121–122, 1971.

Pappas, G.D. and D.P. Purpura. *Structure and Junction of Synapses.* New York: Raven Press, 1972.

Perri, V., et al. Evaluation of the number and distribution of synaptic vesicles at cholinergic nerve-endings after sustained stimulation. *Brain Res.* 39:526–529, 1972.

Phillis, J.W. *The Pharmacology of Synapses.* Oxford, England: Pergamon Press, 1970.

Thesleff, S. Motor end-plate desensitization by repetitive nerve stimuli. *J. Physiol.* 148:659–664, 1959.

CHAPTER SEVEN

Key Concepts

• Muscle spindles are specialized muscle structures that are sensitive to muscle length, tension, and shortening and lengthening velocity.
• The contractile capability of spindles provides a mechanism whereby muscle length and action can be monitored during muscle contraction by maintaining a constant tone on the muscle spindle.
• Spindles adapt morphologically and physiologically to disuse and to overload. Atrophied and hypertrophied muscles are more sensitive to stretch than normal muscles.
• Tendon organs, joint receptors, and Pacinian corpuscles are other sensory receptors that affect the output of alpha-motoneurons and, consequently, can affect movement control.

Sensory components of movement control

Introduction

Approximately one-half of all nerve fibers in the nerve trunks that connect to voluntarily controlled muscles are *afferent* or *sensory*. This fact in itself should alert us to the importance of the sensory information collected by skeletal muscles and transmitted to the CNS. We do not mean to suggest that movements cannot occur without an intact sensory component; however, it has been demonstrated that the utility of a limb is impaired without these receptors.

In addition to our awareness of the well-known senses of pain, pressure, and temperature, we are also aware, perhaps subconsciously, of joint angles or positions, muscle tension, muscle length, and perhaps velocity of movements. We might ask why such information is critical to our ability to run rapidly, jump a hurdle, or walk a straight line. It is important for the CNS to obtain this instantaneous information in order to orchestrate a fluent movement. The sensory data must be collected and integrated by the CNS so that several groups of muscles (thousands of motor units) are contracting and relaxing in a precisely timed manner. In other words, not only must a given muscle contract but also it must contract in relation to what other muscles are doing—and act in such a way that all muscle-fiber contractions are properly coordinated. In addition the CNS must interpret the sensory inflow from the muscles and joints and subsequently determine what is to be done. To summarize, the muscle must receive information from the CNS on: (1) itself, (2) other muscles, both synergists and antagonists, (3) what these other muscles are doing, and (4) what must be done next in order to continue the desired movement.

Muscle Spindles

Muscle spindles are very specialized receptors that are sensitive to muscle length and tension, and probably to velocity of muscular contraction. The muscle spindle provides the initial source of information that simultaneously can lead to an inhibition of some muscle fibers and to a facilitation of others. A common example of its function may be seen in the tendon *stretch reflex*, which is usually tested by tapping the patellar tendon crossing the knee joint and by observing the subsequent extension of the knee. This reaction is eliminated by abolishing the spindle's sensitivity to muscle stretch.

To understand how the stretch reflex occurs, a basic awareness of the morphology of a muscle spindle is required. The relative size and the distribution of muscle spindles are illustrated in Figure 7–1. Note that the largest diameter of a spindle is almost as large as a skeletal muscle fiber. The widespread distribution of spindles within a muscle suggests that the whole muscle (or segment of the muscle) is sensitive to muscle stretch. More slowly contracting

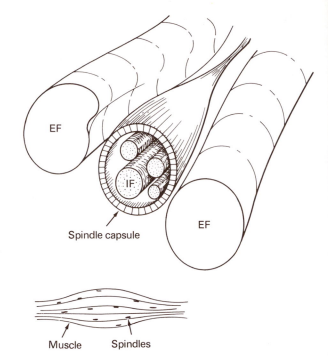

Figure 7–1. Encapsulated muscle spindle fibers (IF) between two extrafusal fibers (EF).

muscles have a greater number of spindles per unit of muscle than the faster muscles.

An examination of the structure of the spindle itself reveals an intricate complexity. Figure 7–2 illustrates a spindle-shaped sheath consisting of connective and *epithelial tissue* (cells that line all surfaces, such as the skin, intestinal wall, inside of blood vessels, tissues, and nerves). Inside the spindle-shaped sheath there are about six muscle fibers that are approximately one-fifth the diameter of muscle fibers outside the spindle. Muscle fibers within the spindle generally are called *intrafusal* (*fusus* is Latin for spindle) as opposed to the *extrafusal* muscle fibers, which, until now, we have referred to as skeletal muscle fibers. When the type of muscle fiber is not specified, we will be referring to the extrafusal fibers.

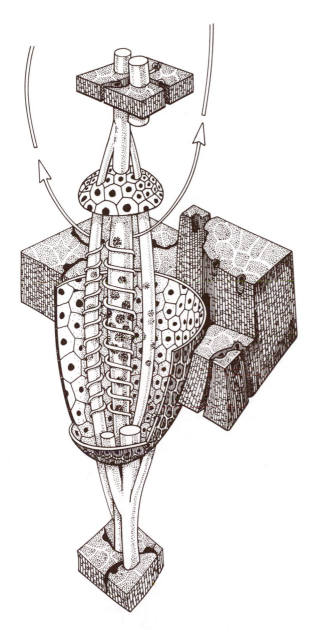

Figure 7-2. Morphological aspects of a muscle spindle, including the epithelial capsule and the primary sensory nerves wrapping around intrafusal fibers.

From J.P. Schadé and D.H. Ford, *Basic Neurology*, 2d ed. Elsevier, Amsterdam 1973, Fig. 164, p. 197.

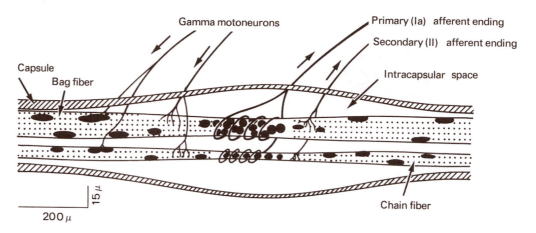

Figure 7–3. Diagram of a muscle spindle showing afferents and efferents.

Courtesy of Alfred Maier, University of Alabama.

There are two kinds of intrafusal fibers. One is relatively large with several nuclei distributed throughout its diameter (called *bag fibers*); the smaller one (*chain fibers*) has nuclei aligned along a single column at the midlength of the intrafusal fiber (Figure 7–3).

At the midlength position of the spindle, there is little, if any, contractile protein. The muscle fiber is multiply wrapped by a nerve fiber that has a large diameter and, consequently, can conduct impulses relatively rapidly. This nerve fiber, known as the Ia or *primary* afferent nerve, is sensory and directs impulses to the spinal cord where it makes direct contact to alpha-motoneurons and interneurons. This sensory nerve ending on the spindle is sensitive to stretch of the midregion (or *equator*) of the intrafusal muscle fiber. Stretch of the intrafusal muscle fiber in this way is caused by a lengthening of the extrafusal muscle.

Another sensory component of the spindle consists of the II or *secondary* afferents. These endings are more localized on the chain than bag fibers. Secondary endings are less sensitive to muscle stretch than primary endings.

The significant difference between primary and secondary endings lies in their CNS connections. When the primary endings are activated, initiated by stretch, and the impulse transmitted up Ia fibers to alpha-motoneurons, they facilitate the same muscle (and selected synergestic muscles) in which the stretch occurred, while antagonistic muscles are inhibited (Figure 7–4). This represents the classical stretch reflex.

It is postulated that secondary endings program the CNS for repetitive movements or perhaps for more long-lasting static contractions, such as during an isometric exercise. Activation of secondary endings by a stretching of the knee extensors inhibits the motoneurons to the extensors and simultaneously facilitates its flexors (Figure 7–5).

From what has been already discussed, two questions present themselves. How can these endings be important for the neural control of movement if they are responsive only to passive muscle stretch—when it is known that movement necessitates active muscle shortening? Second, why does the spindle need intra-

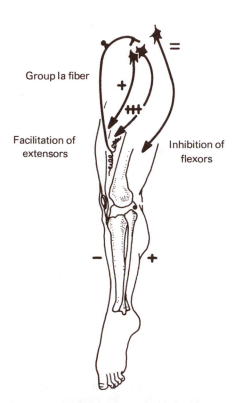

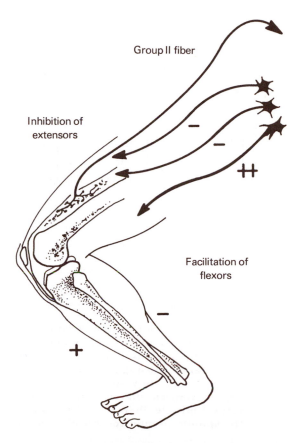

Figure 7-4. Reflex effects of stimulation of group Ia fibers in the vastus intermedius muscle. Strong facilitation of the same muscle and weaker facilitation of synergistic extensors at the same and nearby joints occurs. Concurrently, the antagonist flexors are inhibited.

From E. Eldred, The dual sensory role of muscle spindles. *Journal of American Physical Therapy* 45:290–313, 1965.

Figure 7-5. Effect of knee extension on the discharge of secondary endings. These impulses inhibit the same muscle and synergists while the flexors are facilitated. Thus, the group I and group II afferent nerve fibers have opposite effects.

From E. Eldred, The dual sensory role of muscle spindles. *Journal of American Physical Therapy* 45:290–313, 1965.

fusal muscle fibers that can contract or shorten? Figure 7–6 illustrates what would happen to the tension within an intrafusal muscle fiber if only the extrafusal fibers were to contract actively; obviously this would unload the spindle and shut off any feedback to the CNS concerning muscle tension. Fortunately this does not occur during normal conditions since the

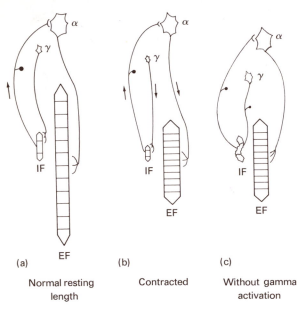

(a) Normal resting length   (b) Contracted   (c) Without gamma activation

Figure 7-6. The function of the gamma (γ) motoneuron is to maintain a constant tension on the intrafusal muscle fibers (IF, spindle) by causing them to contract in response to alpha (α) motoneuron-stimulated contraction of extrafusal muscle fibers (EF). This allows the spindle to monitor muscle length.

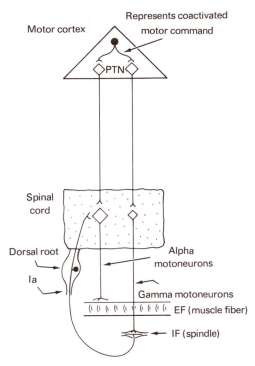

Figure 7-7. The alpha-gamma motoneuronal coactivation by parallel supraspinal influences (PTN: pyramidal tract neuron) is shown.

spindle has motor as well as sensory components. If the intrafusal muscle fibers shorten in proportion to the shortening of the extrafusal ones, the tension on Ia and II endings could be maintained at any stage or degree of extrafusal contraction. Indeed this is what happens.

The intrafusal fibers contain actin and myosin filaments (and probably the identical regulatory proteins of contractions as the extrafusal ones) at both ends of the fibers. The contractile regions are innervated by small motoneurons compared with the size of those that innervate extrafusal fibers (Figure 7-3). These small motoneurons are called *gamma* (γ). It is now known that gamma- and alpha-motoneurons are controlled in such a way that when an ex-

trafusal muscle is instructed to shorten, so are the corresponding intrafusal muscle fibers. This phenomenon is referred to as alpha-gamma *coactivation* (Figure 7-7). Therefore, rather than extrafusal muscle shortening resulting in an unloading of the tension on the spindle sensory nerve endings, as would happen if there were no gamma system, these endings continue to be effective in monitoring muscle tension, length, and velocity of shortening within the whole range of motion of a limb. Actual recordings of Ia impulse frequency, muscle force (*torque*), and EMG (*alpha-motoneuronal output*), resulting from isometric voluntary contractions of the index fingers of humans, are shown in Figure 7-8.

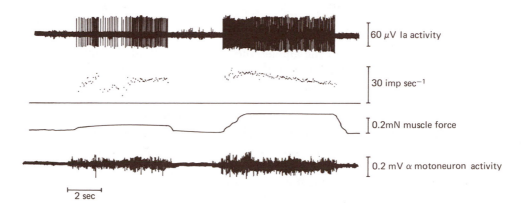

60 µV Ia activity

30 imp sec^{-1}

0.2mN muscle force

0.2 mV α motoneuron activity

2 sec

Figure 7-8. Sensory discharge for two isometric voluntary contractions of muscles to the index finger. The actual impulses, the instantaneous frequency of impulses/sec, the *torque* (muscle force), and the EMG are shown in order from top to bottom.

From A.B. Vallbo, Muscle spindle response at the onset of isometric voluntary contractions in man. Time difference between fusimotor and skeletomotor effects. *Journal of Physiology* 218:405–431, 1971.

As we said earlier in this chapter, human movement is possible without gamma-motoneuronal influence. However, its influence on complex, rapid movements is an important phenomenon: there is a marked impairment of coordinated movements, due to a loss of muscle speed, when the gamma system is blocked pharmacologically. Also the dampening effect of antagonists at the end of a movement (such as throwing) is delayed; that is, the antagonists do not fire at the optimal moment to slow down the arm.

Surgical deafferentation of animals causes conspicuous modifications in their normal gait, but this radical treatment eliminates *all* sensory feedback—not only that from the spindle. Also we can apparently maintain a standing position when the gamma system is blocked; however, the postural sway is magnified—probably due to the loss in sensitivity (hyperpolarizing effect) of alpha-motoneurons ordinarily maintained by tonic (minimal but continuous) gamma activity. When monkeys' forelimbs are deafferented before birth, they do not use their hands in climbing or for picking up objects after they are born. But if deafferentation occurs after adolescence, monkeys do use their forelimbs; thus, we might infer that the movements learned in early life can affect movements as an adult.

The facilitory effect of the *fusimotor system* on alpha-motoneurons is shown by the elevation of the threshold levels of the alpha-motoneurons involved in the tendon reflex after procaine treatment (an anesthetic that can selectively block the gamma effect). In fact, when the gamma system is completely blocked, the stretch reflex, the knee jerk described earlier, cannot be elicited. Thus some gamma bias or influence on the threshold level of the alpha-motoneurons is essential for some normal muscular movements.

The influences of the sensory endings, working in conjunction with the gamma system, probably play a significant role in making a simple walking movement automatic. For example, when the hamstrings are stretched during a walk, they are induced to relax further by inhibitory feedback signals—originating from the sensory secondary endings of the spindle—on the alpha-motoneurons of that muscle.

Neuromuscular Relaxation and the Spindle

We can gain a general impression of a person's state of alertness, tenseness, or relaxation by an analysis of his actual state of neuronal activation. A brisk tendon-tap stretch reflex (knee jerk) is characteristic of a highly tense person. This hypertense behavior can also be caused by hormonal imbalances (such as *hyperthyroidism*). A conscious effort to relax skeletal muscles, or go to sleep, can effectively depress the response to a tendon tap. Relaxation brings about a state in which a person's general musculature is less excitable—and less responsive to unexpected environmental stimuli. Consequently the stretch reflexes are very reliable

measures of the individual's state of relaxation or "mental" alertness.

When "resting" muscle tension or tone is high, the fundamental explanation lies within the gamma system. In the unusually tense or alert individual, gamma bias is exaggerated: the high state of alpha-motoneuron excitability causes greater tension on the spindle, which causes an unusually high number of facilitory impulses to alpha-motoneurons. An individual's neurological state of alertness depends, in part, on motoneuron excitability, which is affected by gamma input, and/or by supraspinal control. This phenomenon is consistent with an individual's inability to induce the stretch reflex when the gamma system is blocked pharmacologically with a chemical such as procaine.

Effect of Exercise on the Stretch Reflex

A single bout of exercise induces a shortening of the time required for the tendon stretch reflex to occur; the same effect occurs with maximal exertion. An explanation for this effect has not been identified, but conceivably it could relate to muscle temperature and its effect on the conduction velocity of axons, on the synapses of the CNS, or on the excitation contraction of the myofibrils.

Differential Utilization of Motor-Unit Types by the Gamma System

Spindle afferent input is greater to those alpha-motoneurons that innervate slow-twitch fibers. Spindles exist in greater concentration in the antigravity, predominantly slow-twitch muscles. The large number of spindles in slow-twitch muscles assure large amounts of sensory output, which, in turn, is used to facilitate the alpha-motoneurons. As explained in Chapter 4, metabolism and circulation of slow-twitch fibers are also better suited for continuous work than fast-twitch fibers. Continuous, partial contractions (isometric) of muscles (desirable in postural maintenance) are most efficiently sustained by slow-twitch muscle fibers.

Since spindles are sensitive to changes in velocity of movement, it is possible that an undesired loss of speed at the onset of muscle shortening could be compensated for by spindle-mediated facilitation in the recruitment of other alpha-motoneurons. These additional alpha-motoneurons would assist in restoring the desired shortening velocity against a given load. It has been shown that this adjustment cannot occur during gamma block. A final fine adjustment may be made in the speed movement by preferential, additional recruitment of either slow- or fast-twitch fibers.

Adaptation of the Spindle to Chronic Disuse and Overload

Disuse

It is well known that disuse of extrafusal skeletal muscle results in atrophy and loss of strength. Intrafusal muscle fibers also atrophy in response to chronic disuse. In addition the spindles' Ia primary endings have a greater firing rate at rest and are more sensitive to muscle stretch in the atrophied than in the normal muscle. Also, at any given increase in muscle length, Ia firing frequency increases more in the atrophied than in the normal muscle; this may be a mechanism whereby the loss of extrafusal muscle strength is compensated for by the greater activating effect of spindle feedback on alpha-motoneurons. For example, for any given stretch or tension on a muscle, which has atrophied by 20%, the spindles of the atrophied muscle stimulate the alpha-motoneurons at a 20% greater frequency, thus compensating for the potential loss of tension (assuming a direct proportional relationship) and thereby achieving the identical amount of final tension.

Overload

Unlike extrafusal muscle fibers, specific intrafusal muscle fibers (chain fibers) selectively atrophy with endurance training. Correspondingly oxidative metabolism seems to decrease slightly in the bag intrafusal fibers. Our understanding of the function of bag and chain intrafusal fibers in normal movements is minimal, making it difficult to determine the significance of the changes noted above.

Other Muscle Sensory Receptors

Tendon Organs

Tendon organs (TO) are specialized projections of collagenous connective tissue surrounded by a thin capsule, which contains sensory nerve endings projecting to the spinal cord. These afferents, classified as Ib (to define their origin as being the TO spindle), send impulses to interneurons, which in turn inhibit alpha-motoneurons in the same or synergestic muscles and facilitate alpha-motoneurons in the antagonist muscles.

The TO distribution is strikingly similar to that of muscle spindles. Contrary to what its name implies, most endings are within the muscle tissue, not the tendon itself. Each TO is attached to about 10–20 extrafusal muscle fibers. Tendon organs are much more plentiful in slow muscles than in fast muscles. They are also found in ligaments of joints and are responsive primarily to joint movement, not position, because they quickly adapt to a static

stretch. In other words, a TO may be activated by the movement, but the impulses will quickly subside once the new position is maintained.

A fundamental difference between muscle spindles and TO is that the spindles are connected in parallel and the TO are connected in series with extrafusal muscle fibers (Figure 7–9). Consequently, when a muscle contracts, the TO tends to fire, whereas the spindle would be unloaded if it were not for the length adjustments made by the gamma-motoneurons innervating the intrafusal muscle fibers.

The sensory output from the tendon organs shows a linear correlation with tension developed by that muscle. The TO is sensitive to rate of tension change, passive or active, with the consequential impulses possibly inhibitory to synergistic alpha-motoneurons. TO are considered the muscle tension sensor; the spindle is considered the length sensor.

Joint Receptors

Awareness of joint positions and movement (*kinesthetic sense* or *proprioception*) is undoubtedly an important mechanism in physical performance of many skills.

Slow-adapting joint receptors are activated within a specific range of from 20° to 30° of joint movement (Figure 7–10). Specific joint receptors may be important in protecting the joint from either extreme flexion or extension—but not both. They probably send signals to the CNS that can be interpreted as either flexion or extension. The slow-adapting receptors also respond to velocity of movement.

Fast-adapting joint receptors are found more rarely. They respond to rapid movements and only fire a single impulse upon activation, although the stimulus induced by the limb movement or position is sustained. Apparently these endings simply tell the CNS that movement is occurring.

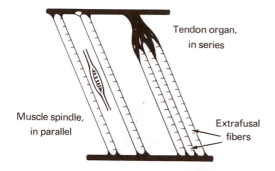

Tendon organ, in series

Muscle spindle, in parallel

Extrafusal fibers

Figure 7-9. Fundamental difference in relations of spindles and tendon organ (TO) to extrafusal muscle fibers. The TO, which attaches at one end to the aponeurosis and receives insertions of extrafusal (EF) fibers at the other, is "in series" with these contractile fibers.

From E. Eldred, The dual sensory role of muscle spindles. *Journal of American Physical Therapy* 45:290–313, 1965.

Pacinian Corpuscles

Pacinian corpuscles (PC) are found extensively in *fascia* (connective tissue sheath) of muscle and in close contact with TO; they may be sensitive to pain when the load becomes excessive. PC are widely distributed and found intermuscularly between membranes connecting bones (e.g., the interosseous in the lower leg, which connects the tibia and fibula), in the *periosteum* or the outer covering of bones, in joint capsules that contain synovial fluids, in tendon sheaths, in the deep layers of sensitive skin areas (like finger tips), and in the mesentery of the gut. They have an onionlike capsule about 500 microns in diameter that serves as the environment for the sensory nerve endings (Figure 7–11). Several PC may be served by a common neuron.

The PC is sensitive to pressure and is very fast adapting, like the TO and some joint receptors. Upon sustained pressure, only a single

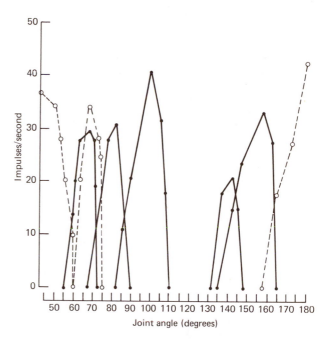

Impulses/second

Joint angle (degrees)

Figure 7-10. The number of impulses/sec for a series of individual joint afferent axons is shown for the knee of a cat in the range of 50° to 180°. Note that a single ending is sensitive to a relatively specific joint angle.

From S. Skoglund, Anatomical and physiological studies of knee joint innervation in the cat. *Acta Physiol. Scand.* 36 Suppl. 124, 1956.

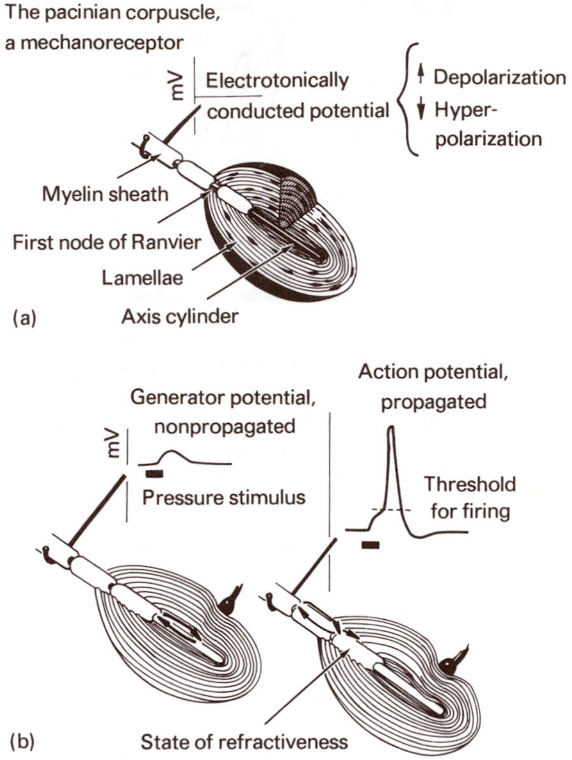

Figure 7-11. a. Structural plan of the pacinian corpuscle. The single, large axon runs through a central core which is relatively undifferentiated in comparison to the concentrically laminated capsule. There is no sheath on the axis cylinder deep within the center of the corpuscle: this is not a functional "nervous conduction." The potential level is steady when no stimulus is being applied, but changes when the capsule is pressed upon. b. Distinction of *generator* and *action potentials* is illustrated by the pacinian corpuscle. The generator potential is a DC shift across the membrane of a sensory axon terminal induced by an "adequate" stimulus—pressure in this case. The spreading currents set by the generator potential also reach the first node of Ranvier, and there shift the membrane potential toward depolarization and instability. When the membrane reaches a certain degree of depolarization, an all-or-none "spike" of total depolarization develops at the node.

From E. Eldred, Peripheral receptors: Their excitation and relaxation to reflex patterns. *American Journal of Physical Medicine* 46:69–87, 1967. © 1967, The Williams & Wilkins Co., Baltimore, Md.

stimulus occurs; another signal is initiated when the pressure is relieved. This type of response is a property of the lamallae surrounding the nerve ending, not the endings themselves. The purpose of this pressure receptor appears to be more of informing the CNS of the change in pressure rather than indicating how much pressure.

Summary

Motor activity is intricately modulated by *sensory feedback*. Whether an alpha-motoneuron is inhibited or facilitated by the action of a sensory nerve depends on the type of sensory receptor excited. We should think of the *muscle spindle* not only as it relates to static postural adjustments and to the classical stretch reflex but also as a critical component in the modulation of all movements.

Gamma-motoneurons, as they alter the load on the muscle spindle, undoubtedly are a very important factor in regulating the output of muscle tension and the speed of skilled movements. Alpha-motoneurons' firing patterns are regulated to a significant degree by the gamma-motoneurons and consequently by muscle spindle output. In fact the gamma influence on the alpha system is an essential element of movement in terms of both speed and precision.

Study Questions

1. Draw and label a muscle spindle.
2. Discuss coactivation.
3. Describe how gamma biasing can facilitate muscle contraction.
4. How is the muscle spindle involved in muscle tension?
5. How is the spindle involved in the stretch reflex?
6. What is the role of the gamma system in motor-unit activation?
7. How does the spindle respond to use and disuse?
8. Contrast the muscle spindle and the tendon organ.
9. How is the joint receptor used?

Review References

Granit, R. *Muscular Afferents and Motor Control.* Stockholm, Sweden: Almqvist and Wiksell, 1966.

Granit, R. *The Basis of Motor Control.* London: Academic Press, 1970.

Maier, A.; E. Eldred; and V.R. Edgerton. The effect on spindles of muscle atrophy and hypertrophy. *Exp. Neurol.* 37:100–123, 1972.

Matthews, P.B.C. *Mammalian Muscle Receptors and Their Central Actions.* Baltimore: Williams and Wilkins Co., 1972.

Smith, J.L.; E.M. Roberts; and E. Atkins. Fusimotor neuron block and voluntary movement in man. *Amer. J. Phys. Med.* 51:225–239,1972.

Swett, J.E. and E. Eldred. Distribution and numbers of stretch receptors in medial gastrocnemius and soleus muscles of the cat. *Anat. Rec.* 137:453–473, 1960.

Wyke, V. The neurology of joint. *Annals of the Royal College of Surgeons, England* 41:25–50, 1967.

Energy support systems of movement and exercise

CHAPTER EIGHT

Key Concepts

• Cardiac output is directly proportional to the intensity of the exercise and is determined by stroke volume and by frequency of the heart beat.

• Cardiac metabolism is oxygen dependent.

• The heart hypertrophies in response to a chronic overload, such as endurance training.

• Hearts of animals adapted to physical training are more resistant to anoxia-induced ischemia than their appropriate controls.

• The electrical characteristics of the heart reveal extensive information about the functional capacity of the heart.

The heart and its vessels

General Features of Cardiac Muscle and Its Vasculature

Heart

The heart consists of a layer of muscle referred to as the *myocardium*. The lining of the inner surface is called the *endocardium*, and the lining of the outer surface, the *epicardium* (Figure 8–1). The thickness of specific walls of the myocardium is a reflection of the stress exerted on them. For example, the walls of the left ventricle are thicker than those of the right ventricle because the pressures in the left ventricle are about five times greater than the pressures in the right ventricle—and up to 50 times greater than those in the atria.

Cardiac muscle is not generally considered to be under voluntary control. However, rats and humans have some degree of conscious control over their heart rates. The heart is innervated by the parasympathetic and sympathetic autonomic nervous systems, with the former acting as a brake and the latter as an excitor of the heart.

Structurally cardiac muscle is unique in that the fibers are connected end to end, separated only by a structure called an *intercalated disc*,

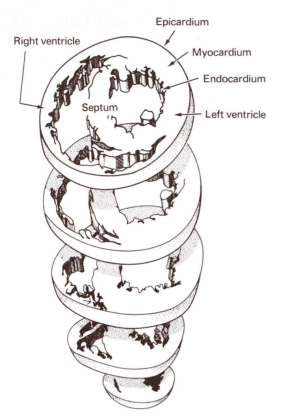

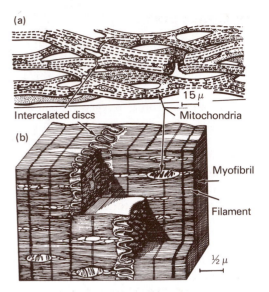

Figure 8-2. Branched nature of cardiac muscle fibers and intercalated discs.

From R. Poche and E. Linden. Untersuchungen zur frage der glanzstreifen des herzmuskelgewebes beim warmbluter und beim kaltbluter. *Z. Zellforsch. Mikro. Anat.* 43:104–120, 1955.

Figure 8-1. Serial section through the rat heart.

From A.F. Grimm; V. Kazimieras; A.K. Stuart; and H.L. Lin. Growth of rat heart. *Growth* 37:189–208,1973.

shown in Figure 8–2. A smooth, progressive contraction from one fiber of the heart to another is facilitated by the lateral branching and interconnections of fibers at the area of the intercalated disk.

Ultrastructurally the heart has T-tubules that are five times larger than those in skeletal muscle. They conduct impulses toward the center of the fiber. The calcium-storing segment of the fiber, *sarcoplasmic reticulum* (SR), is less

extensive in the heart than in the faster-contracting skeletal muscle. The calcium concentration in the fibers is highly correlated to the ability of the fibers to generate tension. The amount of Ca^{2+} released into the cytoplasm at any given time is related to the contractility of the heart. The mechanism for the initiation of shortening by the release of calcium from the SR is similar to that already described for skeletal muscle—except that the Ca^{2+} may enter from outside of the fibers during each contraction rather than only from the SR.

Capillaries

Capillaries in the myocardium are vessels that are about 8 microns thick and consist of a

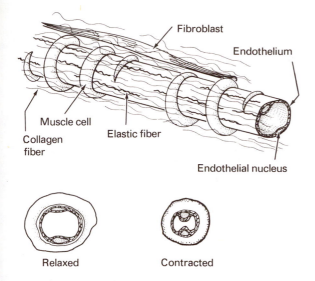

Figure 8-3. Schematic representation of an artery showing the radial nature of muscle fibers which aid in constriction of the vessel.

From H. Elias and J.E. Pauly. *Human Microanatomy.* Philadelphia: F.A. Davis, 1966, Fig. 9–7, p. 117.

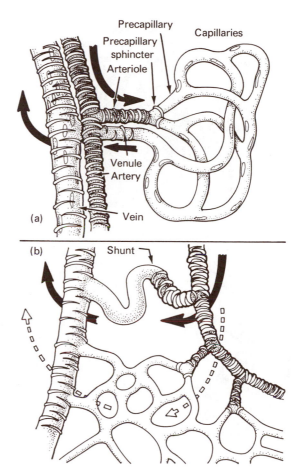

Figure 8-4. Precapillary sphincter regulating capillary flow or shunting.

From H. Elias and J.E. Pauly. *Human Microanatomy.* Philadelphia: F.A. Davis, 1966, pp. 114, 126.

single layer of *endothelium*, continuous with that found in the heart (Figure 8–3). Blood flow in the capillaries is regulated by a variation in the inner diameter of the vessels, particularly at certain points called *precapillary sphincters* (Figure 8–4).

The vascular walls appear to "drink and package" materials into small units called *vesicles;* these can be seen along the surfaces of the endothelial cells (Figure 2–5). Vesicles serve as only one means by which materials are exchanged across capillary membranes.

Arteries and Veins

The larger vessels, the *arteries* and *veins,* consist of varying layers of connective tissue and

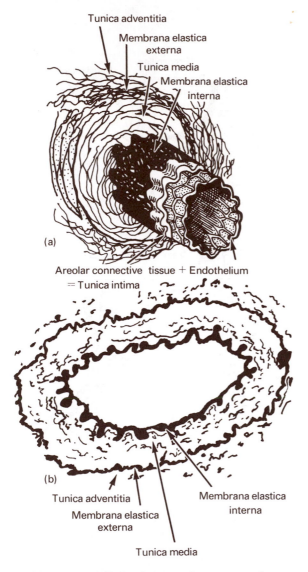

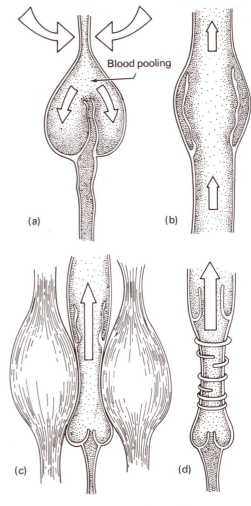

Figure 8-5. Cellular layers of an artery drawn schematically and an actual cross section of an artery.

From H. Elias and J.E. Pauly. *Human Microanatomy.* Philadelphia: F.A. Davis, 1966, Fig. 9-7, p. 117.

Figure 8-6. Valves in the veins (a) prevent returning flow but (b) do not hinder the normal flow very much. Blood can be pushed through veins (c) by nearby active muscle or (d) by smooth muscle.

From H. Elias and J.E. Pauly. *Human Microanatomy.* Philadelphia: F.A. Davis, 1966, Fig. 9-17, p. 126.

smooth muscle (Figure 8-5). The importance of the connective tissue and smooth muscle layers within the vasculature of the circulatory system, in terms of its response to training, has received little attention but may be of considerable significance in our understanding of blood pressure changes that occur with aging and exercise. Larger vessels differ from smaller capillaries primarily in the thickness of the individual layers. The cross section of the artery seen in Figure 8–5 shows the *tunica intima;* the connective tissue portion, which thickens with age, is probably a significant factor in the progressive development of high blood pressure associated with aging. Veins are much thinner than arteries—and consequently are more compliant, have a greater blood-pooling capacity, and lower blood pressures. Figure 8–4 also illustrates an artery-vein shunt, which permits capillary beds of inactive tissues to be bypassed. Figure 8–6 illustrates the presence of one-way valves in the veins, the milking action of muscular contraction, and constriction of the veins, all of which facilitate return of blood to the heart.

Physiology of an Overloaded Heart

Cardiac Output

Cardiac output (Q) is determined by the volume of blood pumped per beat (*stroke volume*) and by *heart rate frequency.* Cardiac output can be greatly increased during exercise; this increased output usually assures adequate blood flow to the tissues. The maximum cardiac output is higher in trained than in nontrained persons (Figure 8–7). Since resting heart rates are lower in trained individuals, it follows that *stroke volume* must be an important compensating factor in maintaining an adequate Q in trained persons. Heart rate frequency is a more important factor in the early exercise response of the heart to enhance Q; however, as the exercise intensity progresses, stroke volume plays a more important part. As exercise bouts approach exhaustion, heart

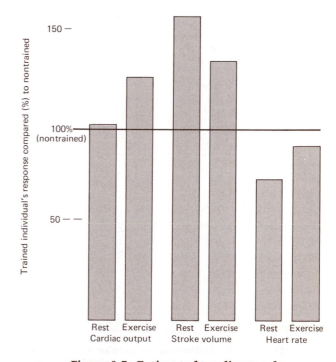

Figure 8-7. Estimated cardiovascular responses to physiologic overload of trained individuals. The numbers represent percent of a nontrained individual's response.

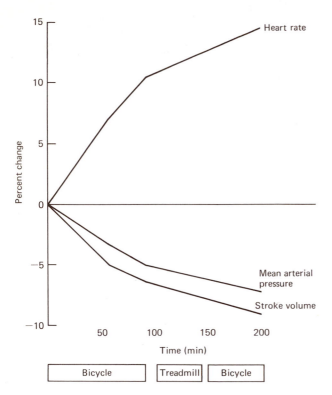

Figure 8-8. Reduction in stroke volume and mean arterial pressure with prolonged exercise (≈70% max $\dot{V}_{O_2}$).

From B. Saltin, Oxygen transport by the circulatory system during exercise in man. In *Limiting Factors of Physical Performance* (J. Keul, ed.). Stuttgart: Georg Thieme, 1973, pp. 235–252.

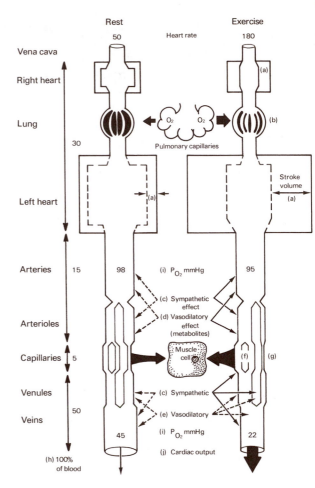

(a) End diastolic and systolic volume

(b) Vasodilation

(c) Decrease vascular capacitance (vascular diameter)

(d,e) Metabolic vasodilatory effect

(f) Vasodilation in active tissue (metabolism)

(g) Vasoconstriction in inactive tissue (sympathetic effect)

(h) Volume of blood in various segments of cardiovascular system

(i) Oxygen partial pressure

(j) Total blood flow (cardiac output)

Figure 8-9. Diagram of cardiovascular responses to exercise throughout the total vascular system.

rate tends to rise gradually as stroke volume is reduced (Figure 8-8).

The interrelationship of heart rate and stroke volume in elevating cardiac output varies under different conditions. For example, in response to a second exercise performed ten minutes after the first exercise, heart rate is much higher than would be expected. Perhaps this is a sign of cardiac fatigue.

The work intensity with which a person can

perform at a given heart rate is greater in trained than in nontrained individuals. Peripheral factors are of unquestionable significance in determining cardiac output and the heart's ability to deliver oxygen to working muscles. Delivery of oxygen is not only determined by cardiac output but also by the capacity of the tissues to utilize it. For example, one important means of increasing the availability of oxygen to muscle fibers is by extracting more oxygen from the blood. The extraction rate is dependent in part on the ability of the muscle to utilize the oxygen. A summary of some of the major cardiovascular adjustments to exercise is diagrammed in Figure 8–9.

Figure 8–10 shows selected cardiovascular adjustments due to changes in postural position as well as to exercise. In the reclining position, note the high stroke volume and the rate of return of venous blood to the right side of the heart, as shown by right ventricular and diastolic volume. This occurs even without exercise as a stimulus.

Blood Pressure

Resting systemic systolic blood pressure in trained individuals is usually the same, or slightly higher, than in nontrained individuals; end diastolic pressure within the heart is usually lower in trained persons. These differences may be due to the greater ease with which the heart can empty as a result of a greater total vascular bed and therefore less resistance to blood flow in the trained individuals. On the other hand, mean pulmonary arterial pressure may be as much as 30% lower in sedentary than in trained persons.

Blood pressure changes in response to a single exercise vary with the nature of the exercise. For example, arm exercise causes a greater rise in aortic mean pressure and diastolic pressure than leg exercise.

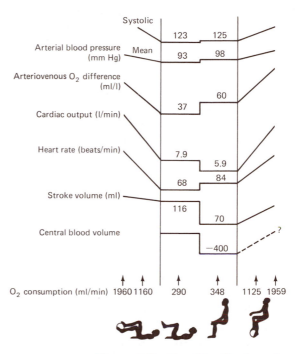

Figure 8–10. Graphic display of cardiovascular responses to exercise.

From P.L. Altman and D.S. Dittmer, eds. *Biological Handbooks: Respiration and Circulation.* Bethesda, Md.: Fed. Amer. Soc. Exp. Biol., 1971, p. 334.

Systolic aortic pressure rises sharply in response to a static (*isometric*) forearm contraction of only 50% of maximum voluntary contraction, while the cardiac output is doubled and heart rate is almost doubled. Venous return to the heart is limited during this type of exercise, which limits any potential adjustments in stroke volume that would tend to relieve the heart of the temporary stress. This situation could be dangerous for marginal cases of cardiovascular disease. The pumping action of the muscles, particularly the legs, in dynamic (*isotonic*) concentric contractions, avoids the stress more characteristic of statically maintained exercises (Figure 8–11).

Role of the Frank-Starling Mechanism in Exercise

The heart has an intrinsic capacity to regulate the strength of its contractions. It does this by being responsive either to ventricular volume or to muscle stretch or length. Greater stretch is induced by greater venous return to the heart, which increases its ventricular volume and consequently muscle length. The greater stretch causes the heart to contract more vigorously, thereby emptying the heart of additional blood that has flowed into the cardiac chamber. By this mechanism the heart is immediately responsive to changes in work demands. This intensification of contraction in response to stretch is called the *Frank-Starling mechanism*. During exercise this mechanism may be partially overridden by the action of the autonomic nervous system. The more strenuous the exercise, the more involved this mechanism is in making the desired cardiac adjustments.

Intrinsic controls of the heart must be quite effective. We know that dogs with their hearts totally denervated can exercise as well as normal dogs can. This suggests that the heart can utilize intrinsic mechanisms to compensate for the apparent absence of extrinsic neural control—or that the extrinsic controls are not that important in normal situations. For example, the parasympathetic system may not play an important role in acute adjustments to cardiovascular demands of exercise since *atropine*, a parasympathetic blocker, has no effect on the maximal heart rate achieved in normal subjects. At the same time the importance of the parasympathetics in mediating some of the classical training effects, such as the reduced resting heart rate in response to endurance training, is evident (see discussion of *Bradycardia and Training* later in this chapter).

Several intrinsic factors play an integral role in regulating the heart's pumping action. The myocardium, in a sense, summarizes a series of physical factors and contracts accordingly. Actual ejection of blood from the ventricles (cardiac output) is modified by: (1) the amount of blood in the heart before contraction begins (*end diastolic volume*), (2) resistance to ejection (*impedance*), (3) intensity or force of contraction, and (4) heart rate.

END DIASTOLIC VOLUME

End diastolic volume refers to that volume of blood in the chambers of the heart when it is relaxed or in diastole. End diastolic volume is dependent on the rate of venous return of blood to the heart and on how much time the blood has to return to the heart between beats. Venous return is increased by an elevation of the legs or by an increase in the pumping action of skeletal muscle contractions. Venous return is enhanced also by a reduction in the amount of blood stored in the venous vasculature, which is generally about 80% of our total volume of blood. A reduction in the amount of blood pooled in the veins is accomplished by a reduction in the size or capacitance of the vessels (*vasoconstriction*). This reduction of pooling could have the same effect as a person's receiving a blood transfusion since it effectively increases the amount of circulating blood.

IMPEDANCE

Impedance is the total resistance to the ejection of blood through the outflow channels. It consists primarily of valve resistance and that resistance to flow along the walls of vessels, particularly arterioles. *Stenotic* (rigid or unpliable) valves or vessels and systemic vasoconstriction elevate impedance, which results in high blood pressure.

There is a marked reduction in resistance to blood flow in skeletal muscles during exercise. The exercise-induced vasodilation results in a decrease in impedance, which may lead to a lower systemic blood pressure—even with an increase in stroke volume.

CONTRACTILE STATE

The *inotropic* properties of the heart refer to the intensity of the contraction or the rate of actomyosin interaction. As in skeletal muscle, the rate of force development is basically controlled by the activity of *myosin ATPase*.

Norepinephrine and epinephrine have a positive inotropic effect on cardiac muscle. These transmitters increase the amount of Ca^{2+} available for the contractile components. Similar effects can be induced by a number of synthetic drugs.

HEART RATE

Heart rate is regulated basically by the autonomic nervous system. Sympathetic nerve endings release the excitatory *catecholamines* (norepinephrine and epinephrine) which

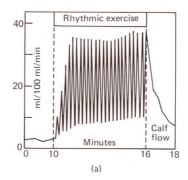

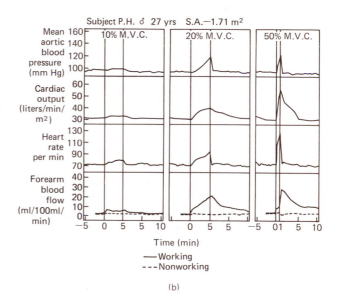

Figure 8-11. Blood flow to muscle during (a) isotonic and (b) isometric contractions.

(a) From H. Barcroft and H.J.C. Swan. *Sympathetic Control of Human Blood Vessels.* London: Edward Arnold, 1953. (b) From A.R. Lind, G.W. McNicol, and K.W. Donald. Circulatory adjustments to sustained (static) muscular activity. In *Physical Activity in Health and Disease* (K. Evang and K.L. Andersen, eds.). Baltimore: Williams & Wilkins 1966, pp. 38–63.

cause an increase in the frequency of the heart beat; conversely, *acetylcholine*, which is released by parasympathetic nerve endings, decreases the heart rate. Epinephrine, released humorally from the *adrenal medulla*, has the same excitatory effect as that which is released neurally—except that its response time is slower and its effect more diffuse. Assuming no decrease in stroke volume, cardiac output is, of course, increased by a higher heart rate.

Cardiac Metabolism

Metabolically the heart is similar to the brain in that both are extremely dependent upon a readily available supply of oxygen, glucose, fatty acids, and lactate. The heart has numerous mitochondria, more than is found in skeletal muscle fibers; in fact the heart has the highest mitochondrial density of all tissues.

Because of the heart's almost complete dependence on a continuous oxygen supply, it is important to recognize the factors that determine the amount of oxygen required. The three major factors are intramyocardial tension (force), contractile state (speed at which the heart develops tension), and heart rate. The stronger and more rapid a contraction, the more oxygen that is required. The oxygen requirement is also directly related to the frequency at which the contractions occur.

Substrate Preferences of the Heart

The relative importance of the various substances withdrawn from the blood and utilized by the human heart is shown in Figure 8-12. Apparently glucose, lactate, and free fatty acids (FFA) are equally important at rest. In trained subjects, resting FFA levels are more important than lactate. As the work load on the heart increases, the relative utilization of lactate may be tripled, while glucose may be reduced by half and FFA by two-thirds in relation to rest

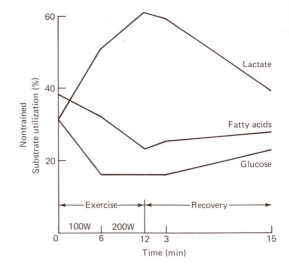

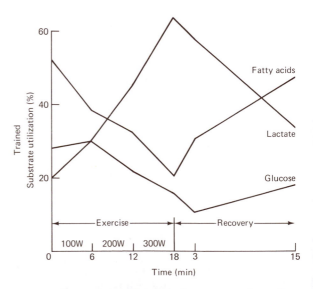

Figure 8-12. The relative proportion of substrate used by the heart at various work loads (w = watts) and during recovery in trained and nontrained subjects.

From J. Keul, E. Doll, and D. Keppler. *Medicine and Sport*, vol. 7, *Energy Metabolism of Human Muscle*. Baltimore: University Park Press, 1972.

values. In absolute terms, utilization of all substrates is enhanced considerably during exercise.

Metabolic Factors in the Ischemic Heart

Glycogen decreases rapidly in *ischemia* (obstructed blood flow), *anoxia* (absence of O_2), and *hypoxia* (low O_2). Enhanced resistance to anoxia has been demonstrated in those hearts that contain elevated glycogen levels. If the 24-hour rhythm in the levels of cardiac glycogen, which has been demonstrated in rats, is shown to be true for humans, there may be some times of the day when ischemia, anoxia, or hypoxia is more critical than at other times. For example, cardiac muscle glycogen may be three times higher at 8:00 A.M. than at 8:00 P.M.

Hearts from endurance-trained rats are more resistant to hypoxia; one reason seems to be an increase in efficiency (oxygen uptake/work performed). Similarly, animals that are adapted to high altitude, compared to nonaltitude-adapted animals, recover their cardiac contractility more completely when exposed to an anoxic environment. Consistent with these findings is the observation that less *necrosis* (cell death) is found when a necrotizing drug is injected in high-altitude-adapted rats than in the corresponding controls. Table 8–1 lists some of the myocardial metabolic shifts that occur in high-altitude adaptation.

Measurement of enzymes in the blood has been an important technique in identifying potential coronary artery patients. *Lactate dehydrogenase* (LDH) is one of the commonly measured enzymes: it is known that LDH activity in the blood increases dramatically upon coronary trauma. Perhaps damage to the heart increases membrane permeability, and LDH from the heart is released into the blood. It has been found that, with cardiac pacing (electrical stimulation of the heart at a desired frequency), lactate production and a depression

Table 8–1. Myocardial Metabolic Shifts During High-Altitude Acclimation

General Category	Specific Category	Change
Glycolysis potential		Increase
High-energy phosphates	ATP (resting)	Decrease
	CP	Decrease
ATPase	Nonpurified	Increase
	Purified	Increase
LDH system (heart to muscle shift)	In young animals	Increase
Myoglobin		Increase
Mitochondrial content		No change

Adapted from O. Poupa, Heart under unusual conditions: Effects of gravitation and chronic hypoxia. In *Recent Advances in Studies on Cardiac Structure and Metabolism*, vol. 1, *Myocardiology*, pp. 779–802. © 1972 University Park Press, Baltimore.

in the ST segment of the ECG occur before the onset of angina. This technique may prove to be superior to the standard exercise test used in detecting potential coronaries, since the rise in LDH efflux from skeletal muscle, also induced by exercise, can be differentiated from the LDH from cardiac muscle. Interestingly, the LDH and lactate responses to exercise are less prominent in trained animals.

Chronic Metabolic Overload

Metabolic pathways are not drastically affected by chronic exercise (training) overload. The relative concentration of *cytochrome c* (found only in mitochondria) and of myoglobin in the hearts of severely trained rats is not changed; either is the activity of the terminal enzyme, *cytochrome oxidase*, in oxidative metabolism. These observations may be misleading since some of these same mitochondrial constituents as well as capillaries increase proportionately. Further research will be necessary before a total understanding of this area is possible.

Chronic hypertension reduces the mitochondria-to-myofibril ratio; this, obviously, makes the heart more susceptible to cardiac insufficiency. On the other hand, long-term training results in a proportional increase in mitochondrial enzymes over myofibrillar substance. Clearly the heart of the trained individual is quite different from the pathologically-induced enlarged heart.

It is debatable whether or not cardiac mitochondrial membranes undergo disruption during an exhaustive exercise. Ultrastructural alterations that have been reported to occur were almost completely reversed within 24 hours after an exhaustive exercise. The possible mitochondrial alterations have been reported to be less prominent in trained than in nontrained animals.

Control Mechanisms of the Heart

Neural Control Mechanisms

The heart is regulated by the autonomic nervous system. It is also particularly responsive to catecholamines released from the adrenal medulla. The autonomic nervous system, consisting of the sympathetic (*adrenergic*) and parasympathetic (*cholinergic*) nervous systems, forms a network of *ganglia* (neurons outside the cord and brain) and nerves near the spinal cord and heart. These nerves finally terminate rather diffusely in the walls of the heart. The sympathetic nerves tend to terminate at the base of the ventricles; the atrial walls receive most of the parasympathetic nerve fibers.

The sympathetic nerve endings release epinephrine and norepinephrine (predominantly the latter), both of which accelerate the heart. The sympathetic nervous system communicates to the heart via receptors called *alpha* and *beta receptors*. Most of the alpha receptors mediate vasoconstriction, whereas beta-receptor activation induces arteriolar vasodilation. Beta receptors augment heart rate and the contractility of each beat. For example, if beta adrenergic receptors are activated, resistance decreases; and heart rate (HR) and contraction speed increase. With alpha adrenergic receptor activation, resistance to blood flow and blood pressure increases. A balance in the activation of alpha and beta receptors is necessary for the heart to execute its total function properly.

Monoamine oxidase, an enzyme that metabolizes epinephrine and norephinephrine, also plays an important role in regulating endogenous amine levels. Its activity is increased in chronically overloaded hearts, suggesting a greater capacity to inactivate the catecholamines. A decrease in *tryosine hydroxylase* (a rate-limiting enzyme essential for catechol-amine synthesis) has been implicated as a factor in determining the net catecholamine levels in nerves during exercise.

Opposing the sympathetic effect are the para-sympathetic nerves, which release a cardiac in-hibitor, *acetylcholine*. Parasympathetic nerves originate in the medulla as *cranial nerve X* (also called the *vagus*). Activation of the vagus nerve in dogs not only markedly reduces heart rate and force of contraction—but also reduces coronary artery resistance by more than half; this facilitates coronary blood flow during vagal dominance. *Vagal dominance* (a term that refers to the situation when the braking effect is greater than the accelerating effect) has been proposed as a mechanism for the bradycardial effect of endurance training.

Intrinsic Regulation of the Heart

The heart has unique intrinsic impulse-con-ducting properties. Located in the atrial wall is a pacemaker known as the *sino-atrial* (S-A) *node*. Each cardiac contraction is initiated at this point. A depolarizing wave spreads across the atria and towards the ventricles via the *atrioventricular* (A-V) node and *Purkinje fibers* in the septum, which separates the right and left halves of the heart. The large, special-ized Purkinje fibers are less densely packed with myofibrils than regular cardiac fibers and conduct impulses about six times faster than ordinary cardiac fibers, permitting almost si-multaneous depolarization of the ventricles when an impulse is initiated.

Extrinsic Regulation of the Heart

Even though the heart has its intrinsic contrac-tile rhythmicity, its external innervation can regulate heart rate, intensity of contraction, and vascular resistance. This modulation originates from the cardiovascular centers in the medulla, which receive peripheral sensory signals from the cardiac depressor nerve (*glossopharyngeal*, IX). The impulses along the cardiac depressor nerve originate from the carotid sinus and from the aortic arch in response to elevated blood pressure. The effect is inhibition of the cardiovascular excitatory center and, therefore, reduction of the number of excitatory sympathetic impulses transmitted to the heart. The depressant effect is overcome during exercise.

A significant portion of the rise in heart rate, in response to muscular contractions, must be mediated centrally. An increase in heart rate and blood pressure in response to an expected contraction—but without the actual contraction of a muscle (neuromuscularly blocked with the drug *succinylcholine*)—demonstrates this central influence. Thus we can hypothesize that the physical shortening of the muscle is not necessary for the central nervous system (CNS) to know that voluntary muscles are going to or have already contracted. However, other mechanisms—such as one including the muscle spindle and other receptors—also have been implicated as having a regulatory effect on the cardiovascular system.

Isometric contractions can initiate an immediate increment in heart rate (within the next heart beat if the contraction occurs early in the pulse interval) and in aortic blood pressure. Apparently the size of the muscle group activated is also a factor; the larger the muscle mass, the greater the response.

Hormonal Control Mechanisms

Hormonal regulators are released into the circulatory system and carried throughout the body to exert a generalized effect. Neural control is considered more localized to a specific tissue and considerably quicker acting.

CATECHOLAMINES

Epinephrine is released by the cells of the adrenal medulla (*chromaffin cells*) and has excitatory effects on the heart similar to the *norepinephrine* released from the sympathetic nerve terminals in the heart. Catecholamines reduce the total resistance to blood flow in large vascular beds—such as in skeletal muscle and in the coronary arteries—by causing vasodilation. However, vascular resistance is increased in the kidneys, lungs, skin, gut, and perhaps the liver because of vasoconstriction induced by catecholamines. The cerebral blood flow is relatively unresponsive to catecholamines.

The indirect effects (metabolic or beta) and the direct effects (neural or alpha) of epinephrine on vascular resistance are, in part, antagonistic. Epinephrine indirectly induces vasodilation by increasing metabolism, while the direct effect on most vascular beds is vasoconstriction (alpha-effect). A marked elevation in metabolism (beta-effect), such as occurs during exercise, camouflages the direct vasoconstriction (alpha-effect) on the active tissue beds. The vasoconstriction will predominate in the inactive tissues.

ACETYLCHOLINE

The transmitter *acetylcholine* released from the parasympathetic vagus nerve slows the heart rate, thereby indirectly decreasing aortic pressure and increasing coronary blood flow through a vasodilatory effect. The bradycardial effect of endurance training probably reflects a type of vagal dominance.

Significance of Electrocardiography

The electrocardiogram (ECG) is a recording of the electrical signals originating from the heart and received by electrodes attached to the sur-

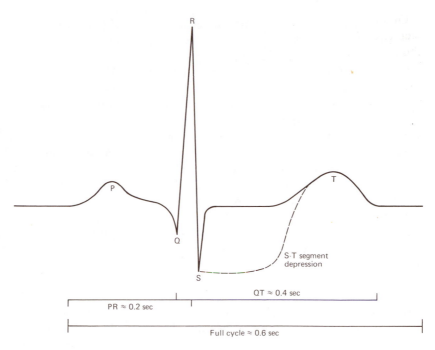

Figure 8-13. The "shape" of one heart-beat cycle as registered on an electrocardiograph. The dotted trace is an example of an abnormal electrical cardiac cycle.

face of the skin. It is the electrical sum of all cardiac fibers in the process of depolarizing and repolarizing. Each heart beat produces a very characteristic electrical wave form (Figure 8–13). The P-wave is caused by depolarization of the atria; the QRS interval reflects depolarization of the ventricles and repolarization of the atria; and the T-wave is the repolarization of the ventricles.

Because of the sensitivity of the muscle-fiber membrane potential, the ECG is an excellent tool for detecting abnormalities of the heart. For example, *ischemia* (or occluded blood flow) changes the rate and magnitude of depolarization and repolarization—and therefore changes the characteristics of the ECG. Other cardiac abnormalities cause specific changes in the normal wave form of the ECG.

Of known physiological importance is the fact that the *subendocardial* (inner layers of the cardiac wall) region of the myocardium has less oxygen supply and produces more lactate than the *subepicardial* (outer layers of the cardiac wall) region of the myocardium. This suggests that inner layers of the myocardium are more prone to ischemia and subsequent myocardial damage.

To exercise physiologists an ECG recording is an extremely important tool for detecting individuals with potential cardiac problems. Abnormalities are much more likely to manifest themselves if the heart is under physiological stress such as during a treadmill or bicycle exercise test. For example, about half the changes observed during a supramaximal exercise load will go undetected if the exercise does not exceed about 85% of the maximal heart rate. Subjects who exhibit a positive ECG usually have high levels of enzymes (e.g., *creatine phosphate kinase*, CPK) in their blood, which come presumably from the heart.

Significance of Echocardiography

The relatively new technique of *echocardiography* should prove of considerable usefulness in working out some of the problems related to cardiac hypertrophy. Its obvious advantage is that heart size in humans can be determined with acceptable reliability. Ultrasonic impulses are transmitted between the costals and received by an ultrasound transducer in a pattern that permits the determination of the thickness of the septum and ventricle thickness. This research tool is very useful to cardiologists and exercise biologists, who have specialized interests in the cardiovascular system.

Hypertrophy in an Overloaded Heart

Although an extensive amount of work has been done on chronically overloaded hearts, most of it has been during conditions which reflect pathological-type stresses; that is, systemic vascular resistance was simulated by partial constriction of the aorta or by some cardiovascular pathological condition. In these pathological hearts, vasoconstriction is present throughout the complete cardiac cycle of every heart beat throughout the day. Under these overload conditions, there is a hypertrophy progression that occurs in three stages of biochemical and physiological changes. In the third stage, the heart hypertrophies but seems to reach a stage of exhaustion. It begins to weaken rather than strengthen as it hypertrophies. The heart will eventually cease to function as it becomes weaker and weaker. The failure may be due, in part, to the chronic stretching of cardiac fibers well beyond the optimal length for maximum tension during a contraction.

Overloading the normal heart by endurance-exercise training clearly is different from any pathological overload that has been studied. First, in endurance training, the overload occurs in intervals so that the stress on the heart is not continuous throughout the entire day. Apparently the intermittent rest periods are essential in permitting the cardiac muscle to maintain and elevate its rate of protein synthesis, as required. The stimuli to adapt occur during the exercise, but most of the actual adaptation takes place between exercise sessions. If the heart is continuously overloaded, there is not time for adaptation.

There is no evidence that the relatively large hearts found in most endurance-trained athletes are pathologic or ever reach a stage similar to that third stage reached by pathologically overloaded hearts, as described above. In fact, all evidence suggests that the normal, trained heart is more efficient in many ways than the normal, nontrained heart. For example, as the heart enlarges in response to training, there is an increase in the number of capillaries, in the amount of blood flow, and in the mitochondrial content and their related enzymes.

Cardiac hypertrophy does not seem to be a necessary factor in the trained individual's ability to endure long bouts of exercise. The so-called "modern spartans of Mexico" are reported to chase deer for several days—or until the deer drops of exhaustion—but these people show no obvious signs of cardiac hypertrophy. Peripheral biochemical and vascular adaptations apparently are important. Hemoglobin levels (17.2 g%) and hematocrit (51%) are reasonable for the altitude (2400 meters above sea level) where they live. Their exercise and postexercise systolic and diastolic blood pressures are relatively low, reflecting a low vascular resistance in these trained men.

Most of the work characterizing the nature of cardiac hypertrophy has been done on hearts in the relatively early stages of pathological overload. Some of the changes that occur in the hearts of endurance-trained individuals are listed in Table 8–2. It can be seen that fiber size, not number, increases. Total DNA content increases, but about 99% of the new nuclei are of nonmuscle origin, such as those found in the cells of capillaries and interstitial connective tissue cells (*fibroblasts*). Cell size increases in diameter because the myofibrils enlarge to about double their usual size and then split longitudinally; the splitting results in more myofibrils per muscle fiber. Mitochondria probably enlarge and/or increase in number; amino acid uptake into mitochondria is increased; the sarcoplasmic reticulum becomes distended, and the Golgi membranes seem to proliferate.

Table 8-2. Summary of Cellular Aspects of a Physiologically Trained Hypertrophied Heart

Increases	No Change
Capillary number	Capillary density
DNA/organ	DNA/nucleus
Fiber diameter	Mitochondria/g muscle
Golgi apparatus enlarged	Myoglobin concentration
Mitochondrial size	Sarcoplasmic reticulum
Myofibril/fiber	
Myosin synthesis	
Nuclei/heart	
Nuclei/muscle cell	
Polyribosomes	
Protein synthesis	
RNA and RNA polymerase	

Mechanism of Cardiac Hypertrophy

The actual triggering mechanism involved in the induction of cardiac hypertrophy is unknown. Although RNA synthesis occurs in response to overload, it appears that some protein synthesis is elevated before the newly synthesized RNA could be effective in augmenting protein synthesis. However, it is probable that the elevated protein and nucleic acid content of hypertrophied hearts is permitted through the normal *repressor-depressor system* of DNA. This would lead to the enhancement of messenger, ribosomal, and transfer RNA content, and of course to increased protein synthesis.

Hypertrophy and Hormones

Pituitary hormones are not necessary for skeletal muscle hypertrophy. Can, however, cardiac hypertrophy occur without hormonal influence? *Testosterone* and *growth hormone* substantially elevate RNA and protein synthesis of skeletal muscle. The fact that castration and hypophysectomy delay cardiac hypertrophy in response to aortic constriction suggests a positive influence of at least one hormone on this compensatory phenomenon. *Thyroid hormone* can induce *cardiomegaly* directly—but also indirectly by elevating the total metabolic rate of the body. Lack of *insulin* lowers RNA synthesis in cardiac muscle. These hormones may play only a permissive role in cardiac hypertrophy rather than an enhancing one.

Bradycardia

Bradycardia and Training

It has been known for years that training results in a reduced resting heart rate (bradycardia) and in an increase in the stroke volume, which are in part a reflection of cardiac hypertrophy and greater vagal dominance. In regard to vagal dominance, *atropine* (which antagonizes the vagal effect and is therefore effectively adrenergic) has a greater effect on the heart rate in nontrained than in trained rats. This can be interpreted to mean that trained rats have more nonneural acetylcholine to compete for the atropine receptor sites or more cholinergic receptor sites. Atrial acetylcholine is more concentrated in trained than nontrained animals and supports atropine effects.

Other factors are at least as important in the bradycardial response to exercise. In fact the bradycardia experienced in trained individuals is influenced by local changes in skeletal muscles. Evidence for this is found in the following observation: if a person trains his arms, there will be a bradycardial effect in response to exercise of the trained arm—but not to exercise of the nontrained legs. The same is true when only the legs are trained. Therefore no marked relative bradycardia should be expected during an exercise that does not involve the trained muscles. Enhancement of muscle blood flow and reduced vascular resistance (or altered afferent impulses originating in the working muscles or joints and their blood vessels) would seem to be logical explanations for these results. It would seem to be very significant that increased maximal $\dot{V}_{O_2}$ can be shown when the trained leg is exercised in relation to when the nontrained control leg is exercised. This points to the importance of peripheral factors in determining maximal $\dot{V}_{O_2}$.

Diving Bradycardia

Immersion of the face in water causes a significant reduction in heart rate. Bradycardia is not related specifically to the metabolic intensity of the amount of work performed, and the level of physical fitness seems not to affect the degree of bradycardia. Bradycardial effect is greater during dynamic exercise than during static exercise or rest. Since blood pressure remains normal during bradycardia, vasoconstriction apparently compensates for the reduced heart rate. Interestingly simple face immersion induces as great a bradycardial response as that experienced during diving—and also lowers blood flow.

Cardiac Ischemia and Warm-up

The warm-up effect of repeated stimulation (known as *treppe*) on skeletal muscle seems to be present in cardiac muscle. It now appears that the heart can function more safely if given some type of warm-up. It has been shown that more than 50% of healthy adults experience a type of ischemic response of the heart for a few seconds after intense exercise is initiated without some type of warm-up (ST segment depression, Figure 8–13). Angina is not usually associated with this warm-up effect. Temporary ischemia is tolerable for most individuals; however, if someone is a potential cardiac disease subject or if the heart is already taxed in terms of oxygen supply, the effect of intense exercise without a warm-up period could be critical.

Beneficial effects can also be realized with a "warm-down" (two-minute walk after exercise), presumably to prevent venous pooling of the blood that occurs when strenuous exercise is completed—and all movements minimized immediately after the exercise.

Summary

Cardiac muscle is similar in many respects to skeletal muscles. It differs physiologically in that it contracts slowly compared to fast-twitch skeletal muscle fibers. Metabolically the heart is more dependent on a constant supply of oxygen and substrates from the blood. The major substrates for the heart are glucose, lactate, and fatty acids. At rest the heart is dependent primarily on fatty acids, but shifts to lactate as the major energy source during exercise.

Cardiac output (ml/min) is determined by the product of stroke volume (ml/beat) and frequency of heart beat (beats/min). Endurance training results in a greater exercising cardiac output and stroke volume.

Controls that are *intrinsic* to the heart are the Frank-Starling effect and the rhythmic firing of the S-A node within the atrium. *Extrinsic* controls involve neural reflexes that lead to inhibition or excitation of the sympathetic and/or parasympathetic nerves to the heart.

The ECG provides information about the electrical characteristics of the heart. From an analysis of the ECG, it is possible to detect abnormalities in cardiac function.

Adaptations to endurance training increase the efficiency of the heart, increase the capillary-to-fiber ratio, and increase the mitochondrial activity.

The bradycardial effect of training can be explained in part by an enhanced effectiveness of the parasympathetic system and/or decreased sympathetic tone of the heart.

Study Questions

1. What are several distinguishing features of the myocardium?
2. How do arteries and veins regulate blood flow?
3. How are frequency and stroke volume related to cardiac output?
4. What happens to blood pressure during exercise?
5. What intrinsic controls influence the heart during exercise?
6. How is heart rate controlled during exercise?
7. What are the energy fuel preferences of the heart?
8. How do hypertension and training affect the heart?
9. What are the parasympathetic and sympathetic influences on the heart?
10. How do hormones affect the heart?
11. What is the significance of the ECG?
12. What is the time course of the RNA and protein response to training?

Review References

Abbasi, A.S., et al. Left ventricular hypertrophy diagnosed by echocardiography. *New Eng. J. Med.* 289:118–121, 1973.

Banister, E.W.; R.J. Tomanek; and N. Cvorkov. Ultrastructural modifications in rat heart: responses to exercise and training. *Amer. J. Physiol.* 220:1935–1940, 1971.

Barnard, J.R., et al. Cardiovascular responses to sudden strenuous exercise-heart rate, blood pressure, and ECG. *J. Appl. Physiol.* 34:833–837, 1973.

Cumming, G.R. Yield of ischemic exercise electrocardiograms in relation to exercise intensity in a normal population. *Brit. Heart J.* 34:919–923, 1972.

Freyschuss, U. Cardiovascular adjustment to somatomotor activation. *Acta Physiol. Scand.* Suppl. 342:1463, 1970.

Goldstone, B.W. and M.J. Silberstein. The role of the vagus nerve in cardiac adaptation to exercise. *Pflugers Arch.* 325:113–124, 1971.

Higgins, C.B.; S.F. Vatner; and E. Braunwald. Parasympathetic control of the heart. *Pharmacol. Rev.* 25:119–155, 1973.

King, D.W. and P.D. Gollnick. Ultrastructure of rat heart and liver after exhaustive exercise. *Amer. J. Physiol.* 8:151–159, 1970.

Tipton, C.M. and B. Taylor. Influence of atropine on heart rates of rats. *Amer. J. Physiol.* 208:480–484, 1965.

Wahren, J. Quantitative aspects of blood flow and oxygen uptake in the human forearm during rhythmic exercise. *Acta Physiol. Scand.* 67:5–92, 1966.

CHAPTER NINE

Key Concepts

• Erythrocyte or red blood cell (RBC) number and cell volume remain constant, but blood volume is reduced during a single bout of exercise.

• Endurance training elevates RBC number and blood volume.

• RBC metabolism is an important factor in determining hemoglobin-oxygen affinity and thus regulates oxygen delivery to peripheral tissues.

• High 2,3-diphosphoglycerate, temperature, and P_{CO_2} and low pH reduce hemoglobin-oxygen affinity—and therefore facilitate hemoglobin-oxygen dissociation in muscular tissue.

• Autoregulation is an important circulatory phenomenon that regulates blood flow in localized vascular beds. The local vascular beds are very sensitive to changes in P_{O_2}, pH, P_{CO_2}, adenosine, K+, and other metabolically related factors.

• Physical training may be an important therapeutic tool in lowering blood pressure in chronic hypertensive patients.

Circulatory factors in exercise

Hematology at Rest, During Exercise, and During Recovery

Body fluids are distributed within a number of compartments. These fluids are located in the vascular, intercellular, intracellular, and cerebrospinal spaces. Although all the compartments are in some way interconnected, the composition of their fluids varies. The intravascular compartment (whole blood) is of major interest to the exercise physiologist since it is the means by which O_2, CO_2, and other essential substrates are transported to and from the various tissues. Blood volume, as well as the concentration of its components, affects the efficiency with which O_2, CO_2, substrates, and metabolites are transported.

Plasma

Whole blood can be separated into *plasma* and a number of cellular components. Plasma consists of: (1) proteins such as antibodies, albumin, and hemoglobin; (2) substrates such as glucose and fatty acids; (3) ions such as Na+, Cl−, and K+; and (4) other metabolites.

The amount of plasma volume and its proper distribution into the various compartments is essential to life. Furthermore special adjustments in plasma volume are essential during exercise, or the appropriate volume of fluid will not be maintained within each compartment. The volume of fluid that is contained within the vasculature is indirectly related to intravascular blood pressure. In order for all tissues to be perfused adequately, but not under- or over-perfused, blood pressure must be kept relatively constant.

Corpuscular Portion of Blood

RED BLOOD CELLS OR ERYTHROCYTES

The cellular component of the blood tends to remain in the intravascular compartment, although red blood cells (RBCs), which are as large as the diameter of a capillary, can squeeze through the capillary wall in a process called *diapedesis* (Figure 9–1). In addition to playing a significant role in the transport of CO_2, RBCs are responsible for almost all of the oxygen-carrying capacity of the blood.

RBCs are less than 0.4 micron in diameter—but vary in size and shape. Each cell has a biconcave disc shape, is relatively thin centrally, and thickens around the periphery of the cell. The biconcavity of the RBC has been shown to be an optimal shape for oxygen uptake. A number of variables, such as aging, blood pH, and certain types of anemia (but probably not training), can alter the physical appearance of the RBC. If the concentration of substances dissolved in the plasma (*osmolarity*) is increased, RBCs lose fluid and decrease in size.

Contrary to the general conception of the RBC being a rather simple cell, it has an extremely intricate cytoplasm and cell membrane. When the normal interrelationships of the compact hemoglobin (Hb) within the cell are disrupted, its oxygen-binding behavior is likewise altered. For example Hb packaged within the RBC membrane takes up oxygen 20 times faster than the same amount of Hb in a noncellular solution, although the Hb concentration is unchanged.

The number of RBCs per unit volume of blood is directly related to the oxygen-carrying capacity of the blood. The greater the number of normal RBCs, the greater the capacity to deliver oxygen until the concentration of RBCs is so great that the blood becomes too viscous; at a certain point the disadvantage of increased viscosity (reduced blood flow) becomes greater than the augmented oxygen-carrying capacity. The size of the RBCs also is directly proportional to the oxygen-loading capacity of the blood. Smaller RBCs transport less oxygen because they have less total Hb, since Hb concentration per RBC tends to remain constant.

Composition of RBCs. The RBC consists of about 63.2% water, 33.5% hemoglobin (Hb), 0.5% lipid, 0.8% sugars, 0.7% salts, and 1.3% other proteins.

Metabolism of RBCs and O_2 Delivery. Metabolic pathways within RBCs are similar to other cells—but have some rather significant differences. *Glycolysis* is the major metabolic pathway for RBCs. A physiologically important "side branch" to the glycolytic pathway in RBCs is the formation of *2,3-diphosphoglycerate* (2,3-DPG) as well as *3-phosphoglycerate* from 1,3-DPG. The importance of 2,3-DPG is

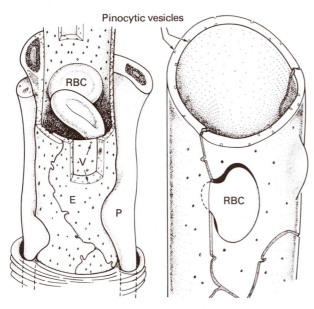

Pinocytic vesicles

Figure 9-1. Diagram (left) of a capillary containing red blood cells (RBC) and (right) of a capillary through which an RBC is escaping into the perivascular space through a process called *diapedesis*. Notice the numerous pinocytic vesicles and the nucleus of the endothelial cells (E). The capillary is a tubular structure made up of numerous endothelial cells and surrounded by special cells called pericytes (P).

(left) From L.V. Leak. Frozen-fractured images of blood capillaries in heart tissue. *Ultrastructure Research* 35:127–146, 1971.

that, along with ATP, it is involved in reducing the affinity of O_2 to Hb. This loss of attraction between Hb and O_2 provides a means for a rapid adaptation to a *hypoxic* state such as that arising from blood loss, high altitude, or perhaps exercise. The reduced affinity of O_2 to Hb facilitates the release of O_2 from Hb when the blood reaches the muscles or any other tissues. In the RBC, 2,3-DPG may also function as an energy store for protection of RBCs against low glucose conditions. DPG also has a significant pH buffering capacity.

Hemoglobin. Hemoglobin (Hb) carries about 98% of the oxygen that is transported in the blood. Consequently Hb would be expected to correlate quite highly with oxygen-carrying capacity over a wide range of Hb concentration values. Hb concentration in whole blood is normally reflected very accurately in the packed cell volume relative to whole blood (*hematocrit*, HCT), which is a reflection of RBC size and RBC number. This relationship is shown by the similarities of Hb concentration and HCT in Figure 9–2.

Blood pH. The pH of the blood markedly affects how much oxygen is delivered to the various tissues. For example a pH shift from 7.4 to 7.2 could increase by 15% the amount of O_2 that is released from Hb. At an O_2 pressure of 27 mm Hg (pH 7.4), about half of the Hb is associated with O_2. It takes a higher O_2 pressure to saturate half the Hb with O_2 if the temperature is elevated or the pH lowered—or if the level of 2,3-DPG in the RBC is elevated. The higher O_2 pressure is a result of the reduced attraction of O_2 to Hb.

Age of RBCs. The normal life span for an RBC is 120 days. The biochemical and biophysical changes associated with RBC aging are probably responsible for its ultimate fragility and lysis. In older cells the capacity for glycolysis progressively decreases as a result of alterations in the membrane or in the Hb molecule itself. Since glycolysis is slower, 2,3-DPG is also lower; so Hb effectiveness in O_2 delivery to the peripheral tissues is reduced. Also older RBCs do not become as saturated with O_2 as younger cells.

RBCs During Exercise. In acute bouts of exercise RBCs are exposed to more physical trauma than in periods of rest because of the forceful bombardment of the cells against the vascular walls. Also, if the exercise involves repeated contact with a hard surface, as in running or marching, RBCs are apparently ruptured—releasing the cellular contents, hemoglobin, and enzymes. If the rupturing is excessive, it results in *hemoglobinuria* and/or *proteinuria.* In addition the *acidosis* that occurs in response to some exercises would also tend to make the RBC more susceptible to lysis.

Effectively, there are an increased number of RBCs available to active tissues during exercise despite the fact that the absolute number within the vasculature is unchanged. This effective increase is due to the elevated blood flow rate. The higher rate at which the RBC traverses a capillary bed during exercise does not generally affect the amount of O_2 or CO_2 that can be picked up or delivered. The tissue and RBC P_{O_2} and P_{CO_2} have reached equilibrium well before the RBC has completed its route through the capillary bed. At rest this equilibrium has occurred by the time the RBC is one-third of the distance through the capillary.

Exercise increases the difference between the P_{O_2} (oxygen pressure) of the arteries and veins. A greater arteriovenous oxygen difference during exercise can be explained by two factors. Venous P_{O_2} will be decreased because of a faster rate of O_2 utilization by exercising muscles. Also, the reduced Hb-O_2 affinity causes a greater unloading of O_2 from Hb to the surrounding tissues. As noted earlier, the Hb-O_2 dissociation curve is shifted to the right by elevated temperature and lowered pH. Arterial P_{O_2} tends to stay at a constant level—except during the most intensive exercises.

The RBC Response to Training. Well-trained athletes usually have high RBC counts—but not always. Male Olympic athletes generally have counts of 16 gm/100 ml, but the count may be lower than 13 gm/100 ml or as high as 18.6 gm/100 ml. Only about 1% of the general population has a Hb concentration greater than 16 gm/100 ml (Figure 9–2). The generally greater Hb level in well-trained individuals is due to a greater number of RBCs—with little or no change in cell size.

Increased trauma to RBCs, perhaps due to the rapid blood flow during exercise, may counteract the general stimulus to synthesize more Hb. It may be the RBC destruction itself that activates the erythropoietic system to augment its rate of RBC production just enough to compensate for the physical disruption of RBCs caused by exercise.

Daily exercise bouts stimulate a more rapid rate of RBC development, perhaps via the hormone *erythropoietin*. Erythropoietin is probably produced by the *juxtaglomerular cells* of the kidney and is responsive to the balance between tissue oxygen supply and demand (see Chapter 12). The role of erythropoietin in mediating circulatory adaptations to exercise is unknown, but it does seem to be a factor in stimulating RBC synthesis. Anabolic steroids also seem to have a stimulating effect on RBC synthesis.

LEUKOCYTES (WHITE BLOOD CELLS)

Under normal circumstances the role of white blood cells (WBC) in serving a special function during an acute bout of exercise is insignificant. When the body is in the process of combating a bacterial infection, however, this line of defense cannot be ignored. Physical activity should be minimized during this time: it tends to facilitate the spread of a localized infection throughout the body.

The number of WBCs gradually rises towards a peak in the late afternoon and then drops to a low in the early morning. Exercise elevates the number of WBCs and causes changes in the population frequencies of specific types of WBCs; that is, of the different kinds of WBCs identified according to their morphology, *eosinophils* are fewer in number. *Eosinopenia*

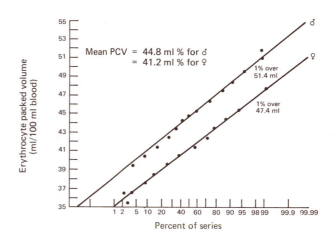

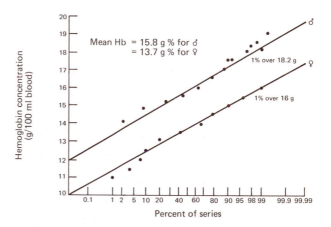

Figure 9–2. Percentile ranks of packed cell volume and hemoglobin with population means.

Adapted from P.L. Altman and D.S. Dittmer, eds. *Biological Handbooks: Blood and Other Body Fluids.* Bethesda, Md.: Fed. Amer. Soc. Exp. Biol., 1971.

(lack of eosinophils) is a well-recognized sign of chronic stress. The reason for this characteristic response to exercise and stress is unknown.

Blood Clotting Elements During Exercise

The clotting process is initiated by the release of chemicals from traumatized tissues and platelets. It involves the formation of a protein called *thromboplastin*, which, along with Ca^{2+}, stimulates the formation of another protein called *thrombin* from the substance *prothrombin*. Thrombin, in turn, activates the formation of *fibrin* from its parent substance *fibrinogen*. The fibrin acts like a fine mesh fishing net, collecting platelets, RBCs, and WBCs to form a clot (Figure 9–3). This complex of reactions is influenced by exercise. Whole blood clotting time decreases and thromboplastin increases about 20% with a single bout of exercise. The effect seems to occur in the initial clotting steps. It appears that the enhancement of factors related to the formation of clots may remain high for up to eight hours after exercise. Also clotting times are shorter in the afternoon than in the morning—another interesting diurnal variation.

The effect of exercise on the rate of clot formation can be accurately evaluated only if the rate of clot destruction is also taken into account. The clot dissolution process, *thrombolysis* and *fibrinolysis*, is controlled by several enzymes called *kinases*. The rate of clot dissolution after a single bout of exercise increases approximately 70%. Thus the net effect of a single bout of exercise may be to more than double the ratio of clot breakdown to clot formation. If these findings are substantiated by further experimentation, we can surmise that a *thrombus* (blood clot) is less likely to become an *embolus* (thrombus blocking a vessel) after an exercise session. The degree of physical ac-

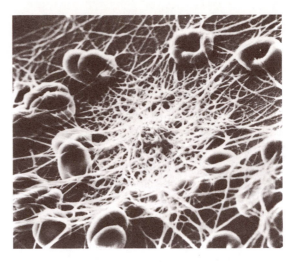

Figure 9–3. Scanning electron micrograph of fibrin and RBCs lying on a surface of polyvinylchloride. The blood clot is of normal human blood.

Courtesy of Nathaniel F. Rodman, M.D., University of Iowa.

tivity during a normal work day does not seem to have any effect on clot dissolution time.

There is evidence that platelet count increases in light exercise. The adhesiveness of the platelets and of other clotting elements may also be altered during exercise. Less is known about the effect of long-term training on the process of blood clotting; there are reports that blood clotting times tend to be longer in men with more active occupations—even though a single exercise bout has the opposite effect.

Blood Volume

ACUTE EXERCISE EFFECTS

During the initial stages of an exercise, blood volume decreases because of blood pressure elevation, which forces the fluid extravascularly. As much as a 20% hemoconcentration may occur in some capillary beds within the first ten minutes of an exercise. The percent of change in plasma volume during exercise can be determined with considerable accuracy by measuring changes in HCT and by plotting the results on the nomogram shown in Figure 9–4. This plasma efflux is counteracted relatively early because of a buildup of pressure in the extravascular compartments. The increased extravascular pressure is caused by the initial net flux of fluid into this volume of limited space. Most of the excess extravascular fluid is quickly returned to the circulation (intravascular space) after exercise.

In cross-country skiers, plasma volume may increase more than 10% after an 85 km race, although the total body weight decreases about 5%; this occurs when the skiers drink several liters of glucose water during the race. Plasma volumetric changes of as much as 25% have been observed without affecting the body's ability to take up oxygen. Cardiovascular adjustments in the redistribution of blood to the more immediately essential vascular beds are sufficient to compensate for these changes. As we pointed out in Chapter 8, a general vasoconstriction of the capacitance vessels (veins) increases the amount of blood available for distribution to the active tissues.

TRAINING EFFECTS

Long-term training may increase blood volume by up to 30%. Similarly, moving from sea level to an altitude of almost 4200 meters causes more than a 25% increase in blood volume over a period of several months. The greater volume of intravascular fluid can be easily accommodated by the compliance of the vasculature, particularly the venous segment. Endurance training or living at high altitudes increases the vascularity of muscles, which provides more space to store the greater blood volume of trained subjects.

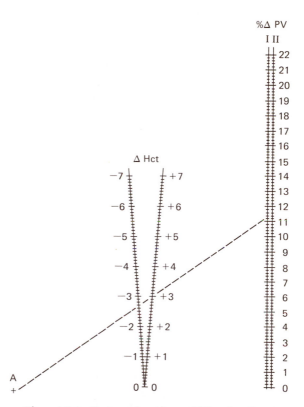

Figure 9–4. Determination of blood volume loss by measuring changes in hematocrit can be made from this nomogram. For example, an increase of 3 + 3 in the hematocrit would be indicative of an 11% decrease in plasma volume. A decrease in hematocrit is represented by the negative Δ Hct. For example, a −2.6 change equals an 11% increase in % Δ PV. Use line I if the initial Hct is greater than 45; use line II if the initial Hct is between 40 and 45. In either case, the final Hct must be less than 55.

From W. Van Beaumont. Evaluation of hemoconcentration from hemotocrit measurements. *Journal of Applied Physiology* 32:712–713, 1972.

Vasoregulatory Mechanisms Related to Blood Distributions

The engorgement of a muscle with blood in response to intensive exercise is a common phenomenon. How does this occur and does it serve a useful function? Blood flow (F), vascular resistance (R), and blood pressure (P) are the major factors that determine blood distribution and the transporting capacity of the blood. There are numerous ways each of these factors are influenced during exercise, recovery, and adaptation to training. Flow is directly related to blood pressure and indirectly related to resistance:

$$F = \frac{P}{R}$$

Pressure is often elevated during various forms of exercise; resistance is the more regulatory factor. Vasoregulatory mechanisms involved in the selective distribution of blood to the appropriate tissue involve neural, hormonal, and metabolic factors.

NEURAL

The sympathetic segment of the autonomic nervous system innervates vasoregulatory structures called *precapillary sphincters* (see Chapter 8). Sympathetic nerve endings release predominantly norepinephrine, a potent vasoconstrictor. (Norepinephrine is vasoconstrictory to most arteries—except to the coronary arteries, which dilate, and to the cerebral vessels, which are not markedly affected by sympathetic stimulation.) The immediate response to sympathetic stimulation is a reduction in the diameter of precapillary sphincters, which increase resistance to blood flow to the capillaries. A secondary and overriding response to norepinephrine is vasodilation. This action is described under the section on metabolic factors.

It is believed that some vascular sympathetic nerves release *acetylcholine* in skeletal muscle and skin (rather than norepinephrine). Acetylcholine causes vasodilation. Teleologically, this explanation seems convenient since sympathetic activity is increased during exercise; and vasodilation, rather than vasoconstriction, would be the more advantageous vasoregulatory maneuver in the active tissues to resist the elements of fatigue.

HORMONAL

Epinephrine is released by the adrenal medulla into the systemic circulation. In most vascular beds this hormone induces vasoconstriction, but epinephrine, like norepinephrine, causes vasodilation in the coronary vascular bed.

Histamine, a vasoactive hormone released from specialized cells throughout the body, reduces systemic resistance to blood flow by inducing relaxation of the vascular smooth muscle. It reduces blood-flow resistance in the muscle, skin, stomach, intestine, and kidney. Blood flow in human muscle can be increased threefold by histamine; however, resistance in the pulmonary veins is elevated by histamine.

Serotonin, another vasoactive chemical that is a known neurotransmitter at some synapses, is effective in inducing vasodilation in skeletal muscle. Histamine and serotonin are released from specialized cells of connective tissue (*mast cells*); the role of these vasoactive agents by themselves in exercise *hyperemia* is probably not of great significance.

METABOLIC

Local vasoregulatory mechanisms are important in regulating blood flow. Vasodilation occurs within a second after the onset of a single muscular contraction and lasts up to 20 seconds. A number of metabolically related chemicals can induce this response, but which factors are the most important has not yet been determined. Low pH, high K^+, Mg^{2+}, lactate, ADP, AMP, adenosine, and *osmolarity* (concentration of particles in solution), all direct by-products of elevated metabolism, are potent vasodilators.

Blood flow is closely linked to oxygen availability in relation to oxygen demand. When the oxygen demand increases, the availability increases by means of an increased blood flow. Physiologically all of the metabolic factors are integrated so that if one of these change, all change accordingly. Consequently exercise hyperemia or active hypermia (elevated blood flow) in skeletal muscles results from a multitude of chemical agents. The vasoregulatory factor that is most important in one tissue may not be important in another tissue.

Changes in pH and Mg^{2+} concentration are most important in the regulatory response of the coronary circulation to stimulation. In the kidney, low venous P_{O_2} seems to be the more important autoregulatory element.

Metabolic influences override the general vasoconstricting effect of the catecholamines, particularly in more metabolically active tissues such as working skeletal muscles. In fact catecholamines indirectly induce a vasodilatory effect after an initial transient vasoconstricting action. The indirectly induced vasodilatory action is mediated via the adrenergic beta-receptors, which, when activated, increase metabolism—and therefore the production of metabolites override the vasoconstrictive effect of catecholamines.

Blood Distribution at Rest and During Exercise

There is only a limited blood volume available to supply all of the tissues of the body. Since all tissues are not normally functioning simultaneously at their maximum rate, it is possible for the body to meet specific metabolic demands—assuming that the blood volume can be redistributed on the basis of need for each tissue bed.

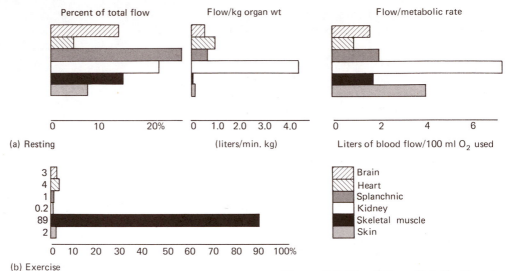

Percent of total flow   Flow/kg organ wt   Flow/metabolic rate

(a) Resting   (liters/min. kg)   Liters of blood flow/100 ml O_2 used

(b) Exercise

**Figure 9-5. Blood flow distribution to
various organs.**

Based on data from K.L. Andersen, The cardiovascular
system in exercise. In *Exercise Physiology* (H. Falls,
ed.). New York: Academic Press, 1968, p. 102. And
from P. Bard, ed. *Medical Physiology*, 10th ed.
St. Louis: C.V. Mosby, 1956, p. 221.

Figure 9-5 illustrates the relative proportions of the cardiac output flowing to the various organs. Also shown is the blood flow to each organ, expressed relative to the metabolic rate and per gram of tissue. It is evident that some relatively small and metabolically inactive organs such as the kidney receive large supplies of blood. Blood flow per metabolic rate shows that the kidney is the most overperfused tissue in the body—with the skin a close second. These ratios are consistent with the fact that the blood flow to each of these two tissues is only secondarily related to the metabolic rate of these tissues. The kidney's primary function is to regulate plasma concentrations by filtering, secreting, and reabsorbing various metabolic by-products, ions, and fluids. To maintain a constant filtering action, blood flow must be relatively high and constant. The primary function of the high blood flow to the skin is to regulate body temperature; the regulation of body temperature is important for its effect on work capacity, particularly in a hot environment where additional blood flow is needed for the skin as well as for the working muscles.

Although muscles receive almost 90% of the blood during vigorous exercise, they are not overperfused, since the increase in blood flow is proportional to the increase in oxygen utilization by the working muscles (Figure 9-6). Blood flow in a working muscle may increase eightfold; oxygen utilization rate may increase sixfold; and total oxygen uptake in the working muscle, thirtyfold.

Muscle contractions can become so intense that blood flow to that muscle is decreased. In Figure 9-7 we can see that a linear relationship exists between work intensity and blood flow up to a peak—at which point the muscle

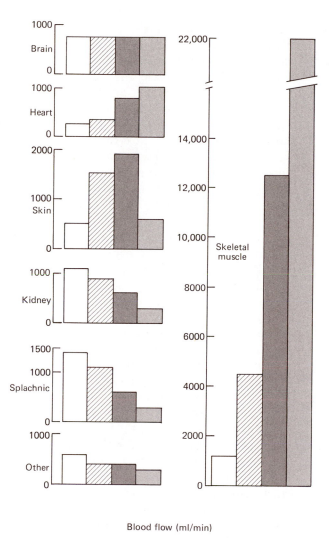

Blood flow (ml/min)

Figure 9-6. Redistribution of blood with increasing intensity of work.

Based on data from K.L. Andersen, The cardiovascular system in exercise. In *Exercise Physiology* (H. Falls, ed.). New York: Academic Press, 1968, p. 102.

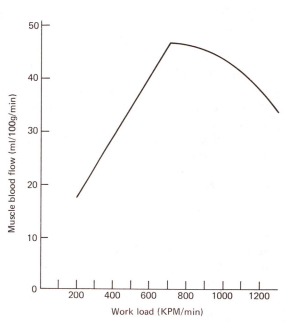

Figure 9-7. Proportionate increase in blood flow to the vastus lateralis with increasing work intensity on a bicycle ergometer (pedal frequency is kept constant). At high work loads, the contractions appear to be prolonged, resulting in reduced blood flow.

From J.P. Clausen, Muscle blood flow during exercise and its significance for maximal performance. In *Limiting Factors in Physical Performance* (J. Keul, ed.). Stuttgart: Georg Thieme, 1973, pp. 253–266.

contraction on the bicycle ergometer decreases blood flow even though pedal frequency is kept constant. In effect, not enough time between contractions was permitted for the vasculature to fill before the onset of the subsequent contraction. Blood flow to working muscles does not decrease at high work intensities when the duration of contractions is short enough and the duration of relaxation is long enough.

The pumping action of contracting muscle is very effective in increasing blood flow to the heart, particularly when running or similar exercises are involved. In fact the skeletal muscle pump is like the myocardium: both propel the blood through the vasculature. The contracting skeletal muscle simulates the systolic phase of the heart; when there is contraction, with subsequent ejection, there is very little blood inflow to the muscle. During muscle relaxation, as in the diastolic phase of the heart, there is a high inflow without the ejection of blood. The skeletal muscle "pump" can account for about 30% of the estimated energy requirement for promoting the return of blood to the heart for recirculation.

Figure 9–6 demonstrates the redistribution of blood resulting from different work loads. Note the following specific points:

1. Although total blood flow to the brain actually remains constant during exercise, regional blood flow in the areas of the brain involved in the movement patterns is increased more than 50% in relation to resting levels. For example, when blood flow was measured in subjects given a visual memory test, blood flow was elevated in the occipital region, decreased in the sensory-motor region, and did not change in the frontal region of the brain. During a reasoning test, increase in blood flow was found in these regions, especially in the occipito-temporal and precentral, but not in the sensory-motor and prefrontal cortex.

2. The heart has a high resting metabolic rate and blood flow; but during long vigorous exercise the flow may increase by fourfold. In contrast to skeletal muscle, the greater oxygen demand of the heart is met principally by increased flow and not by greater oxygen extraction—because in the heart the venous blood is relatively desaturated of oxygen even at rest.

3. The greatest proportional increase in blood flow is in the working muscles. All other systems support the working muscle during exercise. When we consider that the augmented flow represents the flow to *whole* muscles and that specific types of muscle fibers are selectively active depending on the nature of the exercise, the increase in blood flow must be even more dramatically elevated around the individual fibers that are actively contracting. The intensity and type of exercise also determine the nature of the exercise-enhanced flow.

4. Local vasoregulation in the skin capillaries primarily is involved in temperature regulation. During exercise the vasoactive vessels of the skin constrict because of sympathetic stimulation. If the exercise is long-lasting, the vessels dilate in response to the increase in metabolic products or to some unknown neural signals. Although skin blood flow may be reduced during short-term severe work, it is elevated upon cessation of work.

5. Renal and splanchnic blood flow is reduced during exercise. Even with an 80% decrease in blood flow to the liver, in most circumstances blood flow is sufficient to maintain an adequate output of glucose into the blood.

Splanchnic or visceral flow is reduced even more dramatically when exercise takes place at high environmental temperatures. In some cases of strenuous exercise in the heat, the liver of a normal individual cannot maintain an adequate supply of glucose to the blood to meet the metabolic demands of the CNS.

In response to exercise, redistribution of blood flow within a limb may occur without an overall change in blood flow. For example total blood flow to the forearm is unchanged during moderately severe leg exercise, but flow to the forearm skin increases at the expense of forearm muscular flow.

Blood Flow in the Trained Muscle

Significant vascular changes within skeletal muscle occur in response to endurance training. There is an increase in the number of capillaries per muscle fiber with training. The actual density (capillaries per unit of cross-sectional area) also increases—since, with endurance training, little change in fiber size takes place. For example, in Figure 9–8, the blood flow to a *maximally* working muscle group (*vastus lateralis*) increased 9%, while blood flow during a given *submaximal* work level was 15% lower after training than prior to training. This effect was found in middle-aged coronary patients after only six to nine weeks of training. The lower muscle blood flow per relative work load in trained muscle may be due to a smaller buildup of metabolites and/or a reduced effect of exercise on muscle pH or P_{O_2}. Each of these factors is known to contribute to an increase in blood flow in muscles. The greater *maximal* blood flow may be accounted for by a greater total cardiac output and more efficient distribution of more blood to active tissues. Both of these factors would result in greater muscle flow.

Exercise and Training Effects on Blood Pressure

Upon the initiation of muscular contractions, blood pressure rises immediately. The effect is mediated via the sympathetic nervous system and as a result elevates stroke volume,

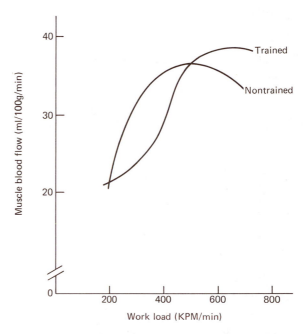

Figure 9–8. Blood flow to the vastus lateralis at a given work load is less in the trained than in the nontrained muscle when the load is submaximal. The maximal rate of flow is greater in the trained muscle.

From J.P. Clausen, Muscle blood flow during exercise and its significance for maximal performance. In *Limiting Factors in Physical Performance* (J. Keul, ed.). Stuttgart: Georg Thieme, 1973, pp. 253–266.

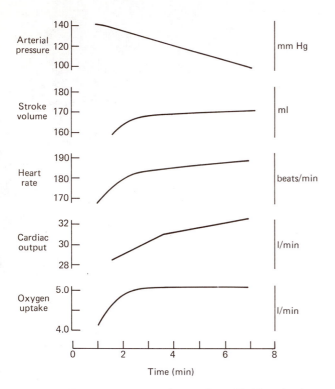

Figure 9-9. Hemodynamic variables during eight minutes of maximal cycling.

From B. Saltin, Oxygen transport by the circulatory system during exercise in man. In *Limiting Factors of Physical Performance* (J. Keul, ed.). Stuttgart: Georg Thieme, 1973, pp. 235–252.

heart rate, and consequently cardiac output. The initial, elevated mean arterial pressure at heavy loads subsides in six to nine minutes. The peripheral vasculature gradually vasodilates sufficiently to counteract the tendency for blood pressure to rise because of the increased cardiac output (Figure 9–9). Static isometric contractions can actually reduce cardiac output, as compared to the onset of dynamic muscular contractions, by reducing the return of blood to the heart. Static arm exercise elevates blood pressure more than leg exercise.

Training also alters blood pressure; the direction of the change is dependent on the initial blood pressure prior to training. In general a normal or below-normal resting blood pressure, prior to training, tends to increase with training or remains constant. Athletes tend to have slightly higher mean arterial pressure than untrained individuals (116 mm Hg as opposed to 112 mm Hg).

Training has proven to be a very effective therapy for lowering the blood pressure of some chronically hypertensive patients, for whom pharmacological attempts were unsuccessful. Three 6-second maximal isometric contractions of the neck, extremities, abdomen, and buttocks each day—for as little as five weeks—have been used to lower systolic and diastolic pressure by 16–42 mm Hg and 2–24 mm Hg, respectively. The mechanism for this effect is unknown. Similar effects have been observed in hypertensive animals subjected to endurance isotonic training.

Other Hematological Factors in Oxygen Delivery

Viscosity of blood is directly related to the number of RBCs per unit volume of blood and indirectly related to flow rate. Under most normal conditions, the viscosity of blood is not a major consideration during exercise. A hematocrit (percent packed cell volume of blood) of 50% has about a 20% greater viscosity than blood with a hematocrit of 40%. Men living at altitudes of about 1000 m usually have hematocrit levels in the high 50s; the normal value at sea level is near 45%. During exercise an increase in the hematocrit of 5% would not significantly increase the blood viscosity.

Summary

The blood consists of *plasma, erythrocytes, leukocytes* and *platelets*. The major constituents of erythrocytes, or red blood cells (RBC), are hemoglobin (Hb) and water. Hemoglobin makes up 95% of the dry weight of an RBC.

The oxygen-carrying capacity of blood is determined by the number of RBCs. RBC metabolism regulates, to some degree, the delivery of oxygen to peripheral tissues by varying the Hb-O_2 affinity. The 2,3-diphosphoglycerate (2,3-DPG) concentration, a special product of glycolysis in RBCs, reduces the affinity of oxygen to hemoglobin. Low blood pH and high temperatures have similar effects on the affinity of oxygen to hemoglobin.

Erythropoiesis, or RBC synthesis, is stimulated by endurance training and by exposure to high-altitude conditions. The hormone, *erythropoietin*, is released from the kidney in response to greater oxygen needs and perhaps low P_{O_2}.

Blood clotting time is shortened by physical activity. The rate of lysis of blood clots may also be more rapid after exercise. These findings indicate a relationship between exercise and the probabilities of developing a thrombus—with consequent vascular blockage.

Blood volume is elevated by training and decreased by a single bout of exercise. Blood volume changes can be estimated by the changes in *hematocrit*.

Vasoregulation within muscle refers to the local control of vascular resistance, which allows blood flow to remain in proportion to metabolic needs. Vasoregulation is under neural, humoral, and metabolic control. Through these effectors, blood is efficiently directed to the more active tissues (muscles) and away from the less active tissues (the splanchnic, renal, and dermal vasculature) during exercise. Low P_{O_2}, pH, and high P_{CO_2}, lactate, adenosine, inorganic phosphate, ADP and K^+ are some of the more important contributors to the vasodilatory response. In strenuous exercise skeletal muscle may receive almost 90% of the body's total cardiac output.

By lowering both systolic and diastolic blood pressures, endurance training appears to have a beneficial effect on chronic hypertension. The viscosity of blood is not usually a critical factor in blood flow.

Study Questions

1. Discuss the body compartments important for body fluid distribution.
2. What factors influence the oxygen-carrying capacity of the RBC?
3. What is important about the hemoglobin concentration?
4. Why do RBCs become more effective as deliverers of oxygen during exercise?
5. How is the blood-clotting mechanism altered during training?
6. What are the effects of exercise and training on blood volume?
7. What is the relationship between blood flow, blood pressure, and capillary resistance?
8. How is the blood flow in capillaries regulated?
9. What is the relative importance of metabolism in regulating blood flow?
10. How is blood distribution altered during exercise?
11. How does training affect blood flow?
12. What is the effect of exercise and training on blood pressure?

Review References

Barnard, J.R., et al. Cardiovascular responses to sudden strenuous exercise—heart rate, blood pressure, and ECG. *J. Appl. Physiol.* 34:833–837, 1973.

Donald, D.E.; D.J. Rowlands; and D.A. Ferguson. Similarity of blood flow in the normal and the sympathectomized dog hind limb during graded exercise. *Circ. Res.* 26:185–199, 1970.

Groom, D. Cardiovascular observations on Tarahumara Indian runners—the modern Spartans. *Amer. Heart J.* 81:394–414, 1971.

Higgins, C.B.; S.F. Vatner; and E. Braunwald. Parasympathetic control of the heart. *Pharmacol. Rev.* 25:119–155, 1973.

Hudlicka, O. *Circulation in Skeletal Muscle.* Oxford, England: Pergamon Press, 1968.

Kannel, W.B. and T. Gordon. *The Framingham Study: An Epidemiological Investigation of Cardiovascular Disease.* Section 24: Diet and the regulation of serum cholesterol. Washington, D.C.: U.S. Dept. of Health, Education and Welfare, Public Health Service, NIH.

Korsan-Bengtsen, K.; L. Wilhemsen; and G. Tibblin. Blood coagulation and fibrinolysis in relation to degree of physical activity during work and leisure time. *Acta Med. Scand.* 193:73–77, 1973.

Tipton, C.M. and B. Taylor. Influence of atropine on heart rates of rats. *Amer. J. Physiol.* 208:480–484, 1965.

Wahren, J. Quantitative aspects of blood flow and oxygen uptake in the human forearm during rhythmic exercise. *Acta Physiol. Scand.* 67:5–92, 1966.

Wintrobe, M. *Clinical Hematology.* Philadelphia: Lea and Febiger, 1967.

CHAPTER TEN

Key Concepts

• Ventilatory responses to exercise are controlled by neural and chemical mechanisms.
• Respiration is responsive to metabolic demands.
• Enhanced ventilation and cardiac output increase oxygen and carbon dioxide exchange in the lungs.
• Oxygen uptake is related to cardiovascular factors and metabolic demands.
• Oxygen exchange is determined by pressure gradient differentials and hemoglobin saturation.
• Oxygen-utilization rate during exercise is determined by muscle metabolism.
• Ventilatory adaptations to training are minimal.

Respiratory factors in exercise

Introduction

The functions of the respiratory system are: to exchange air in the lungs (*ventilation*), to extract oxygen from the air, to cooperate with the vascular system in the delivery of oxygen to the tissues, to provide for the intracellular utilization of oxygen in the mitochondria, to extract carbon dioxide from the tissues, to cooperate with the vascular system in the delivery of carbon dioxide to the lungs, and finally to deliver the carbon dioxide to the atmosphere. In this chapter we will trace the path of oxygen from the atmosphere to the site of utilization (the mitochondrion) and to trace the path of carbon dioxide from the site of production in the cell to its delivery to the atmosphere. These pathways involve the ventilation of air by the lungs, the diffusion of oxygen and carbon dioxide in the alveoli, the transport of the gases via the cardiovascular system, and the diffusion of gases at the capillary-cell border.

Functional Organization

Anatomical Considerations

As the air enters the body through the nose and mouth, it is warmed and humidified. The colder or dryer the air, the more important is this process for transforming the incoming air to the temperature and humidity of the lungs. The mouth, nasal passages, and throat are well prepared for this task because of their high concentration of blood vessels.

As the air leaves the trachea and passes into the lungs, it continues to pass through increasingly smaller airway passages until it arrives at the terminal *alveoli* sacs (Figures 10–1 and 10–2). It has been estimated that there are approximately 250–300 million alveoli sacs in the lungs of adult humans. The passage of the air through these increasingly smaller diameter tubes encounters an airflow resistance. To help reduce this resistance, the alveolar sacs are coated with a slick lipoprotein called *surfactant*.

The nourishing blood supply enters the lungs through the bronchial artery; usually the metabolic requirement forms about 1% of the resting cardiac output. During times of high exercise stress, this supply may exceed 10% of the cardiac output.

The *circulating* blood supply, perfusing the alveoli for the purpose of reoxygenation, arrives at the lungs via the pulmonary artery. At rest the total blood volume, which is approximately 6 liters, passes through the lungs approximately every minute.

There are nearly 300 billion capillaries within the lungs; the capillary-alveoli surface contact area is nearly 70 square meters. When the resting cardiac output is about 6 l/min, an average red blood cell spends less than one second in contact with an alveolar sac. During strenuous exercise, when the cardiac output is increased to around 30 l/min, the time spent in contact with the alveoli is much less; how-

ever, in normal lungs the exchange of gases is still sufficient to permit near normal oxygen saturation of hemoglobin and carbon dioxide exchange. It is highly unlikely that the maximal potential of the diffusion of gases within the lungs is reached during exercise.

Mechanics of Breathing

The actual mechanics of air ventilation (*breathing*) involve a periodic alteration of the pressures within the lungs in response to an alteration of the volume of the thoracic cavity. This volume change is accomplished by the diaphragm, the external and internal intercostal muscles, and the abdominal muscles.

The diaphragm muscle is attached anatomically in such a way that, upon contraction, the volume within the lung cavity is increased. The contraction of the external intercostal muscles also increases the volume in the lung cavity. The increased volume in the thoracic cavity decreases the air pressure and allows air from the atmosphere to enter the lung tissue. A relaxation of the diaphragm, a contraction of the abdominal muscles, and a contraction of the internal intercostal muscles cause a decrease in the volume of the thoracic cavity. This decreased volume increases the intrathoracic pressure and thus forces air out of the lungs into the atmosphere.

During normal inspiration and expiration, the pressures within the lungs vary from −8 mm Hg to −3 mm Hg with respect to the atmospheric pressure (approximately 760 mm Hg). The lungs are elastic to the extent of about 5 mm Hg. This elastic force, in combination with the expiration pressure of −3 mm Hg, gives a +2 mm Hg pressure above that of the atmosphere. This positive pressure allows the lungs to force out air during expiration.

During exercise, the inspiratory and expiratory pressures within the thoracic cavity may range from −80 mm Hg to +100 mm Hg with

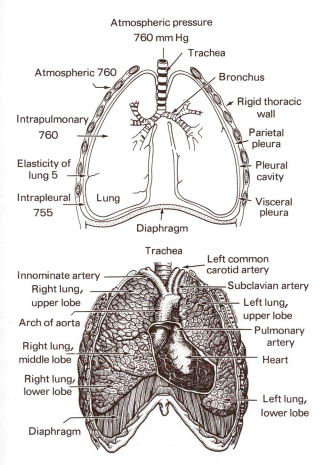

Figure 10-1. Anatomical considerations of the lung, thoracic cavity, and heart.

From Byron A. Schottelius and Dorothy D. Schottelius. *Textbook of Physiology*, 17th ed. St. Louis: C.V. Mosby, 1973.

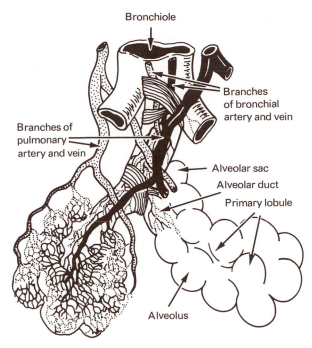

Figure 10-2. The capillary-alveoli junction.

From Byron A. Schottelius and Dorothy D. Schottelius. *Textbook of Physiology*, 17th ed. St. Louis: C.V. Mosby, 1973.

respect to atmospheric conditions, thus creating absolute values of 680 mm Hg to 860 mm Hg. The 680 value would allow air to rush into the lungs; and the 860 value would force air out at a rapid pace—since the atmospheric pressure has not changed from its value of 760 mm Hg. These pressures are generated in response to the rapidity with which the respiratory muscles can contract.

Respiratory Volumes

The amount of air within the lungs can be divided into four major parts. Under normal conditions the respiratory volumes in females are approximately 65–70% of those values found in males; in the following discussion, typical values found in males are used.

Under normal resting conditions, a healthy young male adult exchanges about 500 ml of air with each breath. This type of breathing requires neither the maximum amount of inhalation nor the maximum exhalation. This normal exchange is termed the *tidal volume*.

If after a normal inhalation he continues to inhale to his maximum, this increased value (approximately 3000 ml) is referred to as the *inspiratory reserve volume*.

If after a normal expiration he continues to expire, he would test his *expiratory reserve volume*. This volume is estimated to be about 1200 ml in the young adult male. The air remaining in the lungs after the complete expiratory effort is referred to as the *residual volume*, which averages 1200 ml.

The sum of these four volumes represents the *total lung capacity for air*, approximately 6000 ml (Figure 10–3). Without structural alteration, this total lung capacity cannot change during exercise; however, the individual lung volumes can and do change during exercise. As we would expect, the tidal volume increases; the inspiratory reserve and the expiratory reserve volume decrease. The residual volume of the lungs does not change.

It is convenient to define other functional volumes. The first of these is the *vital capacity*, defined as that maximum amount of air that can be exhaled after a maximum inhalation. Vital capacity can be calculated from the sum of the expiratory reserve volume, the tidal volume, and the inspiratory reserve volume—and equals approximately 4800 ml. The *inspiratory capacity* is that amount of air that can be inhaled after a normal exhalation. This volume represents about 3500 ml in the normal young adult male.

The *functional residual capacity* is that amount of air remaining in the lungs after a normal expiration (2400 ml). The functional residual capacity represents that amount of air available for the transfer of oxygen and carbon dioxide during the period of expiration, prior to the next inspiration.

There is an amount of air that can be classified as *dead space*. This volume represents that air contained in the airway passages from the mouth and nose to the alveoli (estimated at nearly 150 ml in the healthy young adult). In the normal person the *functional tidal volume* is 500 ml minus the dead space volume (150 ml), which equals 350 ml per breath. In pathological conditions where some of the alveoli are nonfunctional, the volume of air in these alveoli should be considered as dead air space.

In the laboratory or classroom, these lung capacities or volumes can be measured by the use of a *spirometer*. The spirometer records the amount of air exhaled and, under the conditions described above, you can measure all of the lung volumes except the residual.

Breathing Frequency

Under resting conditions, the normal tidal volume is nearly 500 ml per breath. The breathing frequency ranges from 10 to 15 breaths per minute (an average of 12). Thus the total air ventilated per minute would be 6000 ml. Under usual exercising conditions, the maximum sustained tidal volume is about one-half of the vital capacity. This rate and depth of breathing would amount to an exchange of air of 120 l/min (50 breaths per minute times a tidal volume of 2400 ml). During maximal short-term exercise, the respiratory rate may rise to 50–60 breaths per minute, and the tidal volume may approach that of vital capacity.

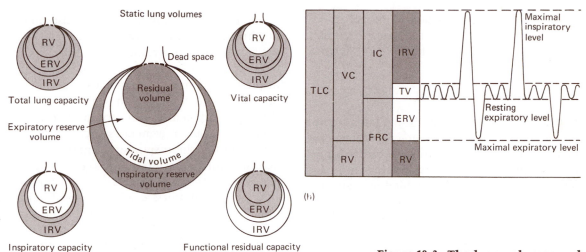

(a)

(b)

It is possible, during short periods of maximal exercise, to ventilate 150–220 liters of air per minute.

Functional Capacities

Two tests of the functional capacity of the ventilatory mechanisms are the determination of the forced expiratory volume and of the maximal voluntary ventilation. The *forced expiratory volume* is determined by the amount of air a person can expire in one second after a maximal inspiration. This value should be about 80% of the total vital capacity; values less than this may indicate increased air resistance or blocked air passages.

The determination of the *maximal voluntary ventilation* rate can be made by measuring the amount of air ventilated in a specified time interval (usually 15 seconds). Values in the range of 100–180 l/min are normal for adult males; values of 70–120 l/min normal for females. Values less than these figures may

Figure 10–3. The lung volumes and capacities. (a) The large central diagram illustrates the four primary lung volumes and their approximate magnitudes. These are: (1) *tidal volume,* **the volume of gas inspired or expired during each respiratory cycle; (2)** *inspiratory reserve volume,* **the maximal amount of gas that can be inspired from the end-inspiratory position; (3)** *expiratory reserve volume,* **the maximal volume of gas that can be expired from the end-expiratory level; (4)** *residual volume,* **the volume of gas remaining in the lungs at the end of the maximal expiration. The surrounding diagrams show the four lung capacities (shaded areas). These are: (1)** *total lung capacity,* **the amount of gas contained in the lung at the end of a maximal inspiration; (2)** *vital capacity,* **the maximal volume of gas that can be expelled from the lungs by forced effort following a maximal inspiration; (3)** *inspiratory capacity,* **the maximal volume of gas that can be inspired from the resting expired level; (4)** *functional residual capacity,* **the volume of gas remaining in the lungs at the resting expired level. (b) Lung volumes as they appear in spirographic tracings. The first vertical bar to the left of the tracings corresponds to that in the central diagram in part (a).**

Adapted from *The Lung,* 2d ed., by J.H. Comroe, Jr., et al. Copyright © 1962 by Yearbook Medical Publishers, Inc., Chicago. Used by permission.

indicate weakened mechanical properties of the lungs, chest cavity, or respiratory muscles, increased air resistance, or blocked air passages.

Ventilatory Control

Ventilation is controlled by neural and humoral mechanisms. The neural control mechanisms seem to be more prominent during normal inspiration and expiration and are reflexive in nature. The humoral mechanisms exert relatively more influence when blood chemicals exceed their normal concentrations, a common condition during exercise.

Neural Control

Since the varying thoracic pressure, which regulates air inflow and outflow, is the result of contraction and relaxation of muscles, it must be under neural control. Therefore we should investigate what stimulus initiates the neural impulse that in turn initiates the muscular contraction.

In essence the neural controlling mechanisms for ventilation involve the inspiratory, the pneumotaxic, and the expiratory centers, the stretch receptors in the lungs, and major influences from the higher centers of the brain. The relationship between these controlling mechanisms, under normal conditions, is relatively straightforward.

There are stretch receptors within the lungs that signal the degree of distension of the lungs. These stretch receptors are stimulated when the lungs are expanded and inhibited when the lungs are relatively collapsed. The neural signals from these receptors are transmitted via the vagus nerve to the respiratory center, located in the *medulla oblongata*. This area of the midbrain consists of an *inspiratory* and an *expiratory* center (Figure 10–4).

The *inspiratory center* sends impulses to the diaphragm and external intercostal muscles via the spinal cord; as a result, these muscles contract and thereby initiate respiration. The inspiratory center simultaneously stimulates the *pneumotaxic center* to send impulses to the expiratory center. The stimulation of the *expiratory center* generates neural signals, which inhibit the inspiratory center. Simultaneously during inspiration, as the lung expands, the *stretch receptors* in the lungs are stimulated and initiate additional inhibitory impulses to the inspiratory center. The summation of the impulses from the expiratory center and the stretch receptors sufficiently inhibit the inspiratory center.

When the inspiratory center is inhibited, (1) the inspiratory muscles relax, (2) decreased stimulation is extended to the pneumotaxic center, thereby decreasing stimulation to the expiratory center, and (3) the expiratory center releases its inhibitory influences upon the inspiratory center. With the relaxation of the diaphragm and external intercostal muscles, the pressure within the lungs increases, forcing air out of the lungs and partially deflating the alveoli—thereby relaxing the stretch receptors. The combination of the removal of inhibitory impulses from the expiratory center and from the stretch receptors allows the inspiratory center to begin the next cycle.

Humoral Control

During exercise there is an alteration in the concentrations of critical chemicals in the blood. Among the most critical chemicals for the regulation of ventilation are the partial pressure of carbon dioxide (P_{CO_2}) and the concentration of hydrogen ions (pH). There are at least two anatomical areas where the concentration values of these variables are monitored. The most important monitoring area lies in the *medulla oblongata* near the respiratory centers. The other is two groups of cells, the *carotid* and *aortic bodies*, located at the bifurcation of the carotid artery and the arch of the aorta.

The concentration of CO_2 and hydrogen ions are positively related to the rate and depth of ventilation; that is, by an increase in the concentration of CO_2, which increases the concentration of H^+ (decreasing the pH), ventilation is stimulated. As we will discuss later, these two factors are closely related to the partial pressure of CO_2 through the formation of carbonic acid (H_2CO_3), which ionizes to H^+ and HCO_3^-.

The central and peripheral chemoreceptors respond to the chemical concentrations in the blood and transmit the appropriate information to the brain via sensory nerves. A rise in arterial P_{CO_2} from 40 mm Hg to 41.5 mm Hg stimulates ventilation two to three times. This stimulus alone cannot account for the total increase in ventilation observed during exercise, as exercise may elevate ventilation by as much as 30 times.

A decrease in the blood partial pressure of oxygen (P_{O_2}) is also known to be a stimulus to increase ventilation. Decreasing the P_{O_2} increases the ventilation; however, the amount of stimulation is relatively insignificant until the P_{O_2} becomes very low. Also normal alveolar ventilation can be reduced to almost half before the oxygen saturation in the blood is altered. Since the alveolar ventilation can change up to tenfold without altering the saturation of the oxygen content of the blood, it is not surprising that arterial P_{O_2} is not the most important factor in regulating ventilation. This control mechanism appears to exert very little influence during resting or exercising conditions.

Mechanisms controlling ventilation are both neural and humoral. The neural mechanisms are: (1) pathways irradiating from the motor

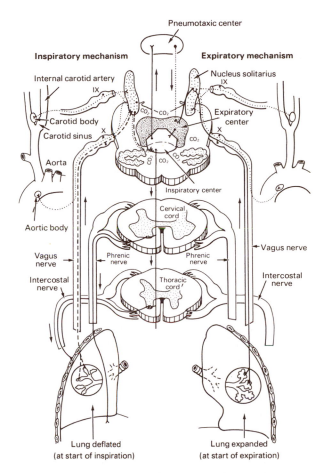

Figure 10-4. Neural control of breathing.

Adapted from an original painting by Frank H. Netter, M.D. from The CIBA Collection of Medical Illustrations copyright by CIBA Pharmaceutical Company, Division of CIBA-GEIGY Corporation. All rights reserved.

cortex to skeletal muscle of the limbs that also send collateral branches to the respiratory center and (2) receptors located within the muscles (spindles), joints, and the walls of the lungs that activate the medullary respiratory center. *Humoral* factors include: (1) *thermoreceptors* that act via the hypothalamus and (2) *chemoreceptors* that act directly on the medullary respiratory center or indirectly through peripheral chemoreceptors.

Evidence of these neural and humoral pathways for activation of the medullary respiratory center is supported by the following observations. *Hyperpnea* (elevated ventilation) is induced at the initiation of passive limb movement (no change in oxygen absorption rate). Hyperpnea during exercise is also induced by chemoreceptor sensitivity to CO_2 production by skeletal muscles. It is of interest that cross-circulation experiments (where the blood from an exercising animal, in this case, is circulated through a resting animal) have shown that there are receptors sensitive to P_{CO_2} since ventilation increases in the resting control animal. Presumably this mechanism assures a steady monitoring of the most critical ventilatory humoral stimulus, blood P_{CO_2}. At higher work intensities other ventilatory stimulants are available such as lactic acid, which lowers the blood pH.

Breathing air containing a high oxygen concentration or pure oxygen during high work intensities reduces the elevation in ventilation. As exercise continues, the metabolic products (CO_2 and lactate) act to adjust the ventilation, via the humoral-neural network, to the necessary increased rate compatible with the severity of the exercise stress.

The mechanisms discussed in this section cannot fully account for the increased ventilation observed during exercise; the possibility remains that some additional unknown mechanisms exist.

Gas Exchange in the Lungs

Alveolar Ventilation

The tidal volume represents the amount of air moved into and out of the lungs per breath. However, as we have discussed before, the effective tidal volume or the *alveolar ventilation* is the tidal volume minus the effective dead air space. With a normal tidal volume of 500 ml and a dead air space of 150 ml, the alveolar ventilation is 350 ml per breath. Given a respiration rate of 12 breaths per minute, the alveolar ventilation per minute is 4200 ml/min.

Alveolar ventilation can be altered by changing the tidal volume and/or the breathing frequency; however, the dead air space will remain the same. During shallow breathing (low tidal volume), more air must be ventilated in order to effect the same alveolar fresh air ventilation. This is because the dead air space occupies a greater percentage of the smaller tidal volume.

The alveolar ventilation volume of 350 ml per breath represents only about one-eighth of the air in the lungs at any one time (the functional residual capacity of the lungs is 2400 ml). If we assume that each expiration removes air other than that air recently inspired, it would take about 40 seconds (eight breaths) to replace all of the old air with new air.

The exchange of air within the lungs allows for a relatively constant gas pressure within the alveolar sacs. By increasing the tidal volume, we alter the partial pressures of the gases within the alveoli and exhale more carbon dioxide. More CO_2 is exhaled since the incoming air has a lower CO_2 concentration than the alveolar air, and the expired air is always a mixture of tidal and alveolar air. The amount of oxygen absorbed will not increase since the oxygen concentration of the blood is already nearly maximum. Only by increasing the blood flow will the rate of oxygen uptake be increased.

During long-term exercise, the tidal volume may approach half of the vital capacity, which allows the lungs to ventilate approximately 2400 ml of air per breath. We cannot assume that all of the old air is replaced by new air during each breath. Due to the behavior of gases, this is obviously not the case, but it does serve to illustrate that by increasing the tidal volume we increase the rate of turnover of air within the alveoli.

The increased tidal volume during exercise results in an increased amount of air in the lungs during the inspiratory phase (a decrease in inspiratory reserve). The increased tidal volume during exercise is accomplished by increasing the depth of inspiration and increasing the depth of expiration.

Ventilation-Oxygen Uptake Ratio

A very useful ratio that can be used to estimate the efficiency of ventilation is the *ventilation-oxygen uptake ratio* ($\dot{V}_E/\dot{V}_{O_2}$). This calculation compares the amount of air ventilated in the lungs to the amount of oxygen taken up by the blood.

During rest and mild activity of young, healthy, male adults, the ratio of air ventilation to oxygen uptake is about 25 (ventilation of 6000 ml/min). With mild exercise the ratio falls to almost 20. With a more strenuous exercise, there is an increase in this ratio. This change occurs when the exercise is such that the oxygen uptake exceeds 3–4 liters of oxygen per minute. (Figure 10–5). If we assume a maximum sustained rate of ventilation of about 130 liters per minute and an oxygen uptake of 5.2 liters per minute, the ratio of $\dot{V}_E/\dot{V}_{O_2}$ is 25. Usually the $\dot{V}_E/\dot{V}_{O_2}$ ratio increases at the high ventilation volumes and may reach a value as high as 30. (Figure 10–5). In well-trained athletes, $\dot{V}_E/\dot{V}_{O_2}$ is less than that which is characteristic of nontrained individuals—

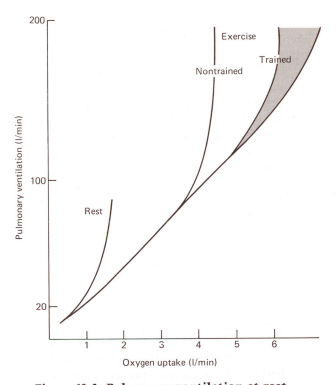

Figure 10-5. Pulmonary ventilation at rest and during exercise of trained and nontrained individuals. Top athletes fall within the shaded area. Note that oxygen uptake does not depend solely upon ventilation. This relationship between oxygen uptake and ventilation would indicate that exercising muscles and cardiac output are the greater determinate of oxygen uptake.

since trained persons can absorb more oxygen per given amount of ventilation.

Ventilation-Perfusion Ratio

Another ratio used to describe the functional status of ventilation is the *ventilation-perfusion ratio*. At rest the normal alveolar ventilation is 4200 ml/min and the blood flow is 6000 ml/min (about 100 ml of blood are in the pulmonary capillaries during any given instant). Given these values, the ratio of the air ventilation to blood flow is 0.7 (4200/6000). During times of strenuous exercise, the alveolar ventilation may rise to 130 l/min and the cardiac output may be up to 25 l/min. These values give a ratio of 5.2.

Oxygen uptake appears to be limited by the maximum cardiac output perfusing the lungs. As the cardiac output nears its maximum and the blood perfusing the lungs continues to be saturated, it is probable that the ratio of air ventilation to oxygen uptake will increase. If at maximum cardiac output the blood is being completely saturated, then any increased ventilation will not increase the oxygen uptake. At this stage the *oxygen cost of breathing* will equal or exceed the additional oxygen uptake as the result of further ventilation (Figure 10–5).

Lung Slices

In many of the calculations in the previous sections, we assumed that the anatomical sections of the lungs were equally ventilated or perfused. It has been shown that this is not the case: there exist portions of the lungs that are either under- or over-perfused and/or ventilated. It is generally accepted that this difference is related to the effect of gravity on the blood volume. The effect of gravity results in a greater volume of blood in the base of

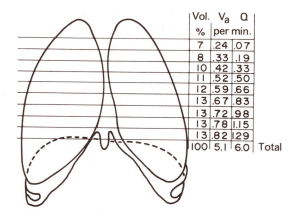

Vol. %	V_a per min.	Q
7	.24	.07
8	.33	.19
10	.42	.33
11	.52	.50
12	.59	.66
13	.67	.83
13	.72	.98
13	.78	1.15
13	.82	1.29
100	5.1	6.0 Total

Figure 10–6. Observed distribution of ventilation and perfusion on regional slices within the lung of a normal man in sitting position. The alveolar ventilation per unit of blood flow may be calculated for each region.

From J.B. West. Regional differences in gas exchange in the lung of erect man. *Journal of Applied Physiology* 17(6):893–898, 1962.

the lungs in comparison with the upper portions. Given this distribution in the blood flow, we would expect differences in the ventilation-perfusion ratios according to the area of the lung under discussion (Figure 10–6). These differences almost completely disappear when measurements are taken in the supine position (where the force of gravity is minimal).

During exercise the blood pressure in the pulmonary vessels increases from resting values near 10 mm Hg to 25 mm Hg during steady-state work and to near 50 mm Hg during maximal work. The increase in the pulmonary blood pressure forces the blood into the under-perfused areas, and the regional differences are almost totally eliminated. Thus the apparent efficiency of the lungs increases with increasing ventilation and perfusion pressures.

Energy Cost of Breathing

Since we have established that the process of breathing requires the operation of inspiratory and expiratory muscles, we know that there must exist energy requirements to operate these muscles. At rest, when the tidal volume is 500 ml/min and the frequency is 12 breaths per minute the energy cost has been estimated to be 6 ml of O_2 (1 ml of oxygen per liter of air ventilated). The 6000 ml of air ventilated per minute result in the uptake of 250 ml of oxygen. Thus the energy cost for the operation of the lungs requires 6 ml of the 250 ml of O_2, which is less than 3% of the total oxygen uptake.

Airway resistance is the source of some energy demands. Analysis of airway resistance indicates that most of this resistance is due to airway turbulence, and less is due to the resistance of the elastic components of the lungs. Therefore, during times of exercise, the resistance to inspiration and expiration increases greatly as the volume of air moving through the airway increases. This increased resistance must be overcome by an increased effort on the part of the respiratory muscles. By the time the exercise is so severe as to require 130–160 liters of air per minute, the oxygen cost of breathing is 520 ml. This cost of breathing is now over 10% of the total oxygen uptake (Figure 10–7).

Respiratory Exchange Index

The ratio of the amount of CO_2 expired to the amount of O_2 absorbed has been used for some time as a partial indication of the energy exchange within the body. This ratio is known as the *respiratory exchange index* (R) or somewhat imprecisely as the respiratory quotient (RQ). The rationale is that O_2 is utilized for energy production and CO_2 is produced from

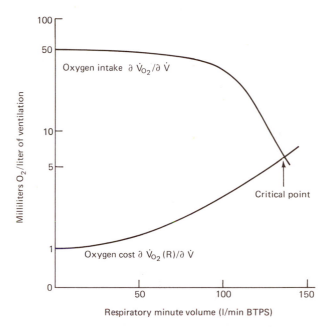

Figure 10–7. Illustration of the concept that the oxygen cost of breathing reaches a critical point with an increase in the respiratory minute volume. Beyond this critical point, the oxygen consumed by the respiratory muscles per liter of ventilation exceeds the oxygen intake achieved by the additional ventilation.

From R.J. Shephard. *Endurance Fitness.* Toronto: University of Toronto Press, 1969, p. 41.

the oxidation of foodstuffs. Under normal resting conditions and with a mixed diet, the R is approximately 0.825.

We should remember that the amount of gas exchange at the lungs is partially a measure of *total* body metabolism. There are some tissues that are underperfused and some that are overperfused; that is, the tissue receives less or more blood than it needs for its metabolic needs. Some resting tissues utilize primarily carbohydrates; others use primarily lipid sources for energy production. Therefore the R, which is determined by an analysis of inspired and expired oxygen and carbon dioxide, represents an average of all active cells in the body.

During short-term moderate exercise, the R value increases to near 1.0. During maximal exercise, the R may increase to as high as 1.20. Although these high values do not represent the type of fuel being oxidized in the cells, they have been interpreted to represent a high degree of nonoxidative metabolism. As we shall see later in this chapter, increased amounts of lactic acid in the blood (resulting from a high rate of nonoxidative metabolism) lower the pH, which aids the diffusion of CO_2 molecules and increases the amount of CO_2 expired. Another source of the excess CO_2, measured during expiration at the high work rates, may arise in the washout of CO_2 from the tissues without the corresponding use of oxygen in these tissues (assuming that the energy production is nonoxidative). The CO_2 may have been residual molecules that were in the cell prior to the shift to completely nonoxidative metabolism. Values of the R above 1.0 observed during recovery may arise from the same considerations affecting the CO_2.

Endurance training results in a lower measured R at a given work level compared with that observed prior to the training. This can be interpreted as representing a greater utiliza-

tion of lipids by the active tissues since pure lipid metabolism would give an R ratio of 0.7 and carbohydrate an R of 1.0.

Other Considerations

HYPERVENTILATION

Hyperventilation is that type of breathing during which the ventilation rate and/or depth is increased over that rate and depth normally required to maintain the metabolic demands of the body. Oxygen uptake will not increase, unless there is a corresponding increase in the cardiac output, since the arterial blood is already nearly 100% saturated. The expiration of carbon dioxide is increased since the partial pressures of carbon dioxide in the lungs and blood are not at an equilibrium under normal circumstances. Since CO_2 is a stimulant to the respiratory center of the brain, a lowered concentration of CO_2 in the blood by hyperventilation would act to lower the breathing rate and to increase breath holding time.

Fainting is known to occur if long-term maximal breath holding follows hyperventilation. The explanation most often given is that the hyperventilation decreases the P_{CO_2} in the blood; then during the breath holding time, the P_{CO_2} rises very slowly. Since we have discussed the critical role that P_{CO_2} plays in the regulation of breathing, you can see that a person may continue to hold his breath until the P_{CO_2} comes up to a level to stimulate continued breathing. In the meantime the P_{O_2} will have decreased to such an extent (since O_2 is being used but not replenished) to be below the critical level for the operation of the nervous system. The low P_{O_2} will result in a fainting point, at which time breathing resumes (Figure 10–8). In this case fainting is a protective mechanism assuring adequate tissue oxygenation. It is highly unlikely that hyperventilation would

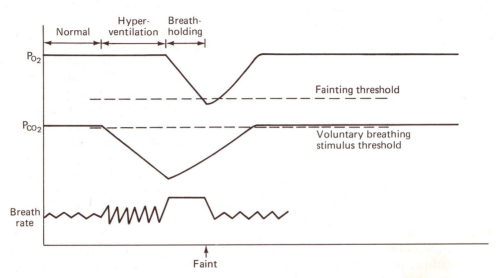

Figure 10–8. Partial pressures of CO_2 and O_2 in blood during normal breathing followed by hypertension and breath-holding. Note that hyperventilation does not increase O_2 content but merely decreases CO_2. During breath-holding, if the fainting threshold of P_{O_2} is reached, fainting occurs with involuntary normal breathing. If fainting does not occur, then onset of breathing is initiated by P_{CO_2}.

aid in athletic events because it is important to maintain a high oxygen pressure. It could prove to be excessively dangerous during underwater swimming.

VALSALVA MANEUVER

The *Valsalva maneuver* is a name applied to a sequence of events that basically involves the attempted forced expiration against a closed windpipe (*glottis*)—following a maximum inspiration. This requires that the expiratory muscles attempt to expire air against a closed exit. If the forces against the thoracic cage are greatly increased and the air cannot escape, the pressures in the cage are greatly increased. These pressures inhibit blood flow and can most effectively reduce the cardiac output. If the normal pressures within the blood vessels in the pulmonary circuit are 80 mm Hg, the Valsalva maneuver can cause an increase to approximately 200 mm Hg. This 200 mm Hg resistance is too much for the right heart to pump against, thus decreasing the cardiac output of the heart by decreasing the rate of return of blood to the heart from the lungs.

The Valsalva maneuver is very common during strength activities such as weight lifting. It is used to stabilize the rib cage for the operation of the powerful muscles of the back and arms.

SECOND WIND

The phenomenon of *second wind* is observed by many people, most often those that are nontrained. It is usually described as an adjustment to exercise. Early in the exercise the breathing pattern is very labored, but as exercise continues there seems to be a point where the breathing pattern becomes easier and the exercise seems lighter.

We feel that the most likely explanation of this observed phenomenon lies in the adjustment of the respiratory, circulatory, and muscular systems to the exercise. We hypothesize that a time lag exists between the onset of the activity and the bodily adjustment to the stress level of the activity. During activity, an increase in muscle temperature, the mobilization of intracellular and extracellular energy sources for the muscle, the mobilization of previously closed capillaries, an increased venous return, and a change in sensitivity of the neural respiratory center to the various stimulants of exercise are only a few of the possible adjustments that must take place at the initiation of an activity. If the nontrained person is slower to make these adjustments, the final completion of the lengthy process of the adjustment may form the basis of the apparent "second wind." As reasonable as these hypothetical considerations appear, very little research confirms the physiological existence of the phenomenon of second wind.

Functional Respiration

Properties of Respiratory Gases

In a mixed gas such as the atmosphere, the pressure exerted by each individual gas is referred to as the *partial pressure* of that gas. A gas mixture containing 21% oxygen and 79% nitrogen at an atmospheric pressure of 760 mm Hg will have partial pressures of 160 mm Hg for oxygen (P_{O_2} = 160 mm Hg) and of 600 mm Hg for nitrogen (P_{N_2} = 600 mm Hg). Partial pressures of gases are altered by the amount of water vapor dissolved in the gas mixture. In the body all gaseous mixtures are saturated with water; this amounts to a partial pressure of 47 mm Hg which is attributed to water vapor.

Table 10–1 lists the expected partial pressures of the respiratory gases found at sea level, in the humidified air as it enters the lungs, in the alveoli, arterial and venous blood, muscle tissue, and in the expired air at rest.

The partial pressure differentials of oxygen and carbon dioxide between the alveoli and blood are not equal; however, both gases diffuse freely. This is accomplished by the higher diffusion coefficient of carbon dioxide—about 20 times greater than oxygen.

Gases diffuse from an area of high partial pressures to areas of low partial pressures. The rate of diffusion is proportional to: (1) the pressure gradient, (2) the distance through which the gas must move, (3) characteristics of the gas (including the solubility coefficient and the molecular weight of the gas), (4) the type of medium through which it must move, (5) the contact surface area, and (6) the temperature of the medium. To calculate the diffusion of gases within body fluids and tissues, the relative diffusion coefficients are: oxygen 1.0, carbon dioxide 20.3, and nitrogen 0.53.

Based upon the alveolar diffusion coefficients and upon the normal laws of gas diffusion, the gases diffuse across the alveoli membranes in accordance with the partial pressure differentials. As oxygen diffuses from the alveoli to the red blood cell, it must cross the alveoli membrane, the interstitial space, the capillary membrane, the plasma volume that exists between the capillary wall and the RBC, and the membrane of the RBC. Net carbon dioxide flux must cover this distance in reverse.

The total amount of gas exchanged in the lungs is dependent upon the blood volume in the pulmonary capillaries and the partial pressure gradient of the gases. Under normal circumstances the uptake of oxygen ($\dot{V}_{O_2}$) is 250 ml/min; the release of carbon dioxide is 210 ml/min. During exercise the pulmonary blood flow increases by the opening of latent capillaries, the dilating of capillaries, and greater cardiac output. The lungs are more completely filled during exercise because of the increased blood volume and increased air ventilation. The increased air volume distends the alveoli membranes, making them thinner and more permeable to gas diffusion. These factors assist in the augmented oxygen uptake during exercise.

PARTIAL PRESSURE GRADIENTS

Gas volumes expand and contract in relation to changes in temperature and pressure. Relative gas concentrations determine, for the most part, the direction of movement of the gas. The exchange of gases between the atmosphere and the tissues is predictable according to the individual partial pressure gradient of each gas. These partial pressures, depicted in Table 10–1, illustrate that the flow of oxygen is from the highest partial pressure (in the atmosphere) to the lowest partial pressure (in the tissue). The flow of carbon dioxide follows a similar pattern of movement from the highest area of concentration (the tissues) to the area of the lowest partial pressure (the atmosphere).

Table 10–1. Approximate Partial Pressure mm Hg of Oxygen and Carbon Dioxide in the Atmosphere and Selected Body Tissue*

	P_{O_2}	P_{CO_2}
Atmosphere	160	0.05
Lung alveoli	149	40
Pulmonary veins	100	40
Large arteries	100	40
Capillaries (arterial)	60	40
Muscle cell	10–30	50–80
Capillaries (venous)	30	48
Mixed venous blood	40	46
Pulmonary arteries	40	46
Pulmonary veins	100	40

*Circular flow of blood through the body, from the lungs to the heart to the tissue to the heart and back to the lungs.

The diffusion of gases in the lungs is responsive to their pressure gradients. The pressure differential for oxygen is nearly 60 mm Hg (partial pressure in the alveoli is 100 mm Hg; partial pressure in the mixed venous blood is 40 mm Hg at rest). During exercise the oxygen pressure differential increases because of the lowering of the mixed venous blood partial pressure for oxygen. Carbon dioxide does not show the same magnitude of pressure gradient: the pressure differential is only about 5 mm Hg at rest but may be as high as 40 mm Hg during heavy exercise. However, the diffusion coefficient for carbon dioxide is about 20 times greater than that for oxygen. During exercise these pressure gradients increase, thus facilitating the exchange of gases at the lungs.

The diffusion of gases at the tissue sites follows a diffusion pressure gradient, but in the case of oxygen diffusion may be facilitated by the presence of myoglobin. Having a much higher diffusion coefficient, carbon dioxide is in less need of facilitated transport. The partial pressure gradient for oxygen from the capillary (40 mm Hg) to the mitochondria (1–10 mm Hg) is less than that of the lungs, but with the help of the myoglobin shuttle system described below the diffusion is easily accomplished.

If oxygen is utilized rapidly in the mitochondrion, the oxygen partial pressure will fall to near zero. The oxygen associated with the nearest myoglobin will move towards the lower concentration area within the mitochondrion. The next myoglobin molecule transfers its oxygen to the first—and so on until the myoglobin nearest the capillary unloads its oxygen molecules and can absorb the oxygen diffusing from the hemoglobin molecule in the capillary. The above shuttle system is a proposed model—and one that has not been totally substantiated. However, it does help to explain the increased oxygen availability in working trained muscles (trained muscles have increased amounts of myoglobin).

Oxygen Uptake and Transport

We know that during resting conditions cardiac output is near 6000 ml/min (given a heart rate frequency of 60 and stroke volume of 100 ml), and the ventilation is about 6000 ml of air (breathing frequency of 12/min and a tidal volume of 500 ml). During exercise the cardiac output can increase to about 25 l/min and the ventilation to near 130 l/min with an oxygen uptake of about 5 l/min.

Oxygen uptake can be measured by an examination of the ventilated gases or by multiplying the cardiac output by the difference in the oxygen saturation in the mixed venous blood and in the arterial blood. If we assume the hemoglobin concentration is 0.15 mg/ml of blood (each mg of hemoglobin binds 1.34 ml of oxygen) and the hemoglobin is 100% saturated with oxygen, the oxygen content of the blood is 0.201 ml of oxygen per ml of blood (0.15 × 1.34). In addition 0.003 ml of oxygen is dissolved per ml of plasma. The total amount of oxygen carried by fully saturated blood is then .204 ml/ml of blood. If we assume that the blood leaving the lungs is 100% saturated (98% is a better approximation), we must then determine the oxygen content of the venous blood and the cardiac output to calculate the oxygen uptake. With a resting cardiac output of 6000 ml and an oxygen content of the returning venous blood of 0.162 ml of oxygen per ml of blood, we can account for the resting oxygen uptake: .042 ml of oxygen uptake per ml of blood flow × 6000 = 252 ml oxygen uptake per minute.

During maximal exercise the total body oxygen uptake is increased up to 20 times (approximately 5.0 l/min). The total cardiac output is increased up to five times (up to 30 l/min), and the arterial-venous (a-v) difference in oxygen saturation is increased by up to four times (a-v difference of .168 ml oxygen/ml blood). Thus the increased cardiac output and the increase in the a-v difference account for the increased oxygen uptake.

The exercise-related increased a-v difference in oxygen content is partially accounted for by the shift in the distribution of the cardiac output. About 90% of the blood flow perfuses the active muscles during exercise as contrasted to resting conditions when muscles receive only 15% of cardiac output. Changes in the affinity of oxygen to hemoglobin are also a factor in the delivery of oxygen to deoxygenated tissues —as decreased pH, increased temperature, and increased CO_2 aid in the release of more oxygen. The affinity fluctuations can be explained by alterations in the oxygen-hemoglobin dissociation curve (Figure 10–9) in response to changes in pH, temperature, and P_{CO_2}.

If we assume that blood flow to muscle is 0.1 ml per gram of wet muscle weight per minute and the a-v difference, under resting conditions, is .04 ml oxygen per ml of blood (4 volumes percent), then the total oxygen delivered per gram per minute of muscle would be 4 μl (.004 ml). During maximal exercise the blood flow may increase to 0.8 ml per gram per minute, and the extraction may increase to 18 volumes percent; the oxygen delivery would then be 144 μl per gram per minute of active muscle. (There have been studies that report as high as 200 μl of oxygen uptake per gram per minute of active muscle). If we alter our assumptions to allow the increase in blood flow to 1.1 ml per gram per minute, the calculated oxygen uptake would be 198 μl per gram per minute. The higher values are not unreasonable during maximum exercise.

Given the active muscle mass as 20 kg during strenuous total body activity and oxygen uptake as 200 μl per gram per minute, we can account for an oxygen uptake of 4 l/min by only measuring the oxygen used in active

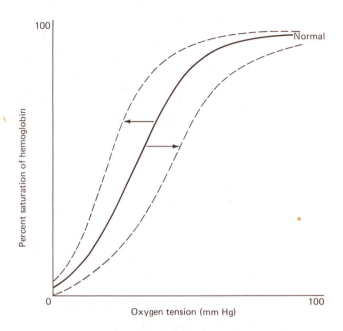

Figure 10-9. Oxygen-hemoglobin dissociation curve under normal conditions (pH = 7.4, P_{CO_2} = 40 mm Hg, and temperature = 38°C). The curve is displaced to the right under exercise conditions (lower pH, increased P_{CO_2}, and increased temperature). The dissociation curve is displaced to the left given an increased pH, decreased P_{CO_2}, decreased temperature, and increased carbon monoxide (P_{CO}).

muscle. Thus the active muscle can be shown to account for nearly 80% of the total oxygen uptake during exercise. These calculations indicate that the ability of the muscles to utilize oxygen determines to a great extent the total oxygen uptake during exercise. At rest the oxygen uptake of the muscle (4 μl per gram per minute) would result in a total oxygen uptake of 80 ml for muscle, which represents 32% of the total oxygen uptake during rest.

MAXIMUM OXYGEN UPTAKE TESTS

Maximum oxygen uptake tests are usually performed in the laboratory, often under the supervision of medical personnel. The tests consist of directly measuring oxygen consumption via a sampling of the inspired and expired air. The exercise protocol calls for a warm-up; after the warm-up, the exercise test begins with an exercise pace estimated to produce approximately 90% of the maximal oxygen uptake. At each succeeding two-minute interval, the work load is increased until the subject decides that the exercise cannot be continued. This "exhaustion" point is highly variable among individuals and leads to intersubject errors in the determination of maximal oxygen uptake values.

In addition to the subject-initiated end point to the exercise, the investigator may stop the exercise test if abnormal reactions of the subject are noticed. Examples of such abnormal reactions are heart rates near 200, abnormal ECG, or respiratory exchange index (R) near 1.2. The high R indicates a continuing washout of CO_2 but a reduction of O_2 utilization. The determination of exercising heart rate, ECG, and R as stress indicators is valuable but requires relatively sophisticated equipment, including *immediate* readouts, usually through the use of digital computers.

The value of the maximum exercise test, for the purpose of measuring *general* cardiovascular fitness, is highly speculative in view of the fact that other acceptable submaximal tests give highly similar results.

SUBMAXIMUM OXYGEN UPTAKE TESTS

Submaximum exercise tests can be divided into direct and indirect oxygen uptake estimations. Direct oxygen measurements unquestionably give the least objectionable results since the oxygen content is measured directly. The primary disadvantage to the direct test is the requirement of oxygen-monitoring equipment. In the laboratory this type of equipment may be readily available, but outside of the laboratory it is relatively rare.

The indirect tests are based upon the direct relationship between heart rate and oxygen uptake. (In Chapter 9 we discussed the relationship between oxygen uptake and cardiac output.) Except at the higher values, the oxygen uptake can be directly determined from knowledge of the heart rate. The objections to this technique are that resting and maximum pulse rates between individuals are variable and that the maximum heart rate varies with age.

Convenient indirect tests that can be used include the bicycle ergometer, the treadmill, the step test, and the running test. The work rate using these tests can be easily controlled. The work rate on the bicycle and treadmill must be set at a predetermined rate; on the step test, the height of the step must be controlled. Running tests, as suggested in Cooper's *Aerobics* (see References, Chapter 19), can be the *time required* to run a given distance or the *total distance* run in a given time.

The determination of heart rate during exercise is preferable to investigation of the heart rate solely during the recovery period. If it is impossible to record the ECG during exercise, the heart rate can be taken by palpation of the carotid artery in the neck, by the ulnar artery on the wrist, or by the use of a stethoscope on the chest. Recovery heart rates immediately following the exercise are closely related to the exercising heart rates and can be used if the exercising heart rates cannot be obtained. The heart rate during recovery returns to normal very rapidly; thus the recovery heart rate must be taken within a well-defined time period. Without some practice, this method will be very inaccurate for determining exercising heart rates. However, there are fitness tests based on the utilization of the recovery heart rate; these tests evaluate the speed at which the heart rate returns to normal following exercise.

Figure 10–10 shows a nomogram that has been widely used in determining oxygen uptake from submaximal work tests. This particular nomogram can be used with a bicycle ergometer or step test. In the illustrated example, a male of 94 kg, with a heart rate of 166 after the step test, will have an oxygen uptake of 3.6 l/min. This oxygen uptake value represents the uptake of 38.3 ml of oxygen per kilogram of body weight. In the second example, a female weighing 61 kg after the 33-cm step test has a heart rate of 156 and projected maximum oxygen uptake of 2.4 liters (39.3 ml/kg).

An even more indirect method of assessing the oxygen uptake capacity of an individual is Cooper's "aerobics" method. This test measures the total distance an individual can cover in 12 minutes or the time required to run 1.5 miles. The effort should be near maximum; that is, one should attempt to run as far as he or she can in the allotted 12 minutes or to run the 1.5 miles in the least amount of time. From the results of these tests an individual can determine his/her own estimated oxygen uptake values, using the established tables.

A WORD OF CAUTION

Physical fitness testing inflicts a stress upon an individual, and this stress is a potential danger

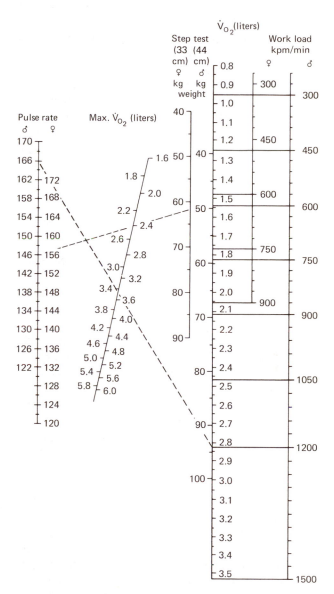

Figure 10–10. Nomogram for estimating aerobic work capacity from submaximal pulse rate.

From I. Astrand. Aerobic work capacity in men and women with special reference to age. *Acta Physiol. Scand.* 49 (Suppl. 169): 159, 1960.

to the body. Extreme caution should be taken, especially when testing an individual for the first time. Unfit individuals may not realize their degree of unfitness, and aged persons may not realize the age-related decrements in performance capabilities. When any doubt exists as to the capability of a subject, one should be instructed to have a medical checkup prior to the activity. Even with medical examinations it may be a wise precautionary move to require a preliminary training period before administrating any fitness test, no matter how easy the test.

SIGNIFICANCE OF RATE OF OXYGEN UPTAKE

Oxygen-uptake rate is traditionally used as a measure of metabolic rate. Consequently resting-body oxygen-uptake values grossly underestimate the metabolically active tissues that make up a small proportion of the total body weight, such as the heart, adrenals, and specific neurons. They likewise fail to reveal the very low rate of oxygen utilization of relative inactive tissues such as bones, cartilage, and tendons.

When an exercise is at a rate such that the oxygen demand is greater than the maximal rate at which oxygen ($\dot{V}_{O_2}$) can be utilized, then the duration of the work becomes markedly limited. Consequently it seems only logical that the rate at which oxygen can be utilized by the body is an important determinant in establishing a limitation to its capacity to endure a high work load. Since $\dot{V}_{O_2}$ is a direct measure of the oxygen absorbed by the lungs, it may be thought that pulmonary function is a limiting factor in work performances. We have shown earlier that maximal $\dot{V}_{O_2}$ is determined, for the most part, by cardiac output; and even during periods of maximal cardiac output the oxygen saturation of arterial blood remains nearly 100%. Therefore it is highly unlikely that ventilatory factors would be considered the limiting factors to oxygen uptake.

In order to identify the factors that are critical in the limitation of oxygen uptake, we need to examine the cardiovascular and skeletal muscle systems. In effect oxygen uptake depends on the individual's capacity to: (1) absorb oxygen, (2) transport oxygen, (3) deliver oxygen, and (4) utilize oxygen.

Absorption of Oxygen. Assuming that there exists an adequate oxygen concentration in the inspired air, there is more than enough alveolar surface area available to absorb oxygen at a rate sufficient to maintain arterial oxygen tension levels at more than 95% saturation even during extremely heavy work.

Transport of Oxygen. The transport of oxygen to the periphery depends upon (1) the amount of blood circulating and (2) the oxygen-carrying capacity of the blood. Cardiac output represents the former; hemoglobin concentration and arteriovenous difference (a-v difference) reflect the latter. Oxygen uptake equals cardiac output (Q) times the a-v difference.

At any blood oxygen-saturation level, the greater the volume of blood pumped by the heart, the greater the oxygen uptake and transport. As discussed in Chapter 9, cardiac output (Q) is determined by a number of factors, including blood volume, stroke volume, and heart rate—none of which are independent of the others. The more blood available for circulation, the greater is the potential oxygen transport. The immediate effect of heavy exercise is for the absolute blood volume to decrease slightly. This effect on oxygen transport is minimal since there is only a slight loss of intravascular plasma; thus hemoglobin is concentrated without changing the total content of hemoglobin.

Endurance training will elevate the blood volume. Conversely a decrease in blood volume occurs with prolonged bedrest (about 7%); most of this is a decrease in plasma unless the bedrest is prolonged. Prolonged bedrest leads to a point of a reduction in RBC volume.

Arteriovenous saturation difference in the lungs is the other factor that actually determines the amount of oxygen transported. This variable is affected by hemoglobin concentration and by the binding capacity of oxygen to hemoglobin. Acidic blood and RBCs with high 2,3-diphosphoglycerate concentrations lessen the oxygen affinity for hemoglobin. This causes a *greater* oxygen delivery to the peripheral tissues since the oxygen is more prone to disassociate from the hemoglobin. Oxygen content is not substantially affected; the alveolar-capillary oxygen gradient is sufficient to load the hemoglobin to about 98% saturation, even with a reduced $Hb-O_2$ affinity.

Delivery of Oxygen. The *oxygen concentration gradient* from the capillaries to the cells partially determines the amount of oxygen delivered to the cells. As pointed out above, more oxygen is liberated by a *reduction* in *$Hb-O_2$ binding affinity*. The amount of oxygen delivered is also dependent on the *diffusion distance* of a tissue (the average distance from the capillary to the center of the cell). Those tissues with a higher *vascularity* or greater numbers of capillaries per unit area (capillaries per fiber) will receive more oxygenation.

Oxygen Utilization. The rate at which oxygen is actually utilized in the cell may be a limiting factor in oxygen uptake. Factors that may act to limit oxygen utilization within a muscle fiber are myoglobin concentration and mitochondrial density and enzyme activity. Mitochondrial enzymes make up a large part of the mitochondrial membranes and increase with endurance training. It is also possible that substrate concentrations (acetyl Co-A, NADH, ADP, Pi, and so forth) could limit oxygen utilization by the mitochondria.

Hemoglobin

Hemoglobin is a protein molecule that binds four heme-containing iron molecules that have an affinity for oxygen. Hemoglobin consists of four chains of amino acids with each chain binding one heme-containing iron molecule, which in turn cooperatively binds one oxygen molecule.

The cooperative binding accounts for the *sigmodidal* shape of the hemoglobin saturation curves. As one of the heme portions of the hemoglobin molecule becomes saturated with oxygen, it brings about a conformational change that increases the possibility of a second heme portion combining with an oxygen molecule. This cooperative binding continues until all four heme components of the hemoglobin molecule are combined with oxygen. As the partial pressure of oxygen is reduced, the hemoglobin molecule gradually gives up its oxygen to the surrounding medium; this reaction accounts for the oxygen binding to the hemoglobin molecule. In the lungs, where the partial pressure of oxygen is 100 mm Hg, the hemoglobin is fully saturated. At the capillary level, where the P_{O_2} of oxygen is 5–30 mm Hg, the hemoglobin tends to release its oxygen load.

Figure 10–9 demonstrates the effect of pH ("Bohr Effect"), P_{CO_2}, temperature, CO and O_2 saturation, and consequently Hb-O_2 affinity. For example venous blood being delivered to the lungs for oxygenation may have a pH of 7.40 and P_{O_2} of 20 mm Hg, which means the hemoglobin is about 35% saturated with oxygen; whereas venous blood at a pH of 7.35 (which could be reached during exercise) would be about 32% saturated. The decreased pH results

in the affinity of oxygen to hemoglobin being reduced by 3%. The reduction in affinity causes greater Hb-O_2 dissociation and consequently more oxygen delivery to the peripheral tissues. The reduction of Hb-O_2 affinity in the lungs is not of concern; the affinity loss is overridden by the surplus of available oxygen at the capillary-alveolar juncture.

Similarly blood that has a P_{CO_2} of 80 mm Hg is 35% less saturated with O_2 (i.e., less Hb-O_2 affinity) than blood that has a P_{CO_2} of 40 mm Hg. During exercise elevated P_{CO_2}, increased temperature, and lowered pH are adaptations that enhance oxygen delivery to active tissues. The P_{CO_2}, temperature, and pH effects are interrelated: high P_{CO_2} adjusts blood pH downward and vice versa.

Similar Hb-O_2 affinity adjustments occur when an individual adapts to high altitude or becomes anemic. In anemics and in individuals adapted to high altitude, an elevated 2,3-diphosphoglycerate/Hb ratio decreases Hb-O_2 affinity and increases oxygen delivery. Although researchers originally thought that 2,3-DPG increased with exercise, subsequent reports have shown that it is not an effective adaptive mechanism in exercise.

The effect of CO on the Hb-O_2 dissociation curve is of special interest in urban societies, in which atmospheric CO levels are elevated due to automobile and industrial pollutants. For example it has been shown that a cardiac patient's tolerance to a treadmill exercise test was reduced by about one-third after he had ridden on a freeway for a few hours. When the subjects were unknowingly breathing purified air, no such detriments in exercise tolerance were noted. This finding clearly indicates that those with marginal cardiac conditions are even more susceptible to a critical cardiovascular overload after CO inhalation. As shown in Figure 10–9, the Hb-O_2 dissociation curve shifts to the left with increasing CO concentrations and indicates less oxygen unloading to the active tissues. The decreased oxygen would be especially critical for the heart patient.

Cellular Considerations

We have already considered cellular topics to some extent in the discussion of the proposed myoglobin shuttle system. The myoglobin molecule, which contains iron, is a single chain of amino acids which attracts oxygen. Myoglobin differs from hemoglobin in that there is no cooperative binding. Compared to hemoglobin, the myoglobin molecule retains its oxygen load until the P_{O_2} falls to relatively low values.

Myoglobin is distributed throughout the muscle cell, and this distribution pattern supports the possibility for the proposed shuttle mechanism. The intracellular distances are such that for oxygen to diffuse adequately to the center of the muscle cell would require a very high diffusion gradient. During exercise it is highly questionable whether this distance could be negotiated by the oxygen molecules in a quantity great enough to supply the metabolic needs of the innermost mitochondria without a shuttle system. Mitochondria are more densely localized near the periphery of muscle fibers, a location that provides the most probable chance of their remaining in constant supply of available oxygen.

Increasing the vascularity of the muscle cell by training is one of the benefits of endurance training. This increased vascularity increases the capillary-to-fiber ratio and thus decreases the intracellular diffusion distances for oxygen and carbon dioxide.

Carbon Dioxide Transport and Expiration

The CO_2 produced in the cell, primarily in the mitochondrion, diffuses from the cell to the interstitial spaces and the capillaries. The

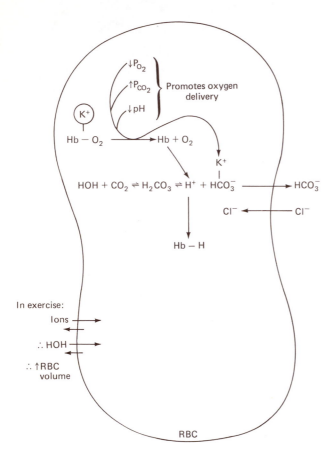

$$\downarrow P_{O_2}$$
$$\uparrow P_{CO_2} \Big\}\ \text{Promotes oxygen delivery}$$
$$\downarrow pH$$

$$K^+$$

$$Hb - O_2 \longrightarrow Hb + O_2$$

$$K^+$$
$$|$$
$$HOH + CO_2 \rightleftharpoons H_2CO_3 \rightleftharpoons H^+ + HCO_3^- \longrightarrow HCO_3^-$$

$$Cl^- \longleftarrow Cl^-$$

$$Hb - H$$

In exercise:

Ions

∴ HOH

∴ ↑RBC volume

RBC

Figure 10–11. Buffering and ionic equilibrium maintenance that is associated with Hb-O₂ association and dissociation.

relative ease of the diffusion is due to the high diffusion coefficient of the molecule.

Under resting conditions the P_{CO_2} in the cell is near 50 mm Hg; the P_{CO_2} in the capillary blood is 40 mm Hg at the arterial end and 46 mm Hg at the venous end of the muscle capillary bed. These partial pressure differentials allow for the diffusion of CO_2 from the cell to the capillary.

Once inside the capillary the CO_2 molecule

radically alters the state of the plasma and red blood cell (Figure 10–11). There are at least three ways by which the CO_2 molecule can be transported from the cell to the lungs for expiration into the atmosphere: (1) by being dissolved in the plasma as CO_2, (2) by binding with Hb at the site vacated by O_2, and (3) by combining with water to form carbonic acid (catalyzed by the enzyme *carbonic anhydrase*), which almost immediately dissociates into the bicarbonate ion and the hydrogen ion. The release of the hydrogen ion, unless buffered, decreases the pH of the blood.

During resting conditions the blood transports about .04 ml of CO_2 per ml of blood. Of the CO_2 transported, nearly 7% is dissolved in the plasma, less than 5% is carried in combination with hemoglobin, and the rest is transported in the form of the bicarbonate ion. The normal high level of CO_2 in the blood is related to the amount of sodium bicarbonate, which acts as a buffering agent to prevent great fluctuations in the pH of the blood.

As we all know hemoglobin is responsible for the transport function of nearly 99% of the oxygen content of the blood. This same hemoglobin, although not directly involved, is indirectly responsible for nearly 90% of the CO_2 transported by the blood.

As the arterial blood enters the capillary, the increased P_{CO_2} and decreased pH influence the oxygenated hemoglobin to release its bound oxygen molecules. Saturated hemoglobin has associated with it potassium ions (K+) so that, when hemoglobin releases its oxygen, the K+ ions are also released. The desaturated hemoglobin then binds hydrogen ions (H+), thus serving as an important buffering system (Figure 10–11).

The CO_2 from the cell readily diffuses into the capillary and further diffuses into the erythrocytes, where CO_2 becomes the substrate for the action of the enzyme *carbonic anhy-*

drase. This enzyme catalyzes the formation of carbonic acid (H_2CO_3). Since carbonic acid is a very weak acid, it nearly immediately dissociates into H^+ and HCO_3^-. The newly formed H^+ is buffered by the desaturated hemoglobin, and the HCO_3^- is neutralized by the liberated K^+. At this stage the concentration of HCO_3^- is high inside and low outside the red blood cell. The ratio of the anions HCO_3^- to Cl^- is not the same in both compartments. There then occurs a shift of HCO_3^- to the plasma and Cl^- to the inside of the cell.

During the above events, the number of ions within the red blood cell increases, and in response to the increased osmotic pressure water molecules enter the erythrocyte. The increased water within the erythrocyte results in an increased volume of the cell, which increases the relative volume of the red blood cell to plasma volume (increased hematocrit). This action partially accounts for the hemoconcentration seen with exercise.

During times of high exercise stress, lactic acid is also released into the blood stream. The H^+ from this acid must be buffered by the buffering capacity of the blood, or the pH decreases (as is usually the case in times of stressful exercise). The lactic acid combines with the sodium bicarbonate ($NaHCO_3$), which dissociates into CO_2 and HOH. (Figure 10–12). As the blood moves back to the lungs the interaction of the CO_2 transport and the lactate transport is in some form of equilibrium. The free CO_2 serves to stimulate ventilation and is also more able to diffuse into the alveoli for expiration into the atmosphere. The CO_2, resulting from excess lactic acid, accounts in part for the increased R during strenuous exercise.

Once the blood enters the pulmonary circulation, the events described above are reversed. The P_{O_2} and the P_{CO_2} gradients are reversed; the flow of oxygen is into the capillary, and the flow of CO_2 is into the alveoli.

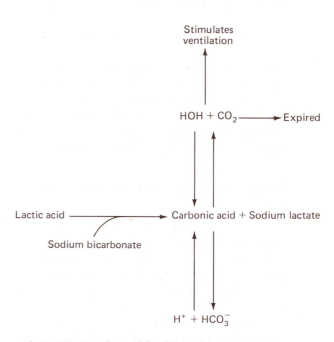

Figure 10–12. The acid effect of lactic acid is buffered by sodium bicarbonate which forms carbonic acid and dissociates into HOH and CO_2. The CO_2 stimulates the neural respiratory centers. Expiration of CO_2 will reduce the acidic effect of the carbonic and lactic acid.

Carbon dioxide moves across the alveoli membrane, thus altering the

$$CO_2 + HOH \longleftrightarrow H^+ + HCO_3^-$$

equilibrium. The removal of the CO_2 pulls the equilibrium in the direction that reduces the bicarbonate concentration. A reduction in the bicarbonate concentration allows for a reverse chloride shift, with the chloride ions moving from the red blood cell to the plasma.

Immediate Adaptations to High Work Loads

Morphological adaptations of the adult lung to endurance training have been difficult to identify. However, studies have indicated that animals exposed to hypoxic environments have lung volumes 20% greater than controls. The greater volume seems to be primarily an enlargement of existing alveoli. It has been reported that physical exercise also causes distension of alveoli, which is followed by some thickening of the alveolar septa with cellular proliferation and later by an increase in number of alveoli. Lung growth is not affected by four weeks of endurance training in laboratory animals.

A reduction in ventilation per minute (decreased $\dot{V}_E$) in response to a standard exercise is the chief pulmonary adjustment to training. The lowered $\dot{V}_E$ is due principally to a decreased breathing frequency, and the greatest adjustment is realized at the higher work loads. This training effect consequently extends the linearity of the $\dot{V}_{O_2}:\dot{V}_E$ relationship, which means that ventilation does not become disproportionately greater than oxygen absorption at the heavier work levels. One explanation might be that the trained individual has developed a greater tissue oxygen gradient, for example greater oxygen utilization by skeletal muscles. Other data also demonstrate a train-

ing effect in ventilation. Inhalation of pure oxygen in trained individuals during exercise results in a less dramatic lowering of the rate of ventilation than in nontrained persons. Similarly training in underwater diving results in lower breathing frequency at rest and less sensitivity to CO_2. Therefore trained individuals may become conditioned to ventilatory stimulation by CO_2.

In supreme exercise efforts in which $\dot{V}_{O_2}$ is very high, elevated diffusion capacity from alveoli to pulmonary capillaries probably occurs. Similarly the pulmonary capillary blood volume and/or distribution may be adaptable in intensively trained athletes.

Summary

The rate and depth of *ventilation* are among the most obvious indicators of exercise stress. The amount of air ventilated per minute is regulated by neural and chemical mechanisms that respond to exercising conditions.

Oxygen uptake in the lungs is related to the cardiac output: those factors that regulate cardiac output also regulate the rate and depth of ventilation. *Oxygen uptake in the tissues* during exercise is primarily related to the specific type of metabolic activity in the active muscles. The oxygen uptake attributed to muscle during exercise can be shown to account for a great majority of the *total* oxygen uptake during exercise.

Oxygen-carrying and carbon dioxide-carrying capacities of the blood are primarily determined by the hemoglobin content of the blood. Other factors such as partial pressures, pH, and temperature alter the gaseous exchange of the respiratory gases in all tissues.

The overall effect of respiration is that oxygen flows from a partial pressure of 160 mm Hg in the atmosphere to a partial pressure area in the tissues equal to about 5 mm Hg. Carbon dioxide flows from a tissue area, where the partial pressure is near 50 mm Hg, to the atmosphere, where the partial pressure of CO_2 is negligible. The transport of O_2 and CO_2 is facilitated, either directly or indirectly, by the hemoglobin molecule contained within the erythrocyte. Myoglobin aids the transport of the oxygen within the tissue from the capillary to the mitochondrion. The alterations in this cycle during exercise are such that the ability of the body to utilize and dispose of the respiratory gases is increased.

Study Questions

1. What are the important respiratory volumes and how do they change during exercise?
2. What are the normal rate and depth of breathing during rest and exercise?
3. What neural and humeral mechanisms control breathing?
4. How does alveolar ventilation change during exercise?
5. How does the ratio of ventilation to oxygen uptake change during exercise?
6. How does the ventilation-perfusion ratio change during exercise?
7. What are the typical values for the respiratory exchange index during exercise?
8. Describe the partial pressures of gases in the several compartments of the body and how these influence gas exchange.
9. What are some possible explanations for second wind?
10. Show how muscle metabolism can account for oxygen uptake during exercise.
11. What are the benefits of the maximal oxygen uptake tests?
12. What are the benefits of the submaximal oxygen uptake tests?
13. How can the indirect submaximal tests be used in a testing situation?
14. How is oxygen utilization related to metabolism?
15. What factors influence the absorption of oxygen in the lungs?
16. What factors influence the transport of oxygen?
17. What is the limitation of oxygen utilization?

18. How is the oxygen-hemoglobin saturation curve altered during exercise?
19. How can myoglobin aid oxygen transport within the muscle fiber?
20. What are the respiratory adaptations to training?

Review References

Assmussen, E. Muscular exercise. In *Handbook of Physiology*, vol. II. Washington, D.C.: American Physiological Society, 1965, pp. 939–978.

Astrand, Per-Olof and Kaare Rodahl. *Textbook of Work Physiology*. New York: McGraw-Hill, 1970.

Balke, Bruno. Variation in altitude and its effects on exercise performance. In *Exercise Physiology* edited by H.B. Falls. New York: Academic Press, 1968.

Dempsey, Jerome A. and John Rankin. Physiologic adaptations of gas transport systems to muscular work in health and disease. *American Journal of Physical Medicine* 46:582–647, 1967.

Grimby, Gunnar. Respiration in exercise. *Medicine and Science in Sports* 1:9–14, 1969.

Otis, A.B. The work of breathing. In *Handbook of Physiology*, vol. I. Washington, D.C.: American Physiological Society, 1964, pp. 463–476.

Schottelius, Byron A. and Dorothy D. Schottelius. *Textbook of Physiology*. St. Louis: C.V. Mosby Co., 1973.

Shephard, R.J. What causes second wind? *The Physician and Sport Medicine* 2:37–42, 1974.

Simmons, R. and R.J. Shephard. Effects of physical conditioning upon the central and peripheral circulatory responses to arm work. *Int. Z. Angew. Physiol.* 30:73–84, 1971.

West, John B. Respiration. In *Annual Reviews of Physiology*, vol. 34, J.G. Comroe, ed. Palo Alto: Annual Reviews, Inc. 1972, pp. 91–116.

Key Concepts

• Hormones are secreted by specialized cells and transported via the blood stream to the target organs.
• Hormonal secretion and regulation are largely under neural control.
• Individual hormones have specific effects on the body.
• Blood glucose levels are largely under the control of hormones.
• Cyclic adenosine monophosphate (cyclic AMP) is the intracellular mediator of many hormonal actions.
• Exercise elicits a unique endocrine response during each stage of physical activity.
• A general (nonspecific) response to stress can be determined.

CHAPTER ELEVEN

Hormonal factors in exercise

Introduction

During exercise the endocrine system, much like the nervous system, exerts a powerful influence over the body's response to specific exercises and helps to coordinate the various mechanisms of the body, including energy production during exercise. The responses of the endocrine system are related to specific exercise variables, involving the physiological and mental states of the individual. These preperformance states determine the degree of stress that the individual will undergo during the actual exercise. The degree of stress on a well-trained individual will be less than the stress imposed upon a nontrained person performing the identical task. The relationship between the individual's level of expectation and his actual "ability to perform" will contribute to the eventual level of stress on the body.

General Considerations

In general terms a *hormone* is synthesized within a well-defined group of cells and is

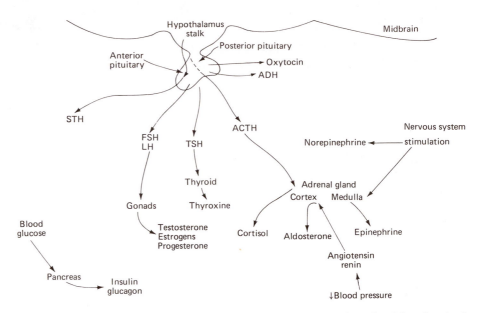

Figure 11-1. Hormones involved in physical activity.

secreted into the blood stream in response to some stimuli. The rate of synthesis of the hormone, as well as the rate of its release, may be controlled by neural innervations and/or by chemical substances. The amount of hormone in the blood stream is dependent upon the rate of release from the endocrine gland, the rate of degradation of the hormone, and the uptake of the hormone by the various target organs. The concentration of the hormone in the blood usually is a good estimate of the stimulation rate and, in part, determines the magnitude of the biological response; in general, there is a positive linear *dose-response curve.*

The release of hormones from the endocrine glands is controlled by a number of mechanisms, a primary one being direct or indirect neural stimulation. The secretion of hormones from the pituitary, for example, is mediated via the neural regulation mechanisms in the

hypothalamus. Figure 11-1 diagrammatically illustrates the hormones secreted by the anterior and posterior lobes of the pituitary as well as the target organs of these hormones. Figure 11-1 also depicts the release of hormones other than those directly controlled by the hypothalamic-pituitary axis. We can see that the pituitary directly or indirectly controls the release of testosterone, estrogen, progesterone, thyroxine, cortisol, growth hormone, antidiuretic hormone, and oxytocin. Epinephrine and norepinephrine (adrenalin and noradrenalin) are controlled by the sympathetic nervous system. Aldosterone, angiotensin, and renin respond to alterations in blood pressure. Insulin and glucagon are secreted by the pancreas in response to blood glucose changes. Parathyroid and calcitonin, secreted by the parathyroid and thyroid glands, respond to blood levels of calcium and phosphorous.

**Table 11-1. Hormones: Their Structure,
Common Form of Stimulation, and their
Primary Metabolic Action**

Hormone	Structure	Primary Stimulus	Metabolic Action
Cortisol		Stress	Increase blood glucose and liver gluconeogenesis
Aldosterone		Decrease in blood pressure	Increase sodium and potassium retention
Estradiol		FSH and LH release from the pituitary (female)	Alteration in lipid metabolism; menstrual cycle
Testosterone		FSH and LH release from the pituitary (male)	Increase amino acid uptake in muscle
Epinephrine		Decrease in blood sugar level; stress	Increase blood glucose; increase blood fatty acids
Norepinephrine		Sympathetic stimulation; stress	Increase blood pressure
Thyroxine		Cold or stress	Increase metabolic rate
Growth hormone (Somatotropic hormone)	Polypeptide (188 amino acids)	Decrease in blood glucose	Increase blood glucose; increase blood fatty acids
Glucagon	Polypeptide (29 amino acids)	Decrease in blood glucose	Increase blood glucose
Insulin	Polypeptide (51 amino acids)	Increase in blood glucose	Decrease blood glucose
Antidiuretic hormone	Polypeptide	Increase in osmotic pressure	Increase water retention

In Table 11-1 we have listed these hormones, their mode of stimulation, and their metabolic actions in the exercise response. The particular mechanisms by which hormones are thought to influence cellular metabolism are included in the following sections, whenever known. In general hormones attach to receptor sites on membranes of cells or pass through the outer membrane to the nucleus of the cell. The receptor sites are probably protein molecules, which become activated by the attachment of the hormone. The activated receptor site then can alter the permeability of the membrane or perform as an enzyme to activate specific metabolic activity within the cell.

Specific Hormones

Functional Mechanisms of the Hypothalamic-Pituitary Axis

The *hypothalamus* exerts control over the secretion of the pituitary hormones. Within the hypothalamus there are clusters of neurons called *nuclei* to which endocrinologists attribute specific functions. These nuclei produce the hormones of the posterior pituitary in addition to specific "releasing factor" hormones, which pass via the "portal" blood stream to the anterior pituitary. Upon arriving in the anterior pituitary, the releasing factor hormones stimulate the specific cells of the anterior pituitary to secrete the respective hormones.

The anterior pituitary is responsible for the secretion of growth hormone, thyroid-stimulating hormone, gonadotropic hormones, and the adrenocorticotrophic hormone. The posterior pituitary is responsible for the secretion of at least two hormones, oxytocin and antidiuretic hormone (*vasopressin*).

The existence of the mechanisms of control exerted by the hypothalamus explains, in part, why emotions influence behavior. The "psychological" approach to exercise is particularly critical in regard to the endocrine system. The hypothalamus receives neural input from all parts of the brain; undoubtedly such pre-exercise emotions as apprehension, confidence, and fear play an important role in the degree of endocrine function during exercise.

Thyroid Hormone

The *thyroid* gland is thought to affect more functions in the body than any other gland—but is not essential for maintaining life. *Thyroxine* is known to alter the activity of more than 100 enzyme systems, including the stimulation of protein synthesis in tissues and the increase in the glucose oxidation rate.

The thyroid affects the basal metabolic rate (BMR), and in *hypothyroid* states the decrease in BMR may be as much as 40%. This hypothyroid state may be manifested by a low energy level and a tendency to overweight. In the case of *hyperthyroidism*, the individual tends to be hyperactive and has very little fat on the body.

The increase in basal metabolic rate seems to be one of the chief clinical indications of increased thyroid activity. This increase in BMR can be partially explained in terms of the effect thyroxine has on mitochondria (with the exception of brain mitochondria). The hormone stimulates the utilization of oxygen without increasing the amount of ADP phosphorylated to ATP. In other words, the hormone uncouples oxidative phosphorylation; this effect could account for part of the increase in oxygen utilization following exercise.

Growth Hormone

Growth hormone (GH) can produce a multitude of metabolic effects including an increase in blood glucose and in the fatty acid levels of the blood; a stimulation of RNA and protein synthesis in bone, muscle, and liver; and an increase in amino acid uptake by the muscles. The release of fatty acids through the stimulation of lipolysis in the adipose tissue may be one of the key exercise effects of the hormone. In studies on the effect of growth hormone on protein synthesis, it was found that treatment with growth hormone retards the weight loss of the denervated soleus and plantaris muscle. In fact, it has been shown that the hormone increased the sizes of the denervated and control muscles proportionately (Figure 11-2).

In terms of exercise, the three main metabolic effects of growth hormone are to increase blood glucose, to increase blood fatty acids, and to increase protein synthesis in muscle.

Adrenocorticotrophic Hormone

Adrenocorticotrophic hormone (ACTH) is released into the circulation and is transported to the adrenal cortex where it stimulates the release of *cortisol.*

The release of cortisol into the bloodstream results in an increase in the release of glucose from the liver, which is accomplished by an increased gluconeogenesis. The increase in gluconeogenesis is at the expense of amino acids and proteins, especially muscle amino acids. Liver glycogen is not decreased in the cortisol-mediated increase in blood sugar—in contrast to the most common mechanism of increasing glucose, glycogenolysis. Cortisol secretion also results in decreased amino acid transport into muscle and decreased protein synthesis, as the amino acids presumably are being used in the liver for gluconeogenesis. In terms of lipid metabolism, cortisol is known to cooperate with GH and epinephrine in stimulating the release of fatty acids from adipose tissue.

High levels of cortisol, maintained over an extended period, induce increased secretion of hydrochloric acid and the two proteolytic

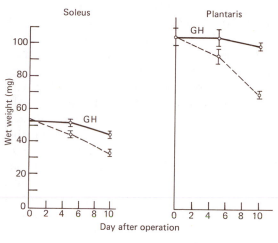

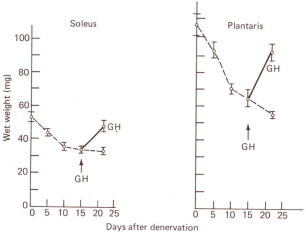

Figure 11-2. Effect of growth hormone (GH) on skeletal muscle immediately and 15 days after denervation.

From A.L. Goldberg and H.M. Goodman. Relationship between growth hormone and muscular work in determining muscle size. *Journal of Physiology* 200:665–666, 1969. By permission of Cambridge University Press.

enzymes, *pepsinogen* and *trypsinogen*. The release of these enzymes may be the basis for precipitation of ulcerative lesions of the gastrointestinal tract in stressed individuals.

Cortisol's antiinflammatory and antiinfection reaction is one of its most well-known actions. Treatment with glucocorticoids for inflammation results in slower healing—but with less inflammation. Antibody formation is decreased; gamma globulins are decreased; and a marked effect on connective tissue metabolism is observed. The cortisol inhibition of connective tissue healing is believed to be accounted for by *hypoemia* (subnormal blood supply), decreased cell migration, inhibition of *fibroblast* (connective tissue cell) formation, and inhibition of collagen deposition.

Gonadal Hormones

The mature female ovary secretes *estradiol* and *progesterone*; the male testes secretes *testosterone*. These hormones are under the influence of the hypothalamic-pituitary axis via several gonadotrophic hormones. The pituitary-regulating hormones are the *follicle stimulating hormone* (FSH) and *leutenizing hormone* (LH). These pituitary hormones are, in turn, controlled by the appropriate releasing factors from the hypothalamus.

TESTOSTERONE

The metabolic action of the *androgens* can be divided into *anabolic* and *androgenic* effects. The anabolic effects are characterized by an increased protein synthesis in the skeletal muscles. The androgenic effects are related to the actions of the androgens upon the reproductive organs.

Testosterone is known to cause increased synthesis of protein, increased incorporation of amino acids into skeletal muscle, and increased synthesis of RNA in skeletal muscle. In its androgenic effect, testosterone is responsible for the hair distribution characteristics of the male and for the complete development of the accessory sex organs (scrotum, seminal vesicles, prostate, and penis).

ACTH is known to have some effect upon the secretion of androgens from the adrenals; this secretion probably accounts for the source of testosterone in females. Either high levels of ACTH (during stressful situations) or an adrenal malfunction or a tumor could lead to increased testosterone in the female—and thus increased masculinization of the female.

ESTRADIOL AND ESTRONE

The metabolic action of the *estrogens* is involved in the regulation of the menstrual cycle, the development of the female secondary sex characteristics (growth of the uterus, vagina, pelvis, breast, and hair distribution) and the characteristic fat distribution of the female.

Antidiuretic Hormone

Antidiuretic hormone (ADH), as the name implies, is related to the maintenance and control of water balance in the body; it is secreted in response to decreased blood pressure. Increased blood levels of this hormone stimulate the kidneys to retain water, increasing the blood volume, which results in an elevation of blood pressure. Emotional stress is known to decrease the levels of ADH. Therefore, with emotional stress, there is an increase in the rate of urine production because of the decreased levels of ADH.

Catecholamines

The *catecholamines* (epinephrine and norepinephrine, also called adrenalin and noradrenalin) are secreted in response to sympathetic nervous system stimulation. Sympathetic nerve endings on the heart, blood vessels, and the adrenal medulla serve as sources for these hormones. The adrenal medulla secretes approximately 80% epinephrine and 20% norepinephrine; the reverse is true at the nerve endings on the heart and blood vessels. Although the adrenal is quite important in the response to certain types of stress, this gland is not necessary for the maintenance of life.

The primary metabolic effects of the catecholamines can be divided into: (1) effects on carbohydrate metabolism, (2) effects on fat metabolism, and (3) effects on cardiovascular dynamics.

The ability of epinephrine to increase carbohydrate metabolism is mediated by the intracellular messenger, cyclic adenosine monophosphate (cyclic AMP), acting upon the phosphorylase system in muscle and liver. This specific metabolic action is discussed in detail, later in this chapter, in the section on cyclic AMP.

Epinephrine is known to increase the release of fatty acids from adipose tissue. As before, the mechanism is dependent upon cyclic AMP, which acts upon a *lipase* (enzymes that split off fatty acids from mono-, di-, and triglycerides).

Catecholamines have extensive effects on cardiovascular dynamics. Norepinephrine, secreted at the sympathetic nerve endings on arterial blood vessels, produces a constricting effect on the vessel that increases systolic and diastolic blood pressure. The norepinephrine secreted in the heart increases the heart rate frequency and strength of the heart beat; these effects, along with the blood vessel constriction, increase the peripheral resistance.

The kinetic data in Figure 11–3 clearly illustrates the dual role of epinephrine, the well-known *calorigenic* (increased metabolism) and the *inotropic* (increased strength of the

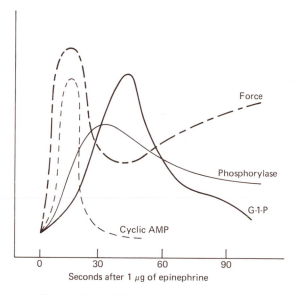

Seconds after 1 μg of epinephrine

Figure 11-3. Effects of epinephrine on the metabolic activity of the heart.

From J.R. Williamson. Kinetic studies of epinephrine effects in the perfused rat heart. *Pharmacological Reviews* 18:205–255, Copyright © 1966, The Williams & Wilkins Co., Baltimore.

heart beat) responses in the heart. This figure shows that the contractile force or inotropic effect on the heart is increased several seconds prior to the increased glycogenolytic effect. These two effects (calorigenic and inotropic) have been labeled as being mediated by beta and alpha receptors, respectively: the beta receptor is the *adenyl cyclase receptor* (responsible for the formation of cyclic AMP); the alpha receptor is the *ATPase receptor* (responsible for the increased contractile force associated with epinephrine infusion). Figure 11–4 reviews the overall role of the catecholamines and their receptors in the heart.

Pancreatic Hormones

INSULIN

Insulin acts as an activator of the key enzymes of glycogen synthesis and glycolysis while simultaneously decreasing the activity of the enzymes of glycogenolysis in the muscle. Insulin is known to increase the percentage of glycogen *synthetase I*, which is ten times as active as glycogen synthetase II.

Insulin increases the transport of glucose into adipose tissue and blocks one of the enzymes of lipid oxidation, thus promoting lipogenesis. These actions lead to an increased synthesis of lipid and increased size of the adipose tissues.

In summary insulin increases the rate of glycolysis in muscle and adipose tissue, increases glycogen deposition in muscle and liver, and increases lipid deposition in adipose tissue while decreasing blood glucose levels.

When there is a decrease in insulin production, as in *diabetes mellitus*, the peripheral cells turn to fatty acids and protein for energy. In this state the build-up of ketone bodies, aceto-acetic acid, acetone, and β-hydroxybutyric acid leads to the metabolic state classified as *ketosis*.

GLUCAGON

The metabolic action of *glucagon* in the liver is mediated by the intracellular messenger, cyclic AMP. In general glucagon activates the adrenyl cyclase system to produce cyclic AMP, which in turn stimulates the phosphorylase system to increase the breakdown of glycogen to glucose-6-phosphate. The G-6-P is then hydrolyzed to the free glucose molecule, and glucose is secreted into the bloodstream to increase the blood level of glucose.

Other Hormones

PROSTAGLANDINS

Prostaglandins are a group of unsaturated cyclic fatty acids, possessing 20 carbon atoms, which were originally found in the prostate gland and hence their name. They have now been identified in all mammalian tissues; they act to lower blood pressure, to inhibit the epinephrine-induced release of glycerol and fatty acids from adipose tissue, and to decrease nervous system excitability. In general it appears that these substances may be inhibitors to many other hormones.

PARATHORMONE AND CALCITONIN

The two hormones, *parathormone* and *calcitonin*, are discussed together as both are involved with calcium and phosphate blood levels. They act in opposition to one another in their effects upon blood calcium and phosphate ion concentrations. The stimulus for their release is the blood level of calcium ion in relation to the phosphate ion concentration. Secretion of parathormone results in an increase in blood levels of calcium and an increase in the urine excretion levels of phosphate ion. The increase in the blood calcium ion arises from the hormone acting on bone, which is the largest single reservoir in the body for calcium. In contrast secretion of calcitonin results in an increased deposit of calcium in bone and an increased retention of phosphorus ion by the kidney.

RENIN, ANGIOTENSIN, AND ALDOSTERONE

The control of blood pressure is one of the high-priority items of the human organism. *Renin* is an enzyme secreted by a special group of epithelial cells located in the wall of the renal artery as it approaches the *glomerulus*. The location of these cells is optimal since kidney filtration is so essential to life. In response to lowered blood pressure, these cells secrete the proteolytic enzyme, renin, which acts upon a blood protein called *angiotensinogen* to form *angiotensin I*. There exists in the blood another enzyme that is capable of converting *angiotensin I* to *angiotensin II*. Angiotensin II is one of the most powerful pressor agents known, and the resulting constriction of blood vessels results in a powerful increase in blood pressure.

Angiotensin II also acts on the adrenal cortex to stimulate the release of *aldosterone*. Aldosterone is transported by the blood and acts on the kidney to increase the amount of sodium reabsorbed; specifically aldosterone influences the kidney to decrease the amount of sodium excreted in the urine, thus raising the osmolarity of the blood by increasing the sodium concentration. Consequently the increased osmotic pressure forces fluid from the extracellular spaces into the blood, increasing blood volume and thus blood pressure.

HISTAMINE AND SEROTONIN

Histamine is found in most mammalian tissue, including lung, liver, muscle, skin, and intestinal tissue. There is some evidence that the histamine-forming capacity of various tissues

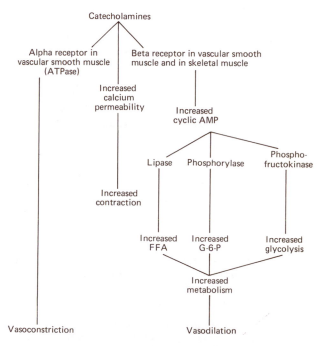

Figure 11-4. Summary of catecholamine effects.

Adapted from L. Lundholm, E. Mohme-Lundholm, and N. Suedmyr. Metabolic effects of catecholamines' physiological interrelationships: Introductory remarks. *Pharmacological Reviews* 18:255–272, Copyright © 1966, The Williams & Wilkins Co., Baltimore.

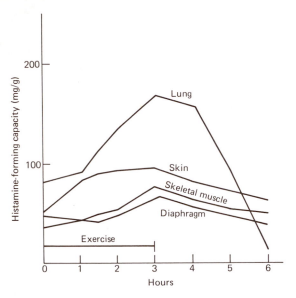

Figure 11-5. Exercise and histamine-forming capacity of various tissues in the mouse, exercising continuously for three hours.

From P. Graham, G. Kahlson, and E. Ressengren. Histamine formation in physical exercise, anoxia, and under the influence of adrenalin and related substances. *Journal of Physiology* 172:174–188, 1964. By permission of Cambridge University Press.

is elevated in response to physical exercise (Figure 11-5). We might postulate that the release of histamine is a vasodilation response, permitting dormant capillaries to open under stress conditions.

Serotonin is a potent vasoconstrictor and has been implicated in neural transmissions. It is found in particularly high concentrations in selected areas of the brain and alimentary tract.

ACETYLCHOLINE

Acetylcholine is the chemical transmitter of the peripheral nervous system and also of some of the actions of the autonomic nervous system. The acetylcholine molecule is either

stored within the mitochondria or released and stored in readily observable vesicles near the end of the axons. Figure 2–4 shows the terminal axon of a peripheral nerve as it forms a typical neuromyo junction. When acetylcholine is released from the axon, it forms a junction on the postsynaptic membrane, thus changing the charge on the receptor site. The altered electrical characteristics result in the development of a wave of negativity that spreads throughout the muscle and eventually stimulates muscular contraction.

Cyclic Adenosine Monophosphate

Cyclic AMP has been implicated as the intracellular mediator of several hormonal actions. After the primary hormone arrives at its specific site of action, it stimulates the formation of cyclic AMP. This secondary messenger in turn stimulates a change in the cellular metabolism. To account for the specificity of this system, it is necessary to accept not only tissue specificity for the primary hormones but also intracellular compartmentation in relation to the activity of cyclic AMP. The ability of cyclic AMP to leak out of one cell and into another must also be controlled.

The cyclic nucleotides are formed by the action of a cyclase enzyme (*adenyl cyclase* in the case of cyclic AMP) as depicted in Figure 11-6. The adenyl cyclase enzyme, which seems to be part of the membrane structure, must be capable of interacting with extracellular space (where the primary hormone comes into contact with the cell) as well as with intracellular space (where the conversion of ATP into cyclic AMP takes place).

The most well-known effect of cyclic AMP is its action in mediating the glycogenolytic effect of glucagon and epinephrine. The overall scheme of these reactions is outlined in Figure

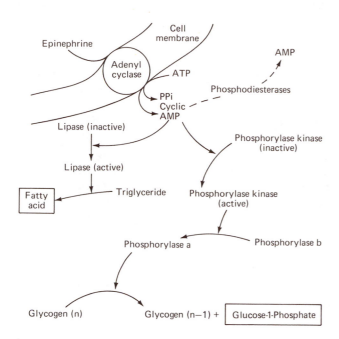

**Figure 11-6. Cyclic adenosine-mono-phosphate
mediated effects of epinephrine.**

11-6. These reactions are a type of biological amplifying system, making it possible for each hormone molecule to activate several adenyl cyclase enzymes, for each adenyl cyclase enzyme to form several cyclic AMP molecules, for each cyclic AMP molecule to activate several kinase enzymes, and so on.

The involvement of cyclic AMP with phosphorylase kinase in skeletal muscle has led to the elucidation of cyclic AMP involvement with other kinases in other tissues. It has been shown that protein kinases (of which phosphorylase kinase is one) effectively phosphorylate histones, which may unmask the DNA molecule; this would, in turn, increase the synthesis of RNA and protein. If this concept is validated, it may uncover the mechanisms that "turn on" previously "turned off" genes.

Table 11-2. Known Metabolic Involvements of Cyclic AMP

Enzyme or Process Affected*	Tissue	Change in Activity or Rate
Phosphorylase	Several	Increased
Glycogen synthetase	Several	Decreased
Protein kinase[†]	Several	Increased
Phosphofructokinase	Flatworm	Increased
Tyrosine transaminase induction	Liver	Increased
PEP carboxykinase induction	Liver	Increased
Serine dehydratase induction	Liver	Increased
β-Galactosidase induction	Bacteria	Increased
Overall protein synthesis	Several	Decreased
Glyconeogenesis	Liver	Increased
Ketogenesis	Liver	Increased
Lipolysis	Adipose	Increased
Steroidogenesis	Several	Increased
Water permeability	Epithelial	Increased
Ion permeability	Several	Increased
Calcium resorption	Bone	Increased
Renin production	Kidney	Increased
Contractility	Cardiac muscle	Increased
Tension	Smooth muscle	Decreased
Membrane potential	Smooth muscle	Hyperpolarized
Amino acid uptake	Liver and uterus	Increased

*PEP signifies phosphoenolpyruvate; ACTH, adrenocorticotropic hormones; TSH, thyroid stimulating hormone; GH, growth hormone; LH, luteinizing hormone; HCl, hydrochloric acid.

†Stimulation of protein kinase is known to be involved in effects of cyclic AMP on phosphorylase and glycogen synthetase systems and may also be involved in many other effects of cyclic AMP.

From E.W. Sutherland, M.D. On the biological role of cyclic AMP. *JAMA* 214(7):1281–1288, Nov. 1970. Copyright 1970, American Medical Association.

Cyclic AMP is known to be involved in over 30 biological processes. Table 11–2 provides a list of known metabolic processes involved with cyclic AMP. Its biological action is very diverse and is known to affect carbohydrate and lipid metabolism, protein activity, permeability, contractility, gland secretions, neurohumoral transmissions, nucleic acid synthesis, and cell growth.

Table 11-2. (*cont.*)

Enzyme or Process Affected*	Tissue	Change in Activity or Rate
Amino acid uptake	Adipose	Decreased
Clearing factor lipase	Adipose	Decreased
Amylase release	Parotid	Increased
Insulin release	Pancreas	Increased
ACTH release	Anterior pituitary	Increased
TSH release	Anterior pituitary	Increased
GH release	Anterior pituitary	Increased
LH release	Anterior pituitary	Increased
Thyroid hormone release	Thyroid	Increased
Calcitonin release	Parafollicular	Increased
Acetylcholine release	Nervous	Increased
Histamine release	Leucocytes	Decreased
HCl secretion	Gastric mucosa	Increased
Fluid secretion	Insect salivary glands	Increased
Discharge frequency	Cerebellar Purkinje	Decreased
Melanocyte dispersion	Skin	Increased
Aggregation	Cellular slime mold	Increased
Aggregation	Platelets	Decreased
RNA synthesis	Bacteria	Increased
DNA synthesis	Thymocytes	Increased
Cell growth	Tumor cells	Decreased

Control of Blood Glucose and Free Fatty Acids

Although insulin is usually thought of as the primary hormone involved in the control of blood glucose, several other hormones are also involved. Why do we need this control over the level of blood glucose? If the level of glucose in the blood increased too much, the organism would lose glucose into the urine. The ability of the kidney to reabsorb glucose is limited to an upper maximum range of 180 milligrams percent. If the level of blood glucose decreased to less than 50 mg %, the

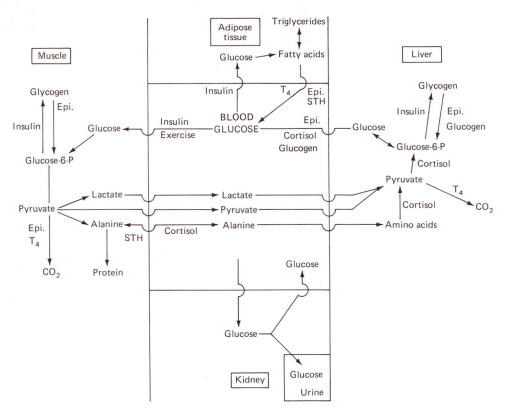

Figure 11-7. Control of blood glucose and free fatty acids (Epi. = epinephrine).

cells of the central nervous system would begin to starve. In general the level of blood glucose concentration is limited to 85 mg %, with a range of 65–160 mg %. Figure 11-7 presents a simplified view of blood glucose and the various hormonal methods of changing its level of concentration.

Most of the hormones act in opposition to insulin by increasing blood levels of glucose, while insulin acts to decrease the level. Epinephrine and glucagon increase blood glucose primarily by increasing glycogenolysis in the liver.

Blood levels of fatty acids are increased at the maximum rate by a combination of thyroxine, epinephrine, and growth hormone.

Syndrome of Physical Activity

At present it is impossible to outline completely endocrine influences on metabolic actions in response to physical activity. Since model building has proven quite useful in molecular biology, it seems reasonable to attempt to build such a model for the "endocrine syndrome of physical activity."

Table 11-3. Hormonal Responses During Physical Activity

Hormone	Anticipatory	Initial		Adaptation		Exhaustion		Recovery		General
	Theory	Theory	Research	Theory	Research	Theory	Research	Theory	Research	Research
ACTH	+	+		−		−				
Cortisol	+	+	−	−	+, −	−	+, −		+, −	0
Epinephrine	+	+		−		−	+, −			+
Norepinephrine	+	+		+			+			+
Histamine		+								+
Extrapancreatic insulin		+								+
Thyroxine				+	+				+	−, 0
Glucagon				+						0
Growth hormone			+	+	+			+	+	+
ADH				+				+		+
Insulin					+, −			+	+	0
Renin			+		+		+	+		+
Testosterone								+		
Aldosterone								+		
LH										0

The theoretical chemical changes in the hormonal levels in the blood are indicated in the theory column within each subdivision. Where data is available, the research is indicated in the appropriate column. Blood levels of hormones do not always indicate the secretion rate but reflect the difference between secretion and degradation of the hormone.

Physical activity represents the attempt of an individual to move from one position to another. An individual's ability to perform a given activity is dependent upon his or her nutritional state, state of training, external resistance—as well as other variables. Furthermore it is well known that physical activity can be (1) of a low, medium, or high level of intensity, (2) of general or specific muscular involvement, and (3) of short or long duration. With these factors in mind, we have attempted to describe a model illustrating the endocrine involvement in physical activity.

Table 11-3 lists the predicted and substantiated directional changes to be found during a physical activity. Many of these endocrine alterations may be elicited by "emotional" stimuli; these are sometimes hard to separate from truly activity-stimulated changes: the perceived rather than actual stress may elicit the response from the endocrine system. For our purposes we have assumed that emotional stimulation contributes to the endocrine response in the performance of a physical activity. This assumption is made whether the performance involves a competitive athletic event or the emotional stimulation of a person exercising for a personal health program.

For the purpose of describing the endocrine response to exercise, we propose to divide physical activity into five stages: (1) a *preactivity stage*, (2) an *initial activity stage*, (3) an *activity stage*, (4) an *exhaustion stage* (occurring only if the activity is severe enough), and (5) a *recovery stage*. In other words an exercise bout consists of anticipation, reaction, adaptation, possibly a breakdown, and recovery.

Whether preparing for athletic competition or for health-related exercise programs, there is an anticipatory time period prior to the activity when the individual realizes that he or she will soon be required to move from the resting steady state to a higher level of energy expenditure. At the beginning of the exercise there is an initial metabolic "rate of change" from the resting state to the exercising condition. This initial rate of change elicits the greatest endocrine response.

As the exercise persists, the individual attempts to adapt the energy-producing reactions to the requirements imposed by the exercise. In attempting to find a new steady state, the body may try shifting energy sources or shifting into a more acute stage of response. At some later time, either the activity will end, or the body will reach the stage of acute exhaustion owing to its inability to maintain the constant elevated level of stress. The point at which this happens will be determined by the individual's stage of training and motivation. At the cessation of activity, whether due to voluntary stoppage or due to acute exhaustion, the individual makes the necessary adjustments to recover to the resting state. As in the initial reaction phase, the initial recovery phase is marked by a rapid rate of change in the body metabolism; during this stage we would expect several hormonal systems to alter their activities rapidly.

In our model, as opposed to the stress theory of Selye (see p. 195), the stimulus for activity must come from *within*; that is, the individual must want to be active in order to stimulate his or her muscles to operate. We must then ask the question, "What factors are produced during muscular work or what neural connections are utilized to stimulate the responses of the endocrine system?" Obviously, if an activity is of *low intensity* in relation to an individual's exercising potential, the stimulus to the endocrine system will differ from the stimulus that arises when the activity calls for a much *greater* percentage of exercising potential. The endocrine responses shown in Table 11–3 are based upon a level of activity that taxes the body to at least 60% of maximum (max $\dot{V}_{O_2}$). We chose this level of exercise since it has been shown that this level is necessary to elicit the maximum adrenal response. Given this level of exercising intensity, 60% of maximum, we can discuss the response of the endocrine system during the arbitrary five stages of activity, although we recognize that the evidence for these responses is incomplete.

Anticipation

The degree of anticipation, or the extent of the emotional involvement, prior to an exercise bout varies from one individual to another. At one extreme is the individual who acknowledges that he or she will shortly engage in physical activity, but is not particularly concerned. At the other end of the scale is the individual who for one reason or another is overexcited, worried, sick, or scared. The endocrine response of the former individual is minimal; the endocrine response of the latter individual is excessive. In general, during the anticipation phase, we can expect a release of ACTH and cortisol owing to a general stress reaction and a release of norepinephrine owing to sympathetic nervous system stimulation. This combination increases blood sugar, increases gluconeogenesis in the liver, decreases protein synthesis in the muscle, increases sweat gland activity, increases heart rate, increases the strength of the heart beat, and increases blood pressure.

Initial Stage of Exercise

During the initial stages of exercise, the body reacts to the great rate of change from the homeostatic conditions. During this time, cortisol secretion is continued, and epinephrine and norepinephrine are released in increasing amounts. Therefore, a situation exists in which: blood sugar is increased due to gluconeogenesis and glycogenolysis in the liver; peripheral resistance is decreased owing to the effect of epinephrine on the muscle cells; inotropic and calorigenic responses are increased in the heart; and the number of open capillaries is increased in active muscles owing to increased metabolic products and perhaps to histamine release.

Adaptation

During the adaptation stage, cortisol secretion probably shuts off because of the absence of any rate of change in the level of activity. Epinephrine and norepinephrine continue to be secreted. Due to the high-activity level, blood glucose probably decreases to some extent; this decrease helps to stimulate the release of growth hormone, thyroxine, and glucagon. During exercises that last over two minutes, some combination of the existing cortisol, epinephrine, growth hormone, glucagon, and thyroxine stimulates the adipose tissue to increase the rate of lypolysis. The increased concentration of fatty acids in the blood stream represents the shift in energy substrates from carbohydrate to lipid which spares the remaining stores of carbohydrate in the body—stores that are essential for central nervous system metabolism.

As the exercise continues, the fluids of the blood are forced into the intercellular spaces, and the osmolarity of the blood is increased. The hemoconcentration is likely to stimulate the release of antidiuretic hormone, which conserves water through the kidney filtration process; this leads to the well-known antidiuretic effect of exercise.

Exhaustion

The events transpiring at the stage of exhaustion are probably the most difficult to propose; there are many diverse speculations for the cause of the exhaustion. Depletion of hormones from the adrenal medulla remains as one possibility and the decreased rate of cortisol secretion is another.

Adrenal or pituitary depletion (or shutoff) due to negative feedback factors would decrease the available energy substrates. Pituitary depletion cannot be the total answer as hypophysectomized animals can also exercise to a small degree, as can thyroidectomized animals. It appears that in the normal person under normal exercise conditions, the endocrine system is not a limiting factor which causes exhaustion.

Recovery

The recovery phase, whether due to voluntary or involuntary cessation of the activity, is the final phase of our exercise model. During this stage the individual returns to the resting steady state. The postexercise levels of the hormones are high; even if the secretion rate declines, the levels remain elevated for some time until all the active molecules are degraded. Sympathetic stimulation ceases, to a great extent, and the heart rate begins its return to normal. Blood pressure decreases because of dilation of muscle blood vessels in response to metabolic products and because of dilation of skin blood vessels owing to temperature regulation. The action of ADH is opposed to this decreased blood pressure, and the renin-angiotensin-aldosterone system may be activated during this stage to maintain the blood pressure within the limits necessary for complete recovery. Since the levels of growth hormone are elevated, it aids in the incorporation of amino acids into muscle cells and stimulates RNA and protein synthesis. Although it seems reasonable to expect that testosterone levels would be increased to aid in protein synthesis, no evidence is available as of this time to suggest or contradict this hypothesis.

Summary

The above "syndrome of physical activity" cannot be documented in its entirety; yet enough information exists for us to propose this type of syndrome in the hope that it will define the concept of endocrine responses to activity. Once the concept is accepted, the remaining unproven specific facts can be approached in a systematic manner. Obviously the endocrine system does not operate independently of other systems—a point that we have stressed repeatedly.

Acute Stress Syndrome

When one discusses *stress*, the name of Selye must be included. Much has been written about the stress syndrome as conceived, formulated, and popularized by the Canadian physician Hans Selye, who describes the development of this concept in *In Vivo*; we recommend it for information on the stress concept or research methods. Selye's concept of stress arose from his observation that there were common symptoms to "just being sick." After several years and many experiments, Selye, in the early 1950s, coined the expression "general adaptation syndrome" to describe the organism's attempt to react to a stress. There are several types of *stressors* (stress-producing agents) and several degrees of stress. Regardless of the type of stressor or the degree of stress, Selye lists the organism's response in terms of: a general alarm reaction, a stage of resistance, and if the stress persists or is overpowering, a stage of exhaustion.

Stressors other than physical activity are known to arouse a response from the endocrine system. Examples of these other stressors are pain, emotional stimuli, innocuous agents, and acute viral infections; for these the initial response seems to be an increase in the output of ACTH, resulting in an increase in the circulating levels of cortisol. The exact mechanism for the release of ACTH is not clear, but it appears to be neurally mediated. Stress can be shown to stimulate the sympathetic nervous system and to increase the electrical activity in several nuclei within the hypothalamus. It has been demonstrated that stress increases the release of *gonadotropins* from the pituitary, indicating that there is an increase in the levels of circulating gonadal hormones. Also the rate of pituitary release of growth hormone and thyroid-stimulating hormone has been shown to increase during stressful situations. Antidiuretic hormone release is elevated in response to most stressors.

If the intensity of the activity and/or the duration of the stress are such that the body can readily adapt to the stress, there are probably not many more alterations than those presented in the previous paragraph. However, if the intensity or duration of the stress is such that the body cannot cope with it, the stage of exhaustion will set in, and the ability to respond to that stress will be lost. In some cases this will mean acute exhaustion; with more severe stressors, it may mean the death of the individual.

In the endocrine system, the imposing of one stress upon another is additive; that is, two different stresses taken together may elicit a state of exhaustion whereas, if taken separately, neither of the stresses would tax the body to exhaustion. This brings us to the concept of "additive stresses". We can ask the question, "Does the experience of one stress have any effect when a second stress is imposed concurrently?" The answer to this appears to be yes, as exemplified by the proverb about "the straw that broke the camel's back". However, it has been shown that pretreatment with mild stressing agents protects and aids the body in responding to the second independent stress. For example exposure to the stress of a mild physical activity would help an individual to respond to a related or unrelated stress, such as pain or emotional stimuli.

Summary

The primary effects of hormones are to regulate local and systemic blood pressure, protein synthesis, blood levels of glucose and fatty acids, and to influence energy metabolism. These main effects are critical to the exercising capacity of an individual.

The endocrine system is largely under the influence of the nervous system; as a result, emotions can readily alter hormonal secretion patterns during exercise. The hormonal response is related to the relative degree of stress that the exercise places upon the body; the more stressful the activity, the more severe the hormonal response.

The total endocrine response to exercise can be divided into preactivity, initial activity, activity, exhaustion, and recovery stages. We propose that hormones are differentially released during these stages of exercise.

Study Questions

1. How does exercise affect the endocrine system?
2. What hormones are most important during exercise?
3. What is the role of the hypothalamus and pituitary during exercise?
4. What is the function of thyroxine during exercise?
5. What is the function of ACTH and cortisol during exercise?
6. What is the function of the growth hormone during exercise?
7. What is the function of the gonadal hormones during exercise?
8. What is the function of the catecholamines during exercise?
9. What role does cyclic AMP have in mediating the effects of hormones?
10. Describe the control of blood glucose.
11. What factors determine the effects of exercise on the hormonal system?
12. How do the different stages of exercise alter hormonal release?
13. How would you describe the stress syndrome?
14. How can stresses be additive?

Review References

Gollnick, P.D.; R.G. Soule; A.W. Taylor; C. Williams; and C.D. Ianuzzo. Exercise-induced glycogenolysis and lipolysis in the rat: Hormonal influence. *Amer. J. Physiol.* 219:729–733, 1970.

Gollnick, P.D., and C.D. Ianuzzo. Acute and

chronic adaptations to exercise in hormone deficient rats. *Med. Sci. Sports* 7:12–19, 1975.

Hartley, L.H. Growth hormone and catecholamine response to exercise in relation to physical training. *Med. Sci. Sports* 7:34–36, 1975.

Irvine, C.G.H. Effect of exercise on thyroxine degradation in athletes and non-athletes. *Journal of Clinical Endocrinology and Metabolism* 28:942–948, 1968.

Killinger, D. Testosterone. *Canad. Med. Assoc. J.* 103:733–735, 1970.

Lamb, D. Androgens and exercise. *Med. Sci. Sports* 7:1–5, 1975.

Morgan, W.P., ed. *Erogenic Aids and Muscular Performance*. New York: Academic Press, 1972.

Prokop, L. Adrenals and sports. *Journal of Sports Medicine and Physical Fitness* 3:117–121, 1963.

Seyle, H. *In Vivo*. New York: Liverright, 1967.

Seyle, H. The evolution of the stress concept. *Amer. J. Cardiol.* 26:289–299, 1970.

Shephard, R.J., and K.H. Sidney. Effects of physical exercise on plasma growth hormone and cortisol levels in human subjects. In *Exercise and Sport Sciences Reviews*, vol. 3, J.H. Wilmore and J.F. Keogh, eds. New York: Academic Press, 1975, pp. 1–30.

Terjung, R.L., and W.W. Winder. Exercise and thyroid function. *Med. Sci. Sports* 7:20–26, 1975.

Tharp, G.D. The role of glucocorticoids in exercise. *Med. Sci. Sports* 7:6–11, 1975.

Vranic, M.; R. Kawamori; and G.A. Wrenshall. The role of insulin and glucagon in regulating glucose turnover in dogs during exercise. *Med. Sci. Sports* 7:27–33, 1975.

CHAPTER TWELVE

Renal and digestive responses to exercise

Key Concepts

- The kidney plays a key role in maintaining a relatively homeostatic whole body acid-base balance.
- Exercise retards digestion.
- Some evidence indicates that physical training has an antiulcerongenic effect.
- The liver plays a supportive role in the timely distribution of metabolic components according to muscular tissue needs.
- Exercise can assist some diabetic conditions by aiding glucose transport.

Renal Function and Exercise

Basic Function

The kidney's primary function is to preserve the composition, volume, and pH of extracellular fluids. The kidney's detoxifying function is related to the partial elimination of the ammonium, urea, and other metabolic by-products that are not easily metabolized or eliminated. It is not that the kidney prevents substances from circulating but that it prevents the accumulation of toxic levels of metabolites.

The kidney is also an endocrine organ. A specialized segment of its tubular complexes (*juxtaglomerular apparatus*) produces a hormone, *renin*, in response to low renal blood pressure. Renin elevates systemic blood pressure through a series of interactions with other hormones from the liver and adrenal cortex (see Chapter 11). The renin system seems to be a very important factor in the etiology of some types of hypertension.

Compared with any other organ, the kidney receives the greatest blood flow per amount of oxygen consumed. Although it accounts for only 0.4% of the body weight, the kidney receives about 25% of the resting cardiac output. This fact suggests that the renal blood flow is important in a way that is independent of supporting the metabolic demands of the kidney.

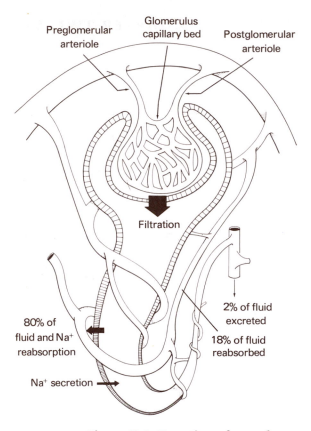

Figure 12-1. Function of a nephron.

This critical function is the maintenance of a normal composition of body fluids.

The kidney performs its main function by a combination of *filtration, reabsorption, secretion,* and simple *diffusion* (Figure 12–1). These processes occur in the cells lining the renal tubules; the relative importance of each of these is different along the various segments of the tubule.

FILTRATION

The process of *filtration* is dependent on the *blood pressure* (or driving force) within the renal vascular system in the tuft of capillaries called the *glomerulus*. These capillaries permit the entry of a relatively large flux of vascular fluid, including electrolytes and occasionally proteins, into the first winding segment of the tubular system called the *proximal convoluted tubule*. In fact the osmolarity of the fluid within the proximal convoluted tubule is very similar to the osmolarity of the blood.

Normal changes in systemic blood pressure do not affect the filtration rate because the kidney can autoregulate blood flow. Within a wide range of blood pressures, renal blood flow remains relatively constant since the kidney can alter its capillary resistance. For example, when systemic blood pressure rises, resistance to renal blood flow increases proportionally; thus a constant blood flow and a normal rate of filtration are maintained.

REABSORPTION

The outer layer of the kidney (the *cortex*) retrieves hexoses, amino acids, electrolytes (Na^+, K^+, Cl^-, NH_4, and others), and a proportional amount of fluid by *reabsorption*, a process essential for life. There is a maximum limit to the amount of each substance that can be retrieved (reabsorbed) from the tubules. This maximum is referred to as the *transport maximum* (T_m) and is expressed as mg/min. For example the T_m for glucose is about 375 mg/min. If any more glucose, per unit of time, flows through the renal tubules, it spills over into the urine, and the body loses an important energy source. An example of this is the occurrence of glucose in urine shortly after the ingestion of a candy bar or soft drink. However, this spillover is short-lived and of a low magnitude.

SECRETION

The process by which the kidney *actively* discards substances, mainly sodium, into the tubular cavity (*lumen*) to be lost in the urine is called *secretion*. This process counteracts reabsorption; however, by working together, the two processes can assure a greater regulatory effectiveness in maintaining proper balances of various electrolytes and fluids, as well as the pH of body fluids.

Composition of Body Fluids and Their Regulation

About 75–80% of the body weight consists of water. The percent of the body weight that is water depends on the relative amount of fat, protein, and carbohydrate that make up the body weight. About 65% of the total water is located intracellularly; 28% is in interstitial spaces; and 7% is associated with the cellular membranes themselves.

Prolonged muscular activity elevates the blood pressure within the capillaries; the elevated blood pressure results in the filtration of plasma fluid extravascularly (Figure 12–2). More indirect changes in the amount of fluid in the various compartments occur with severe fluid loss by sweating. Excessive sweating leads to dehydration and consequently to reduced blood volume.

As much as 20–30% of the body's Na+ can be lost prior to reaching the critical level that results in an individual's death. The Na+ deficiency can be corrected rapidly with a high sodium intake. Military exercises in the desert have resulted in a fluid intake of 15 liters of water within a 24-hour period. Throughout this fluid exchange, an appropriate level of Na+ in the body fluids had to be maintained.

Generally exercise causes an increase in K+ concentration in urine while Na+ and Cl− concentrations decrease. This alteration in ion distribution can partially be explained by the continuous outflux of K+ from muscles while more Na+ passes into the intracellular compartment during exercise. Dehydration also results in an elevated K+ content in the urine. Mild exercise elevates Cl− concentration in the

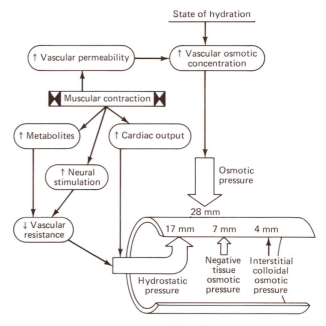

Figure 12-2. Factors affecting the forces which regulate fluid influx and outflux of the capillaries.

Data from Arthur C. Guyton. *Textbook of Medical Physiology*, 4th ed. Philadelphia: W.B. Saunders, 1971.

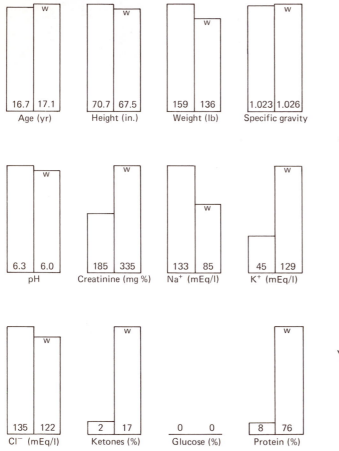

Figure 12-3. Urinary profile of wrestlers (w) and nonwrestlers (mEq = milliequivalent).

Adapted from E.J. Zambraski; C.M. Tipton; H.R. Jordon; W.K. Palmer; and T.K. Tcheng. Iowa wrestling study: Urinary profiles of state finalists prior to competition. *Medicine and Science in Sports* 6:129–132, 1974.

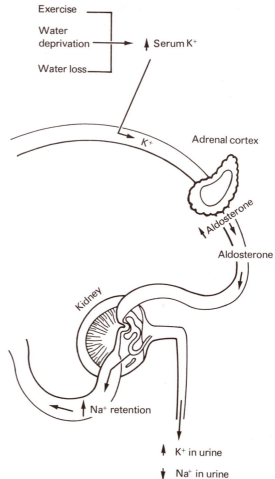

Figure 12-4. Increased K+ in the blood serum stimulates the secretion of aldosterone in the adrenal gland. This, in turn, induces Na+ retention and K+ excretion in the kidneys.

urine; however, exhaustive exercise is characterized by low Cl− content in the urine. The significance of this observation is not clear.

Urinary concentrations of various cellular components of high school wrestlers and nonwrestlers are shown in Figure 12-3. *Crea-* *tinine*, a product of creatine that comes from muscle, is almost twice as high in the urine of wrestlers as it is in nonwrestlers. Since the daily production of creatine is proportional to muscle mass, wrestlers may secrete more creatinine. It is also known that a single bout of exercise elevates urinary creatinine. Sodium levels are decreased, and potassium is more concentrated in the urine of wrestlers. It is possible that chronic exercise and suppression of weight gain, both common among wrestlers, result in high potassium in the urine through the mechanism shown in Figure 12–4.

The sympathetic nervous system provides one control system for determining the rate of urine formation by inducing renal vasoconstriction. If the vasoconstriction is severe enough, a reduced filtration, and thus a reduced urine formation, results (Figure 12–5). This can be induced by severe exercise or by general stresses such as cold, pain, fright, and hemorrhaging. Teleologically, such a mechanism seems reasonable since a reduction in renal blood flow from 1200 ml/min to 200 ml/min would make one more liter of blood available to tissues that are in greater immediate need of blood. But even under radical changes in renal resistance, the kidney has an amazing capacity to function without severe alterations.

Renal Regulation of Fluid Volume and Blood Pressure

The kidney's role in maintaining a constant fluid volume and its effect on blood pressure are shown in Figure 12–6. This diagram illustrates the several mechanisms available to the body to insure the maintenance of an adequate blood pressure—a very critical physiological parameter that assures adequate perfusion of the active vital organs (such as the brain, heart, and kidneys).

Because of the numerous variables during exercise that can cause fluid displacement, the homeostatic mechanisms are placed in even greater stress. Some of these exercise-related factors are: (1) loss of fluid from sweating, (2) elevated sympathetic activity, (3) hemoconcentration due to high arterial blood pressure, resulting in a net extravascular flux of fluid, and (4) redistribution of the blood volume (e.g., from kidney and gut) to the more immediately active tissues (working muscles).

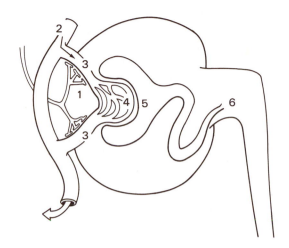

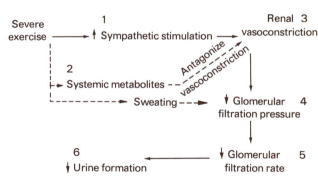

Figure 12-5. Effect of exercise on urine formation.

Regulation of Acid-Base Balance

The acidification of body fluids induced by exercise is partially compensated by the kidney. It regulates the pH of the body fluids by a combination of excretion of acid, secretion of ammonia, secretion of K^+, and reabsorption of bicarbonate ions. Acid-base stability is also facilitated by ventilatory adjustments (see Chapter 10).

Water and Salt Balance

A major function of the kidney is the preservation of a stable osmotic pressure. Exercise induces the loss of water and Na^+ and Cl^- via perspiration. The actions of elevated antidiuretic hormone (ADH) as well as of aldosterone and angiotensin partially compensate for the water loss due to sweating. Body temperature changes can also enhance release of ADH and, therefore, water retention during exercise. As an adaptation to training, the body secretes a sweat that is less concentrated in salt.

Implications for Athletics

Recognition of the existence of exercise-renal interaction is very important for the complete understanding of physical exercise. An awareness of these interactions and their control mechanisms can be important in the design of exercise and training sessions. For example, during early fall football practice, it is usually quite hot and perhaps humid. Fluid and salt loss as well as the complicating factor of temperature regulation—with their potential physiological ramifications—should not be ignored. During practice sessions, salt tablets are frequently taken without recognition of the fact that salt loss is significantly reduced after adaptation to heat has taken place. Reduced work capacity may be evident when as little as one percent of body weight has been lost. This amount of water can be lost within 30 minutes of hard work at a temperature of only 20°C (68°F).

The practice of not permitting athletes to drink water during or after exercise is sometimes too strict because of the erroneous assumption that performance will be adversely affected by ingestion of water. Similarly, extreme measures regarding fluid restriction are taken, not infrequently, by some coaches to assure that a wrestler "makes weight" for his weight class. A more reasonable attitude about salt balance and water intake should be that large amounts of water can only make an athlete uncomfortable, while too much salt or too little water can have serious effects on fluid homeostatic processes.

The thirst mechanism should not be trusted to be the sole regulator of water intake under the stress conditions that often occur in athletics. The thirst mechanism has been shown to be a rather inconsistent and imprecise sensor of the body's need for fluids.

Drinking Fluids with Electrolytes and Glucose While Working

Glucose and electrolytes are absorbed by the intestinal tract during exercise at about the same rate as they are absorbed at rest. The *osmolarity* (concentration of electrolytes and glucose) of the fluid is an important factor in regulating the rate at which the fluid is absorbed into blood. During severe work, when blood flow to the intestinal tract is minimized, up to 1.5 liters of fluid and up to 60 grams of glucose can be absorbed per hour. It appears that an ingested fluid of less than 2.5% glucose, less than 10 mEq of sodium, and less than 5 mEq of potassium at 25°C is most appropriate during exercise. About 0.2 liter should be ingested every ten to twelve minutes during the work. With this intake, the rate of absorption approximates the rate of ingestion, and the glucose absorbed will allow for a maintenance of normal blood sugar concentration. If higher glucose concentrations are ingested, the rate of absorption is depressed; and the advantage of the greater glucose concentration is offset by the lack of gastric emptying.

Electrolyte (e.g., Na^+) replacement of losses also deserves attention. For athletes who participate in prolonged intense competition, it may be advantageous to augment fluid and food intake for several days prior to the event. Long-distance runners are encouraged to ingest 400–500 ml of fluid 10–15 minutes before competition. For the more casual participant, it is recommended that salt tablets be taken in the later stages of prolonged exercise. Of course, water should be ingested throughout prolonged work periods in 0.2 liter lots.

Renal Regulation of Erythropoiesis

Although some controversy remains regarding the role of the kidney in stimulating red blood cell (RBC) synthesis, it appears likely that this organ produces a substance (*erythropoietin*) that has an erythrocyte-stimulating effect. Nephrectomized rats and rabbits, unlike normal animals, fail to produce *polycythemia* (high RBC concentration) in response to low P_{O_2}. The fluid extracted from a homogenized kidney stimulates *erythropoiesis* when injected into normal animals. Similarly blood from hypoxic or anemic animals is a potent source for erythropoietin. The role of this substance and its response to exercise remain unclear at the present time.

The Kidney: A Source of Metabolic Support

The kidney, like the liver and intestine, but unlike skeletal muscle, has the capability to synthesize and then to release glucose into the blood stream. The relative significance of

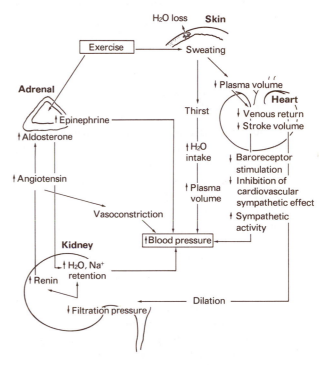

Figure 12-6. Mechanisms that maintain blood volume and blood pressure.

this organ in the maintenance of an adequate blood glucose level during exercise is unknown. It is known that *gluconeogenesis* (synthesis of glucose from amino acid and lactate) in the kidney is increased by exercise and by a low carbohydrate diet.

Urinalysis: Its Use in Assessment of Exercise and Training Effects

In addition to a slightly depressed renal blood flow and urine formation during strenuous exercise, *proteinuria* may also occur. This increased concentration of protein in the urine reflects changes in the serum protein concentration. Consequently it is a potentially valuable tool with which to investigate exercise effects on the whole body.

Elevated serum albumin concentrations account for most of the proteinuric response. Part of this response can be accounted for by the exercise-related hemoconcentration resulting from the flux of fluid from the vascular to extravascular compartments. Increased permeability changes of the glomeruli in the kidney could also cause proteinuria. Strenuous exercise increases glomerular permeability and thus the filtration rate; however, an enhanced reabsorption rate of some low-molecular-weight proteins in the proximal tubules tends to counteract this effect.

Hemoglobinuria and *myoglobinuria* are found in some subjects in response to vigorous exercise and give the urine a dark appearance. The causes of hemoglobinuria and myoglobinuria differ somewhat. Hemoglobinuria seems to be a manifestation of a fragmentation of the red blood cells caused by physical trauma during exercise. It is more severe when the subject runs long distances on hard surfaces. In fact, it has been eliminated in some subjects by their running on softer surfaces. Iron loss, and consequently anemia, might occur over a pro-

longed period due to hemoglobinuria. A similar phenomenon is called "march hemoglobinuria" and has been found in almost 10% of military trainees. Renal necrosis has been reported in subjects with severe hemoglobinuria and myoglobinuria after a 26-mile marathon run. This matter needs to be investigated more thoroughly because of the potential permanent renal damage that has been associated, although not in a "cause and effect" way, with individuals susceptible to hemoglobinuria or myoglobinuria.

Myoglobinuria in response to exercise is probably caused by an increased permeability of the muscle membrane to its proteins. Other proteins, such as lactate dehydrogenase, creatine phosphokinase (cytoplasmic), and serum glutamic oxalacetic transaminase (mitochondrial), also are found in higher concentrations in the blood after exercise.

For some unrecognized reason, certain individuals are more susceptible to proteinuria than others. (Although some protein is filtered through normal glomeruli, any concentration above 20 mg/100 ml of urine is considered abnormal.) As we might expect, a single exercise bout will induce a greater proteinuria effect in an untrained individual than in a trained individual. Also the longer the exercise period, the greater the proteinuria. But the proteinuric effect of exercise should not be overemphasized since, as a rule, the maximal change is only 10–15% different from resting values. If proteinuria persists, kidney disease is implied. Renal ischemia may be a contributing factor to exercise proteinuria.

Direct physical trauma to the kidney may occur in boxers. Many of the urinary symptoms seen in boxers during the week following a match are remarkably like those characteristic of *glomerulonephritis* (inflammation of glomeruli). In a group of fighters studied, 46% had increased urine turbidity; 80% high specific gravity; 68% albuminuria; and 73% had RBCs in the urine. Evidence of calcification was seen in more than 50% of the boxers.

Gastrointestinal and Accessory Digestive Tissues

The gastrointestinal tract (GI) and accessory tissues, such as the salivary glands, liver, gall bladder, and pancreas, may affect performance indirectly in that they serve a metabolically supportive role in assuring a sufficient energy supply for the working muscles and the nervous system. The liver plays an integral role in providing a constant energy supply during rest as well as during demanding exercise. And of course nutritional considerations cannot be isolated from the energy demands of exercise and training.

Gastrointestinal Tract: Effects of Exercise on Digestion

PERISTALSIS

Peristalsis is the wavelike contractions of the GI tract. This action is neurally controlled by the autonomic nervous system. The parasympathetic neural supply comes largely from the vagus nerve, which activates peristalsis, whereas sympathetic activation generally depresses gut motility. Exercise decreases gut motility perhaps through neural pathways, but hormonal factors are also involved.

SECRETIONS

During exercise watery secretions from salivary glands are depressed and replaced by more viscous secretions due to the influence of the sympathetic nervous system. A probable example of this exercise effect is the occurrence of "cotton" or dry mouth with severe exercise. General stress such as pain, anxiety, and fear

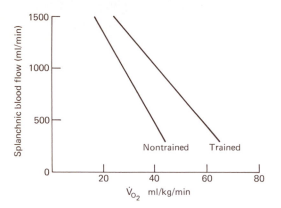

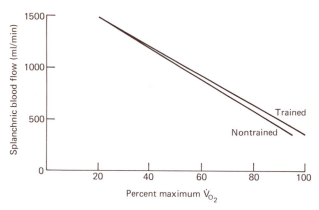

Figure 12-7. Differences of splanchnic blood flow in trained and nontrained individuals. Note the decrease in flow with increased $\dot{V}_{O_2}$ uptake (exercise). When the data are expressed as percent of maximum, the two curves are nearly superimposed, indicating a direct relationship between splanchnic blood flow and maximum $\dot{V}_{O_2}$ uptake, not training.

From L.B. Rowell. Regulation of splanchnic blood flow in man. *The Physiologist* 16(2):127–142, 1973.

can induce similar physiological responses in the salivary glands. In animals it has been shown that gastric secretions are depressed by daily bouts of swimming.

BLOOD FLOW

Directly after meals, blood flow markedly increases in the superior mesenteric artery, one of two major blood vessels to the gut. Within five minutes after the feeding is initiated, blood flow increases; after three hours blood flow may be still 50% higher than at prefeeding time. In dogs, the enhanced blood flow is due largely to a redistribution of the total blood volume since cardiac output changes are negligible. For example, the flow to the gastrointestinal tract may double its share of the cardiac output after feeding. In humans, a 25% increase in cardiac output has been observed after eating.

Splanchnic (stomach, intestines, spleen, and liver) blood flow decreases in proportion to exercise intensity (Figure 12–7). At a work intensity approximately equal to an individual's max $\dot{V}_{O_2}$, splanchnic blood flow may decrease up to 80%. At a work rate requiring an oxygen consumption of 40 ml/kg/min, the splanchnic flow of untrained subjects is less than half that of trained subjects. If the increasing work intensities are expressed as percentage of max $\dot{V}_{O_2}$, the slope of the line is the same. In other words the decrement in flow will be similar in a trained and nontrained subject if both work at 50% of their maximum rate.

The differences between "poorly conditioned" and well-trained subjects are related to such cardiovascular-respiratory adaptations to training as greater blood volume, cardiac output, and hemoglobin concentrations. It can be postulated that with these improvements following training, additional oxygen demands by muscular tissue can be satisfied with less stress on the splanchnic tissue and, therefore, with less effect on splanchnic blood flow. Splanchnic blood flow is similar in trained and nontrained individuals at rest.

Splanchnic blood flow can be predicted with considerable accuracy from changes in heart

rate, a much more readily measurable parameter than oxygen uptake (Figure 12–8). The
slope of the regression line in Figure 12–8 demonstrates that for every increase of three heart
beats per minute, a two percent decrease in
splanchnic flow can be expected. For example
we can predict that an increase in heart rate
from 60 to 180 would cause an 80% decrease
in splanchnic blood flow.

What are the functional considerations of
the drop in blood flow to the splanchnic tissue?
At rest only about 15–20% of the oxygen delivered to this tissue is utilized for metabolism.
During exercise the arteriovenous difference in
oxygen increases, compensating for a decreased
flow, and provides for an adequate supply of
oxygen to the splanchnic tissues. But the gastric emptying rate is reduced when blood flow
is reduced. Normally this would be of no consequence since digestion is only briefly delayed
and is not immediately critical to performance. Diversion of blood from the splanchnic
tissues means as much as 300 ml/min more
oxygen is available for the active tissue. By
combining the effects that occur in the skin
and the kidneys, this figure may be as high
as 600 ml/min.

Maintenance of adequate blood pressure
during exercise is also dependent on blood redistribution. In respect to the massive peripheral vasodilation in response to exercise, blood
redistribution is as essential in maintaining
normal blood pressure as it is in assuring an
adequate oxygen delivery.

One cause of decreased splanchnic flow during exercise is sympathetic activation, which
induces vasoconstriction of the splanchnic
bed. In addition vasoconstriction of the venous vasculature reduces its *capacitance*
(storage space); therefore less blood is "stored"
as a relatively inactive pool. In effect, the muscles are receiving a transfusion as a result of
splanchnic vasoconstriction.

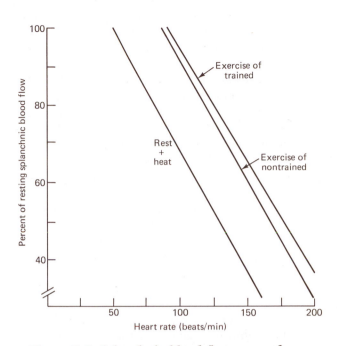

**Figure 12–8. Splanchnic blood flow responds
to increases in heart rate similarly
regardless of treatment.**

From L.B. Rowell. Regulation of splanchnic blood flow
in man. *The Physiologist* 16(2):127–142, 1973.

Stress, Exercise, and Peptic Ulcers

A *peptic ulcer* is a localized loss of tissue from the wall of the digestive tract, including the mucosa, submucosa, and muscular layers. Ulcers occur in the localized regions where acid-pepsinogen gastric juice is secreted. In today's industrialized populations, 10–15% of the males and 5–10% of the females develop peptic ulcers within their lifetimes. This frequency is about six times greater than it was in the mid-1930s. Peptic ulcers are the cause of death in about 6% of the diagnosed cases.

Other variables associated with peptic ulcers are: blood groups, genetic characteristics, serum pepsinogen levels, hypersecretion, and body build. In summary an ideal candidate for a peptic ulcer is lean, generally tense, has blood group O, high serum and urinary pepsinogen—and has a history of peptic ulcers in the family.

A common type of ulcer, which occurs in the beginning segment of the small intestine, is referred to as a *duodenal ulcer*. A commonly accepted explanation for the development of a duodenal ulcer is hypersecretion of hydrochloric acid and of a protein-digesting enzyme, *pepsinogen*, from the gastric wall. Autonomic imbalance, much like that associated with greater vagal tone in inducing the bradycardial response to exercise, could also be related to hypersecretion and formation of peptic ulcers. In fact many ulcer patients are treated by severing the vagal innervation to the lining of the stomach, thereby eliminating vagal dominance.

Vagotonia of ulcer patients may be a manifestation of a personality trait, for example nervous tension. Exercise used as a psychotherapeutic measure in the release of stress for the treatment of vagotonic ulcer patients is a theoretically rehabilitative possibility. It must also be realized that exercise itself is a form of stress—two comments might be pertinent. First, either the exercise intensity or duration (or both) may determine whether or not exercise is a noxious stress. Second, although exercise is commonly equated with other forms of stress, it is unlikely that all types of stress have the same psychogenic and general systemic physiological effects. The heterogeneity of the causal factors of stress ulcers is shown by the fact that at least one type of stress ulcer is related to a sympathetic, rather than a parasympathetic, factor in some laboratory animals.

Ulcers can be produced in animals by various forms of stress such as restraint stress, electrical stimulation, and starvation stress. Starvation stress is associated with hypersecretion of gastric hydrochloric acid, which has ulcer-inducing properties. In restraint stress, gastric erosions are thought to develop because of a loss of RNA, which causes a decline in mucoprotein production. *Mucopolysaccharides*, an essential component of connective tissue, seem to play protective roles in maintaining normal intestinal walls. Mucopolysaccharide synthesis is inhibited by hydrocortisone, an antiinflammatory drug that induces gastric ulcers. On the other hand ulcers can be prevented pharmacologically by enhancing mucopolysaccharide synthesis. It may prove to be very significant that several weeks of training by swimming enhance mucopolysaccharide synthesis in the stomach of rats; this may be the reason that exercise has an antiulcerogenic effect.

There is another antiulcerogenic effect related to training and gastrointestinal secretions. It is known that histamine secretion causes a marked increase in pepsinogen and hydrochloric acid secretions, which make the stomach wall more susceptible to digestion by the hydrochloric acid that it produces (Figure 12–9). The acid secretory response to histamine is depressed in trained rats but is back to normal within two weeks of the cessation of training. These studies suggest some beneficial effects of exercise; however, more extensive research will be necessary before exercise can be recommended positively as a rehabilitative treatment for ulcerogenic patients.

Accessory Digestive Tissues

HEPATIC CELL RESPONSE TO EXERCISE

In spite of the reduction of hepatic blood flow during exercise, the liver's function of providing glucose for skeletal muscle is not normally compromised. Glucose output from the liver can increase at least fivefold during exercise; when the exercise is severe in a hot environment, glucose release can increase up to fourteen times. The elevated oxygen consumption of the liver during exercise is in part a reflection of the metabolic changes essential for releasing glucose. Most of the glucose comes from *glycogenolysis* (the sequential cleavage of single glucose molecules from glycogen). As much as 300 mg/min of glucose can be released by the liver during exercise. During prolonged exercise, plasma glucose contributes about 10–15% of the energy expended by working muscle. Plasma-free fatty acids actually account for as much as 60–90% of the energy needs of skeletal muscles during prolonged work (longer than an hour). The plasma-free fatty acids are not contributed by the liver, which actually takes up FFA during exercise, but by the adipose tissue distributed throughout the body.

Hepatic glucose metabolism is neurally controlled to some extent. In the liver of the rabbit, glycogen deposition is greatly enhanced by vagal stimulation but reduced by sympathetic stimulation. During exercise, sympathetic activity predominates, promoting glycogenolysis and assuring an adequate amount of plasma glucose. On the other hand vagal activity

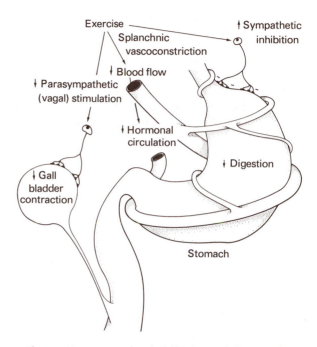

Figure 12-9. Exercise inhibition of digestion.

predominates during digestion when there is a need for glucose storage in the liver.

Physical training augments the liver mitochondrial and ribosomal content. A single bout of exercise apparently can result in a significant structural alteration of hepatic cells and their relationship to the circulatory network. Thus the liver can adapt metabolically and structurally, and these adaptations may be important factors in the general training syndrome related to skeletal and cardiac muscles.

GALL BLADDER

The *gall bladder* is subservient to the liver since it only stores and concentrates the bile salts produced by the liver. The gall bladder is not essential for life; it is frequently removed in humans. The release of bile from the gall bladder to the duodenum via the cystic and common bile duct is controlled neurally and hormonally. The vagus nerve increases the secretion of bile salts.

From the control mechanisms, we can hypothesize why vigorous exercise depresses digestion (Figure 12–9). Blood flow is reduced to the gut, thereby minimizing the hormonal effectiveness. In addition vagal activation is reduced, minimizing the function of the gall bladder during intense exercise. However, the duration and intensity of exercise required to induce this inhibitory effect are unknown.

PANCREAS

The *pancreas* synthesizes molecules that flow into a system of ducts (*exocrine gland*) leading to the duodenum; it also produces molecules that are released into the blood (*endocrine gland*) and distributed throughout the body.

The exocrine secretion of sodium bicarbonate assists digestion by regulating acidity of the gut. The pancreas also produces proteases, amylases, and lipases, which assist digestion.

The endocrine portion of the pancreas, consisting of the randomly distributed groups of cells called *islets of Langerhans*, produces the hormones *glucagon* and *insulin*. Glucagon is synthesized in alpha cells, histologically distinguishable from the insulin-producing beta cells. Since both of these hormones function as regulators of blood glucose, it might be expected that they would play a significant role, particularly during exercise; however, this is not necessarily the case.

Insulin. Insulin and muscular contractions facilitate glucose transport from blood to muscle. Since the two effects are additive, they apparently do not operate via the same mechanism. In fact it is because of this transport facilitation that a controlled exercise program for diabetics potentially can be beneficial.

Physical training increases the sensitivity of cellular membranes of muscle and of adipose tissue to glucose in obese subjects, who normally have low insulin sensitivity. A decreased sensitivity requires a greater insulin production; this could explain the tendency of obese people to develop diabetes. Consequently, exercise should be a very effective therapeutic tool in preventing diabetes in the obese. A single exercise bout of sufficient duration and intensity can induce *hypoglycemia*. These benefits are magnified in light of the increase in incidence of cardiovascular disease in diabetics and of the apparent protective effect of long-term exercise programs against cardiovascular diseases. The increased incidence of cardiovascular diseases is apparently related to an augmented dependence on fats and minimized dependence on glucose owing to its lack of availability in diabetics. This adaptation is demonstrated by the greater lipid mobilization in diabetics during exercise than in nondiabetics and by the tendency towards normalization in response to insulin therapy.

For the diabetic athlete, another factor should be recognized: muscle glycogen in most, if not all forms of diabetes, is low. This low muscle glycogen is critical in selecting the particular event for which a diabetic should train, because muscle glycogen level is an important factor in determining the time of exhaustion in events that require approximately 70% max $\dot{V}_{O_2}$. Insulin treatment causes a rapid increase in muscle glycogen in diabetics, particularly the first few days of therapy.

For the diabetic, all of these factors mean that: (1) exercise can be beneficial from several physiological viewpoints; (2) the degree of exercise and insulin dosages should be coordinated carefully during training; and (3) some special consideration is appropriate in the selection of an athletic event.

Glucagon. Glucagon's target organ is the liver where it has a glycogenolytic action (by activating the glycogenolytic enzyme, *phosphorylase*). Glucagon is released from the alpha cells of the pancreas when blood glucose levels are low. Glucagon also increases blood glucose levels by increasing gluconeogensis. Therefore glucagon acting on the liver tends to rectify hypoglycemia. Glucagon and insulin work antagonistically on glucose mobilization to maintain blood glucose homeostasis.

Summary

Since the kidney has a very effective autoregulatory system that assures normal renal blood flow and pressure, renal function is normally affected little by the cardiovascular adjustments to exercise. The kidney also assists in the homeostasis of blood pH. The kidney has an endocrine function as well, since it is the site of release of *renin*, a hormone that affects blood pressure, and *erythropoietin*, which stimulates RBC synthesis and maturation. The body fluids excreted by the kidney can be examined to test the adequate functioning of the kidney and its response to exercise and training (examining for proteinuria, hemoglobinuria, myoglobinuria and ionic balance).

Exercise can retard the digestion rate by causing a reduction in blood flow and by depressing gastrointestinal motility. There is evidence that exercise also modifies digestive secretions in such a manner that they provide an antiulcerogenic effect.

The digestive accessory organs, the liver and pancreas, are important supportive organs during exercise. The liver helps to maintain available substrate supplies (*glucose*) to working muscles. The pancreas is the source of *insulin* and *glucagon*, which help to maintain a constant blood glucose concentration.

Although a diabetic can participate in physical exercise like anyone else, special considerations should be taken into account in selecting his or her sport activities. Furthermore there are reasons to believe that a rational daily exercise program would be of special benefit to diabetics, and even more beneficial to diabetics with a tendency to obesity.

Study Questions

1. How does the kidney work?
2. Why is the kidney essential for life?
3. What ionic balances are shifted during exercise?
4. How does the renal system influence fluid volumes?
5. How are water loss and salt metabolism related?
6. What limitations should be placed upon fluid and glucose ingestion during exercise?
7. How can the kidney contribute to the metabolic demand during exercise?
8. What effects do proteinuria, hemoglobinuria and myoglobinuria have and how are these altered during training?
9. How does exercise alter the GI tract functions?

10. What are the response and the significance of intestinal and splanchnic blood flow during exercise?
11. What is the source of glucose secreted by the liver during exercise?
12. How can exercise aid the diabetic?

Review References

Bergstrom, J., et al. Diet, muscle glycogen and physical performance. *Acta Physiol. Scand.* 71:140–150, 1967.

Dugas, M.C.; R.L. Hazelwood; and A.L. Lawrence. Influence of pregnancy and/or exercise on intestinal transport of amino acids in rats. *Proc. Soc. Exp. Biol. and Med.* 135:127–131, 1970.

Ezer, E. and L. Szporny. Prevention of experimental gastric ulcer in rats by a substance which increases biosynthesis of acid mucopolysaccharides. *J. Pharm. Pharmacol.* 22:143–145, 1970.

Frenkl, R., et al. Antiulcerogenic effect of exercise in rats. *Acta Physiol. Acad. Sci. Hung.* 25:97–100, 1964.

Ludwig, W.M. and M. Lipkin. Biochemical and cytological alterations in gastric mucosa of guinea pigs under restraint stress. *Gastroenterology* 56:895–902, 1969.

Olsson, K.E. and B. Saltin. Diet and fluid in training and competition. *Scand. J. Rehab. Med.* 3:31–38, 1971.

Pollard, T.D. and I.W. Weiss. Acute tubular necrosis in a patient with march hemoglobinuria. *New Eng. J. Med.* 283:803–804, 1970.

Rowell, L.B. Regulation of splanchnic blood flow in man. *Physiologist* 16:127–142, 1973.

Shimazu, T. and T. Fujimoto. Regulation of glycogen metabolism in liver by the autonomic nervous system. IV. Neural Control of Glycogen Biosynthesis. *Biochim. Biophys. Acta* 252:18–27, 1971.

Applied biology
of physical activity

Key Concepts

• Body growth is the enlargement of existing cells (*hypertrophy*) and/or the formation of new cells (*hyperplasia*).
• Body growth rate is influenced by age, sex, and physical activity.
• The effects of excessive stress and/or lack of activity on growing bones are more likely to result in permanent alterations than the effect of these same stresses on mature bones.
• Muscle growth is regulated by hormones, muscle activity, muscle stretch, diet, and neurotrophic factors.
• Important modifications in tendons and ligaments suggest that connective tissue adapts to hyperactivity as does muscle tissue.
• Neuronal growth and development and motor development are intricately interrelated and can be modified by varying the environment particularly during the phase of rapid brain development.

Introduction

Human growth is the process of assimilating substances into existing cells, which results in an increase in the size of existing cells or in the formation of new cells. In essence we are talking about an increase in cell size (*hypertrophy*) and number (*hyperplasia*), both of which result in body growth. Within a given tissue, either one or both of these processes can be responsible for the enlargement or the expansion of tissues. The process that is dominating at any one time is not necessarily dominating at another.

The terms *hyperplasia* and *hypertrophy* must be examined in greater detail if we are to understand the influence of physical activity on growth. Hyperplasia can be defined as an increase in the number of elements. According to this definition, we must consider

CHAPTER THIRTEEN

Influence of activity on growth

not only an increase of total cell number, but also proliferation in number and size of organelles within a cell. As a further complication, either cellular or organelle hyperplasia can result in hypertrophy of a cell and therefore a tissue or organ. Cell proliferation is far more complex than cell hypertrophy since the former usually necessitates nuclear division (*mitosis*), including replication of DNA, whereas the latter involves greater use of existing DNA in cell growth.

Then what is the nature of exercise-induced muscle hypertrophy? Does it differ from normal growth? In this chapter we will consider exercise-induced hypertrophy of a cell as an enlargement beyond what would be expected during the normal course of growth.

Adaptation and Growth

Adaptation is the cell's response to some genetic or environmental factor that has altered the normal balance of supply and demand within that cell. Exercise is an environmental factor that directly or indirectly places greater demands on most tissues of the body. Due to the complexity of the response to this demand, few if any fundamental mechanisms have been identified that can account for the many known adaptations. The complex adaptations to exercise are such that the effects of a single *adaptagent* (agent inducing adaptation) cannot be determined without considering interrelationships to other simultaneously occurring adaptations.

The adaptations to exercise may be short or long term; that is, the nature of the adaptation is determined by the specificity of, and the elapsed time after, the adaptagent release. Examples of short-term, or acute responses, to an overload (such as exercise) are: activation of enzymes by structural modification, change in ADP/ATP ratio, greater stimulation of existing ribosomes rather than synthesis of new ones, and changes in membrane permeability. Examples of long-term, or chronic changes, are: increased amounts of RNA, ribosomes, enzymic and nonenzymic proteins, mitochondria—and perhaps more nuclei and DNA in cells. All these characteristics of chronic changes are traditionally associated with tissue growth.

Exercise induces a series of sequential tissue imbalances, which are followed by sequential adaptations, which in turn may act as adaptagents. For example ATP used in muscle contraction is converted to ADP and P_i, and the resulting ratio, ADP/ATP, is increased over that at rest—and thus becomes an adaptation. The increased ratio now acts as the adaptagent that turns on the mitochondrial oxidative phosphorylation, resulting in the formation of more ATP. If the exercise stress continues day after day (training), the high ADP/ATP ratio *may* indirectly act as an adaptagent for the production of more oxidative enzymes and more mitochondria.

The chronic response to stress, through the process of training, yields quantitative changes in the tissues responding to that stress. The stress and the response to the stress are generally quite specific as seen in endurance versus strength training. For example muscles respond to acute endurance-exercise stress by synthesizing and storing more glycogen after the exercise bout than was available before the exercise (*supercompensation*). The maximum amount of glycogen supercompensation occurs from one to five days after the end of exercise; then the glycogen levels gradually return to normal (Figure 13–1). In response to the daily exercise stress, the muscles begin to synthesize more glycogen-synthesizing enzyme and more mitochondria for greater energy reserves and for faster, more efficient utilization of each glucose molecule. Other systems begin to change also: the number of capillaries increase, more

oxygen is delivered, and more lipids are utilized as fuel.

Adaptation to training is an example of
growth modification made to meet the exercise
demands on the tissues, cells, organelles, and
other components that are overloaded. A
muscle adapts to weight-lifting training by
hypertrophying—or modifying the usual level
of *anabolism* (constructive phase of metabolism) and *catabolism* (destructive phase of metabolism) of contractile proteins. Connective
tissue hyperplasia in overloaded muscle also
occurs, which provides additional protection
against "damaging" physical stresses.

Another example of how the body adapts to
stress is seen in the formation of a callous.
The abrasive forces between the foot and sock
can induce the underlying germinal epithelial
tissue to proliferate, thereby eventually forming a thickened callous to protect the sensitive
blood-vessel-packed dermis.

General Body Growth

General body growth is affected by exercise in
ways other than through changes in body composition. The effect is particularly dependent
upon the age of the individual and the intensity of the exercise. For example, rats that are
allowed to exercise at will do grow more rapidly
than those rats whose activity is restricted or
those upon whom the demands of exercise, at
least at a prepuberty age, are too strenuous.
In one-month-old rats forced to swim for a 100-
day training period, reductions in weight of the
kidneys, liver, adrenals, and spleen are associated with reductions in cell numbers. The
reduced weights and cell number of these animals provide further evidence of a potential
catabolic effect of exercise on some organs. Although these findings cannot be directly related to humans, they do suggest that we should
be aware that the potential response to exercise

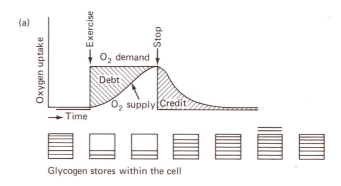

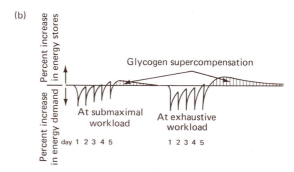

**Figure 13-1. At the onset of exercise, the
demand for energy is immediate, but since
the blood supply of oxygen increases slowly,
the cells must rely on nonoxidative
metabolism of glycogen to lactate (3 ATPs
formed per molecule of G-6-P). After
increased O_2 is supplied to the cell, it uses
glycogen more efficiently by converting it to
CO_2 (39 ATPs per molecule of G-6-P). (a)
After exercise, glycogen stores are repleted.
(b) With repeated bouts of daily exercise,
the repletion phase does not stop at pre-
exercise levels of glycogen, but production
of stored glycogen is increased in
anticipation of future exercise. This process
is called "glycogen supercompensation."**

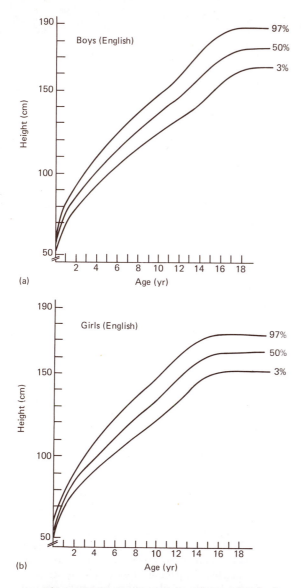

(a)

(b)

Figure 13-2. Cross sectional standards for height attained at each age. The central line represents the mean or 59th percentile. The other centile lines represent 3% of the population below 3% or above 97%.

depends upon the subject's age and the severity of the exercise (intensity and duration). While one tissue may be affected in an anabolic manner, another may be catabolically affected.

On the other hand potentially beneficial effects of exercise on general body growth have been noted in healthy young boys and girls (late teens) who had engaged in strenuous physical exercise during their childhood; they grow taller and heavier with larger chest girths and knee joint widths. There is some evidence that the onset of rapid pubertal growth in girls is accelerated if they have participated in competitive swimming for several years. However, the size at maturity did not differ from that expected from a nonparticipatory population. Other studies have compared the rate of growth of specific athletes with nonathletes and have studied relatively short-term training effects on general body growth. The results of these studies have shown, in general, only changes in body composition and muscular development.

Postnatal Growth in Height

At the age of two years, an individual has reached half of his or her expected adult height. Adult height will be reached at age eighteen years in males and sixteen and one-half in females (Figure 13-2). Further increases in height are only slight. Some of the later growth is due to an increase in the length of the vertebral column—until about age thirty.

A more striking characteristic of growth is the fluctuation in growth rate. The rate of growth in height during the first postnatal year is four times that of the rate at age ten, and double that of the peak rate during the adolescent growth "spurt" (Figure 13-3). Note also the relative age at which the peak growth spurt

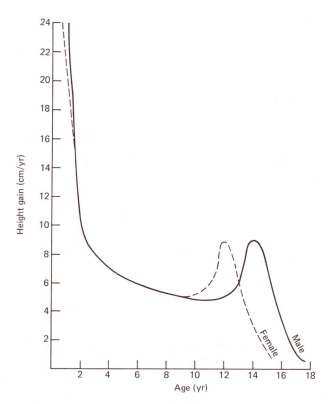

**Figure 13-3. The rate of height gain at each
age for male and female.**

appears for girls (at about twelve years of age)
and for boys (at fourteen years of age). It is
also apparent in Figure 13–3 that the difference
in adult height between the sexes is minimal
until the growth spurt.

The potential of predicting the adult
height using measurements as early as those
available at two years of age has not been fully
realized. For example a girl who at age two
is 87 cm (34.2 in.) tall will be close to 160 cm
(64 in.) by age fifteen. These predictions might
be particularly advantageous in making realis-
tic judgments on the chances of an individual
having a reasonable degree of success in a given
sport or in some other specialized profession.

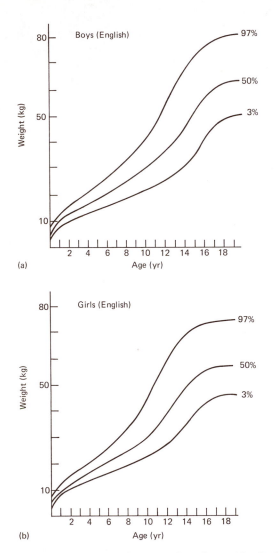

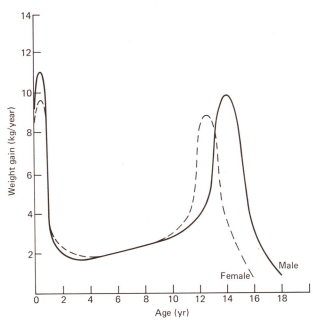

Figure 13-5. Relationship between age and the rate of weight gain for males and females.

Figure 13-4. Cross-sectional standards for weight attained at each age. The central line represents the mean or 50th percentile. The other centile lines represent 3% of the population below 3% or above 97%.

Postnatal Growth in Weight

Weight is much more variable than height because it is apparently influenced by more variables than is height. The variability is reflected in Figure 13–4 (compared to Figure 13–2) and is particularly obvious with increasing age.

Unlike growth in height, an individual can expect to attain half of his or her adult weight about the age of ten years. It is at this age for girls that an accelerated rate of growth begins and peaks at about 12 or 13 years. Boys are about two years behind in growth; however, the accelerated rate of weight gain in boys of this age lasts for slightly longer periods, which is primarily a reflection of height increases (Figure 13–5).

Growth of Bones

Growth of Short Bones

Growth of cartilage precedes formation of short bones (*ossification*), and this forms a model for the eventual shape of the final ossified structure. But cartilage growth (*chrondrogenesis*) and ossification later proceed simultaneously, with ossification eventually overtaking chrondrogenesis (Figure 13–6), at which time the bone size is determined. Consideration of the growth of short bones is particularly important because they make up the carpals and tarsals of the hands and feet. These bones are exposed to considerable stress, particularly during youth when bone growth rate is optimal, and are therefore the most susceptible to being adversely affected by injury.

Growth of Long Bones

Ossification of long bones begins at the periphery of the bone shaft. The bone enlarges and elongates by adding bone to the surface (this process is called *osteoblastic activity*) after removing it from the interior of the shaft (*osteoclastic* or *bone dissolution activity*); the result is a greater shaft diameter. As these events are taking place, *ossification centers* have begun at the ends of the shaft and at the ends of the bone (Figure 13–7). The ossification centers in the shaft and at the ends of the bone form a junction, which is cartilaginous and is called the *epiphyseal plate* (Figure 13–8). As long as this plate exists, bone length can continue to increase; but once this region is ossified, no further increase in bone length can be expected. Ossification of the plate may occur from sixteen to eighteen years of age in females and from eighteen to twenty-one years of age in males. The nature of bone elongation is shown in experiments in which bone markers in the shaft do not change their distance from one

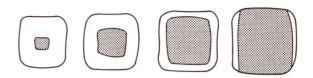

☐ Cartilage deposition

▦ Bone formation

Figure 13–6. Cross section through the center of a bone such as the carpals. The cartilage is replaced by the bone because its growth rate is slower than that of bone.

Adapted from David C. Sinclair. *Human Growth After Birth.* London: Oxford University Press, 1969. Reprinted by permission of the publisher.

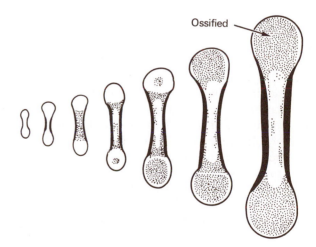

Ossified

Figure 13–7. Longitudinal sections of long bone growth.

Adapted from David C. Sinclair. *Human Growth After Birth.* London: Oxford University Press, 1969. Reprinted by permission of the publisher.

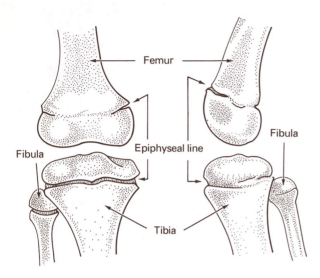

Figure 13-8. Epiphyseal plate.

Adapted from David C. Sinclair. *Human Growth After Birth.* London: Oxford University Press, 1969. Reprinted by permission of the publisher.

another—meaning that the existing bone is not expanding but that the new bone is being added at the ends of the shaft (epiphyseal plates).

One end of the bone grows more rapidly and for a longer period of time than the other. In the leg, the knee end of the tibia and the femur grows faster than the opposite end (hip or ankle). But in the arm, the ends of the long bones that form the elbow grow more slowly than their opposite ends (shoulder and wrist). This is pointed out because the more rapidly growing ends of long bones are more susceptible to serious injury than the less actively growing ends. Injury of the cartilage segments of young bones can easily result in permanent bone malformation.

CONTROL OF BONE GROWTH

The concept of bone being a metabolically static tissue is true to some degree, but actually it is in a constant process of dissolution and deposition of calcium and phosphate. A proper coordination of these two processes is essential during bone growth and for maintenance of mature bones.

HORMONAL CONTROL

Growth hormone (GH) from the anterior pituitary, calcitonin from the thyroid gland, and parathyroid hormone from the parathyroid gland make up the major hormonal sources of control of bone growth. These hormones are influenced by thyroxine, androgens, estrogens, and insulin levels—all of which promote protein synthesis. Growth hormone stimulates proliferation of cartilage, formation of bone, and protein synthesis in the area of the epiphyseal plate. Each of these effects promotes a lengthening of "immature" long bones. Unlike what might be expected, the adolescent growth spurt is not caused by a sudden increase in GH, since GH blood levels seem to be about the same throughout a person's lifetime. Daily fluctuations of GH secretions can be caused by diet adjustments, insulin, and exercise; however, exercise during the day does not affect the high rate of GH release that is characteristic of sleep.

Parathyroid hormone regulates, and is sensitive to, calcium concentration in extracellular fluid. If calcium concentration is low, the parathyroid hormone release is stimulated; it induces higher calcium release from bones, greater calcium absorption from food in the gut, and less calcium loss in the urine (Figure 13-9). Calcitonin has an opposite effect on fluid calcium concentrations. A high calcium concentration in extracellular fluids stimulates release of calcitonin, which inhibits release of calcium from the bone; the effect is that calcitonin lowers extracellular fluid calcium (Figure 13-9).

ENVIRONMENTAL CONTROL

Diet. Improved dietary factors are probably among the major reasons for the greater height of recent generations. However, this increase in height cannot be contributed solely to dietary factors. Also the maximal height is being reached at an earlier age now than in former generations. This earlier growth has been demonstrated in several countries: for example American and West European children between five and seven years old were ten cm (4 in.) taller in 1950 than in 1880. However, it appears that the rate of increase in size of Americans is beginning to level off.

Use and Disuse. The effects of inactivity on bone are better understood than the effects of activity. Limbs inactivated by immobilization or muscular denervation develop bones that are lighter in weight because of mineral loss (*decalcification*). Like muscle, bone "atrophies" internally with disuse. This condition is like *osteoporosis* (bone condition in which there is a loss of Ca^{2+} and a weakening of the bone) since the bone becomes more porous and less dense. In spite of the loss of calcium and bone density, vertebrae may actually grow in the absence of normal stresses of weight bearing.

The *osteoclastic* (demineralization) activity is manifested in elevated rates of calcium excretion. Long-term bed rest has similar effects. Astronauts have experienced similar alterations in bone metabolism: their skeletal systems are confronted with the problems of weightlessness and physical confinement. Since inactivity or the absence of the normal stresses causes bones to become more porous and spongy, they become more susceptible to pathological fractures (Figure 13–10). This deterioration is shown very clearly in Figure 13–11 where bone deformation following immobilization in a cast for sixty days is consider-

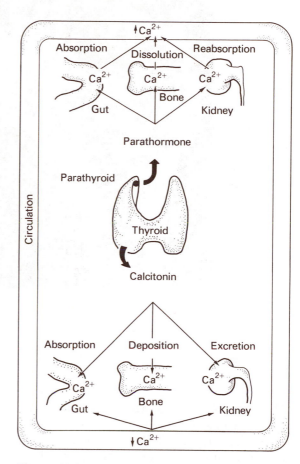

Figure 13-9. Low serum Ca^{2+} yields increased parathormone which increase serum Ca^{2+}. High serum Ca^{2+} yields increased calcitonin which acts to decrease serum Ca^{2+}.

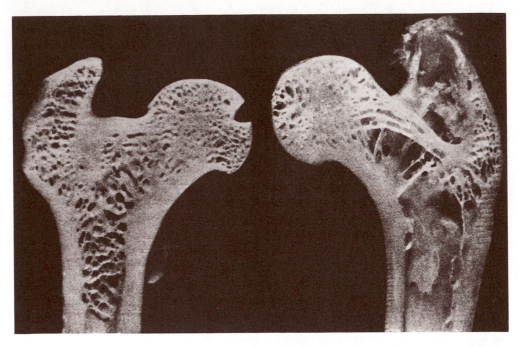

Figure 13-10. Normal and osteoporosed bone.

From L.E. Kazarian and H.E. VonGierke. Bone loss as a result of immobilization and chelation. *Clin. Orthoped. Rel. Res.* 65:67–75, 1969.

ably greater than in normal bone. Reduced activity, due to arthritis, also causes osteoporosis.

Reduced *osteoblastic* (bone-synthesizing) activity may also contribute to bone "atrophy." These effects are localized to the inactive bones, suggesting that they are not precipitated by a hormonal factor such as parathyroid hormone, which would effect all bones.

During immobilization there is a lack of the normal stimuli to maintain bone: both normal tensile and compressive forces seem to be essential for normal bone growth and maintenance. For example, weight of the *psoas muscle* is highly correlated with vertebrae mineral weight. It seems that muscle weight reflects the magnitude of the forces that it exerts on bones to which it is attached; in this way mus-

cle weight is a very important determinant of bone mass.

It appears that bone growth is stimulated by electrical potentials, which are induced within bones when they are exposed to muscular forces. This bone strengthening in response to the stress of muscle force may explain why women's bones are less dense than men's; why the skeletal density of Caucasians is less than that of Negroes; why bone mass is reduced with age (in proportion to loss of muscle mass); why osteoporosis is more common in women than men; and why anabolic steroids prevent or limit deterioration in some osteoporotic patients (causes muscle hypertrophy).

On the other hand excessive muscular forces can lead to resorption of bone, perhaps because of interrupted blood flow whenever the tissue is compressed. Excessive compression and general bone and joint stress are particularly important considerations for normal growth and development until the age of puberty. Physical stress can modify the epiphyseal growth zone so that it seems to adapt in order to protect itself against the stress rather than to function solely as a center for bone elongation. The responsiveness of bone to stresses also has been shown in an experiment in which tooth-to-bone compression stimulated osteoclastic (dissolution) activity, whereas pulling forces activated osteoblastic (deposition) activity.

The avoidance of excessive physical stresses to growing bone is of great importance to those involved in the supervision of highly competitive youthful, community baseball and football teams. A high incidence of epiphyseal injuries in participants of Little League and Pony League Baseball seems to occur—although some evidence to the contrary has been reported. Enlarged epiphysis of the elbow end of the *humerus* and bone demineralization and fragmentation have been found in baseball pitchers in their early teens. Another study of boys, nine to fourteen years old, revealed that 80 out of 160 pitchers of Little League Baseball had discernible degrees of epiphysitis (inflammation) of the medial epicondyl epiphysis (humerus at elbow). This condition in nonpitching players and nonplaying controls was rare. The knee and shoulder are other joints prone to developing epiphysitis. Relief from the trauma-inducing activity is essential for recovery. But some damage may be permanent—thereby restricting joint mobility in later life.

During the adolescent growth spurt (nine to fourteen years of age), the most common osteochondritic condition is *Osgood-Schlatter disease*. High forces exerted by the quadricep musculature through the patellar tendon on the tibial tuberosity can cause bone damage, with consequent discomfort and tenderness. It is usu-

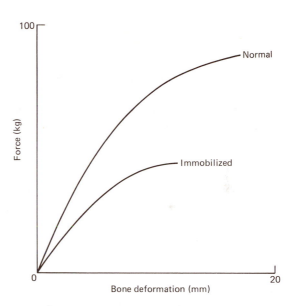

Figure 13-11. Effect of immobilization on force necessary to produce deformation of bone.

Adapted from L.E. Kazarian and H.E. VonGierke. Bone loss as a result of immobilization and chelation. *Clin. Orthoped. Rel. Res.* 65:67–75, 1969.

ally necessary to restrict the individual's athletic involvement and any direct strain on the knee until the active phase of this condition subsides—which may take months. Actually any epiphyseal area may become osteochondritic as described above; physical activity is not encouraged during any manifestation of this condition.

The repeated observation that the bones of the dominant limb of an individual are generally greater in length and width suggests that its greater use is one reason for the difference in size. The relationship may not be one of cause and effect because the larger arm may be a factor in determining the preference for that particular limb. In contrast to an augmentation of bone growth, the bones of chronically trained (running) rats are shorter and thinner than their controls. These data suggest that the effect of activity must be considered with respect to age and severity of exercise. For example, about a 7% drop in total ash (minerals remaining after combustion of the bone) is observed in men every decade after the age of thirty. In women about a 9% drop can be expected. Obviously more research is needed to determine the effects of various kinds and intensities of activity on bones at different ages.

Growth of Skeletal Muscles

The relative rates of *protein synthesis* and *degradation* determine whether there is an increase or decrease in muscle mass. Theoretically an increase in muscle mass should be accounted for by more new cells (hyperplasia), or additional cytoplasmic components for existing cells (hypertrophy).

How do different types of exercise influence these factors? Endurance-type training stimulates a net synthesis of proteins that are important to the muscle fiber in resisting fatigue. Mitochondria increase in number and, to some degree, in size; they also seem to be altered qualitatively in that the relative content of the various mitochondrial proteins (enzymes) that facilitate ATP production is altered with endurance training. The sarcoplasmic reticulum has not been shown to change with endurance training. Similarly the number of true muscle nuclei within muscle fibers is not affected by training. Within the whole muscle, new nuclei are formed, but they are of extrafiber origin associated either with connective tissue or with the satellite cells of muscle fibers.

Between birth and adulthood, muscle-fiber diameter increases at least two to three times. Total muscle mass increases from 25% of the body mass at birth to 45% in adulthood.

Satellite cells produce new DNA and divide by mitosis after which one of the daughter cells is incorporated into a muscle fiber. Satellite cells make up to 10–15% of the apparent adult muscle nuclei as seen with light microscopy. It has been estimated that the population of nuclei within muscle, but outside of the muscle fibers, is elevated by 2–3% per day in rats when the muscles are still maturing and differentiating into fiber types. Satellite cells are the sources of new nuclei in the latter stages of maturation. The verification of migration of satellite nuclei, as a result of training, might prove to be a significant factor in explaining the cell's adaptation to various forms of chronic overload (Figure 13–12).

Growth of Muscle Fiber Cross-Sectional Area

Muscle fiber cross-sectional area increases steadily in most muscle up to the onset of puberty, at which time a sharp increase in fiber size is evident. Muscle fibers continue to enlarge up to the fourth decade and thereafter begin to decrease in size. Factors other than genetic ones can influence muscle growth: re-

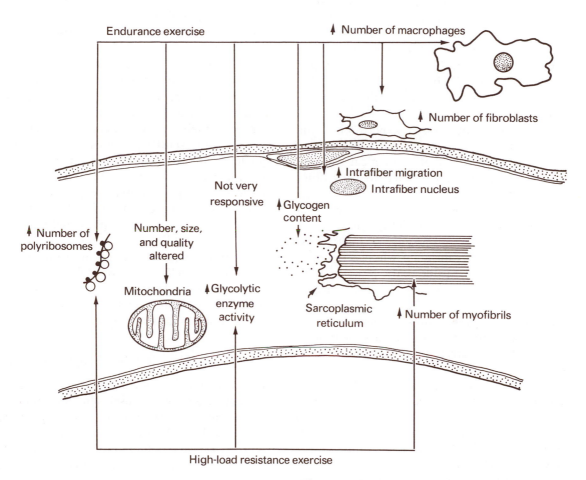

**Figure 13–12. Response of cellular organelles
to two types of exercise.**

petitive increases in muscle-tension overload
(exercise), age, hormonal condition, and muscle stretch are examples of such factors. The
mechanisms involved in inducing accelerated
muscle growth caused by each of these variables would seem to differ, since the nature
of the apparent stimuli are so varied.

In animal experiments in which the tendons
of synergistic muscles are cut (*tenotomized*)

so as to overload the remaining intact muscles, muscle hypertrophy occurs rapidly, reaching a maximum weight within a week after surgery. It is very important to realize, however, that this specific type of muscle hypertrophy is a very poor model of the actual events that may occur in hypertrophy induced by weight lifting or similar exercises. Our skepticism of the model is warranted because: (1) we cannot induce the degree of "muscle hypertrophy" due to synergist removal by normal exercise, and (2) it has been shown that practically all the new RNA synthesis is related to cells other than the muscle fiber itself.

As we have noted, the responsiveness of a muscle to overload (*trainability*) depends on age. At prepuberty and puberty rat muscles hypertrophy at the greatest rate, whereas in young adult rats, the magnitude of the response is reduced 10–15%. In much older rats the magnitude and rate of muscle response to overload are considerably less. Similar effects have been observed in humans at various ages who have been on a strength-training program (Figure 13–13).

The type of overload is a critical factor in determining the nature of the adaptive response of muscle. High-resistant, low-repetitive exercises tend to stimulate the synthesis of the contractile elements of the muscle fiber as opposed to the increase in the energy-related proteins of fibers following endurance training (Figure 13–12). The muscle enlargement following high-resistant exercise is due to the increased size of the myofibrils and to an increase in the number of myofibrils per muscle fiber (Figure 13–14). However, there is evidence that more muscular individuals have a greater number of muscle fibers than less muscular ones. This difference may be *genetically* determined since evidence inferring that muscles hypertrophy by an increase in the number of muscle fibers (hyperplasia) is extremely weak.

Some "hyperplasia" occurs by fiber splitting in muscles of normal adult laboratory animals and in diseased human muscles. The frequency of occurrence of fiber splitting is increased by chronic training in rats. But the effect due to a moderate training intensity seems to be localized to the slow-twitch soleus muscle of the leg, a muscle that makes up only a small percentage of the muscle mass of the leg. Extremely severe exercise in animals can induce some splitting in muscles that consist predominantly of fast-twitch fibers. These findings suggest that the number of fibers may increase, or appear to increase, under very specific and unusual conditions, but the general growth of adult muscles (hypertrophy) is due primarily to additional myofibrils in existing cells—and not to the formation of new myofibrils for new cells. A study of the satellite cell may provide answers to many of our current uncertainties about skeletal muscle maintenance and adaptation to exercise.

Growth of Muscle Fiber Length

It is apparent that as bone length increases, so does muscle length. Muscle length is increased by the addition of new sarcomeres, predominantly at the ends of fibers (at the *musculotendinous junctions*). Immobilization reduces the number of sarcomeres developed, but subsequent mobilization is followed by a rapid rate of synthesis of new sarcomeres. Incorporation of satellite cell nuclei into an elongated muscle would seem to assure adequate nuclear control of the new cytoplasmic components.

Sex Differences in Growth of Muscles

There are some fundamental differences in male and female muscle tissue. For example a seventeen-year-old female has about two-thirds of the muscle nuclei of a male. The fewer

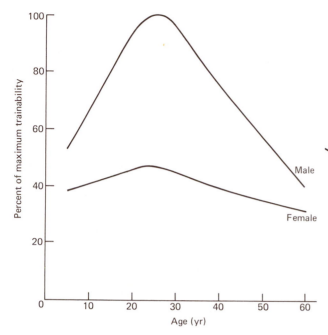

Figure 13-13. Variation of trainability with age; that is, the rate of muscle strength increases during training.

Adapted from Theodor Hettinger. *Isometrisches Muskeltraining.* Stuttgart: Georg Thieme, 1964.

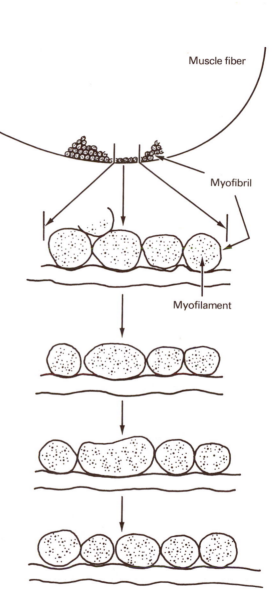

Figure 13-14. A myofibril enlarges, and eventually divides, in response to high-resistant overload by increases in the number of myofilaments. The overall result is hypertrophy of the muscle fiber.

nuclei mean that a nucleus, in female muscle, has a more demanding function since it must maintain more muscle mass and protein.

The size of the actual regulating unit of muscle (the amount of cytoplasm per nucleus) is different in males and females. Although strength per muscle mass is the same in males and females, females have less total muscle mass and a lower nucleus-to-cytoplasm ratio. This low nucleus-to-cytoplasm ratio may be a factor in explaining the apparent difference in trainability of female and male muscle—because training necessitates greater synthesis of those proteins related to energy production and/or contractile components. Increased trainability capacity would seem to require more nuclear DNA involvement. This difference in trainability was shown in Figure 13–13. DNA may not be a factor, however, if muscle nuclear substance is not functioning near its capacity or if sufficient satellite cells are available for integration into the muscle. These extra nuclei would then be able to accept the additional protein-synthesizing demands.

Control of Muscle Growth

ANDROGENIC AND GLUCOGENIC HORMONES

Growth Hormone. Growth hormone (GH) facilitates muscular growth by stimulating protein synthesis in muscles. The elevated GH secretions that are induced by exercise, however, appear to be primarily a mechanism to maintain vascular fatty acids as substrates for working muscles rather than a mechanism to facilitate muscular growth.

Sex Hormones. Testosterone is the most potent of the natural androgenic sex hormones (Table 13–1). It stimulates protein synthesis by increasing amino acid incorporation into muscle and by increasing the amount of polyribosomes. Since males have greater testosterone secretions, they would seem to have a greater protein-synthesizing capacity. Castration of male rats before puberty completely arrests myonuclear multiplication and retards growth of contractile materials. Testosterone injections correct this depression. Apparently testosterone induces hyperplasia of nuclei and hypertrophy of muscle fibers while castration causes aplasia and atrophy.

The question immediately arises as to which hormones in females control these muscle growth parameters. Unfortunately little is known about this point. We can speculate that the testosterone output from the female adrenal and the estrogen hormones, which are slightly androgenic, may be sufficient to support the normal growth process. The female hormone, *progesterone*, facilitates protein catalysis.

Testosterone is also involved in the phenomenon of glycogen supercompensation after an exhaustive exercise. Without testosterone, there is a slower rate of repletion of muscle glycogen following an exhaustive exercise and a final lower resting level of glycogen in muscles. Much less is known about which hormone would be most important in females; *glucocorticoids* may be sufficient to promote post-exercise glycogen supercompensation.

Ingestion of *synthetic androgens* among male and female athletes is not uncommon today. Performance by weight lifters is improved by androgen supplements; performance by swimmers apparently is not. Anabolic steroids do stimulate RBC synthesis. We might expect some beneficial effects of androgen supplements because of the enhancing effect of testosterone on muscle glycogen; glycogen levels in muscles are known to be related to resistance to fatigue. Due to the androgenic effect of an increase in the amount of contractile proteins, we can expect a greater gain in strength training if an anabolic steroid is taken concomitantly with the training.

Possible side effects of anabolic steroids are increased appetite, alteration of secondary male characteristics, and depression of the interstitial cells of the testis (testicular atrophy), which normally produce testosterone. The adrenal cortex atrophies and loses lipids; the total adrenal weight decreases, and the overall adrenal secretory capacity is impaired (in male rats after testosterone treatment). Liver disease, including cancer, has also been implicated.

Androgens can induce sexual sterility in laboratory animals; however, the dosage level and duration at which this occurs in humans is unknown. This uncertainty, in itself, should be sufficient warning to those who take anabolic steroids for the sole purpose of enhancing muscle mass and performance. In one study, 25% of androgen-treated female mice were sterile while others had difficulty in conception after the first insemination. Total reproductive capacity was reduced by half, the fertility span was shortened—and there was evidence of early aging of the uterus.

Growth retardation and delayed puberty have been reported as effects of androgen treatments. Supraphysiologic dosages of testosterone can cause premature calcification in bones. However, used in a proper therapeutic manner, anabolic steroids have great value. One anabolic steroid, *oxandrolone*, a growth-promoting agent, has been administered to previously growth-stunted children, doubling their growth rate without compromising their ultimate adult height.

Insulin. Compensatory hypertrophy of muscle induced by removal of synergists, as described earlier, is not dependent on insulin levels, even though insulin facilitates protein synthesis in muscle. Because insulin can increase the permeability of the muscle membrane to amino acids and glucose, it is possible that insulin is a factor in the normal exercise-induced hypertrophy.

Other Hormones. The hormones, epinephrine, prostaglandins, thyroxine, and others, may play some role in inducing muscle hypertrophy; if they do, their effects probably are induced secondarily by altering the fiber's energy metabolism so that protein metabolism is enhanced.

MUSCLE STRETCH

Muscle stretch is an effective stimulus to induce muscular hypertrophy. In fact even a denervated muscle can hypertrophy if it is sufficiently stretched. This stretch phenomenon

Table 13-1. Effects of Supplemental Testosterone Administration

Performance Enhancement	Other Effects
Increased protein synthesis	Increased appetite
Increased muscle glycogen	Gonadal atrophy
Increased muscular strength	Development of sterility or shortened fertility period
Increased maximum rate of O_2 uptake	Increased early aging of uterus
Increased growth of connective tissue	Premature ossification of epiphysis and cessation of growth
	Development of male secondary sexual characteristics in females
	Adrenal cortex atrophy
	Decreased adrenal secretory capacity
	Decreased lipids in adrenal cortex

also applies to immobilized muscles; muscles that are immobilized in a stretched position atrophy less than those fixed at shorter lengths. This stretching effect has considerable utility in prevention of atrophy and in general muscular rehabilitative procedures.

EXERCISE-INDUCED HYPERTROPHY

The specific stimulus needed to activate the protein-synthesizing machinery of a muscle fiber is unknown. Only in a general way do we know the nature of the stimuli that induces muscle hypertrophy. For example, high-resistant, low-repetitive exercises (heavy weights usually lifted between two and ten times) performed a few times a week can induce hypertrophy in a relatively short period of time. Isometric exercise (in which there is a minimal change in muscle length) is known for its strength-yielding effects. The increased strength has been claimed to occur without muscle hypertrophy. A long-distance runner's muscles, which are not loaded with high resistances, do not enlarge, although the muscles are stretched during running. In formulating the best exercise program for the normal, rehabilitating, or training individual, it will be extremely useful to discover the specific stimulus that induces the appropriate effects to modify performance in the desired way.

Neural Control of Muscle Growth

Growth of skeletal muscle is modulated to a great degree by its own motoneurons. The motoneuron determines directly or indirectly not only the type of proteins synthesized but also the rate of synthesis—and probably of degradation. These effects are mediated by the specific frequency or pattern of impulses transmitted to the muscle or by some influences presumably induced by secretion of some neuronal

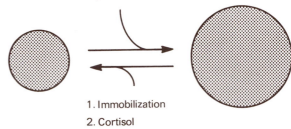

1. Weight lifting
2. Hormones (testosterone)
3. Muscle stretch
4. Neural trophism

1. Immobilization
2. Cortisol
3. Progesterone

Figure 13–15. Some of the factors which cause hypertrophy or atrophy of muscle fibers.

substance (neurotrophic substance) to the muscle. The neurotrophic substance does not seem to be the neuromuscular transmitter, *acetylcholine.* Many muscular diseases are known to be of neural origin; in fact it has been suggested that there are no muscular diseases that originate in the muscle. Since the type and amount of protein produced are under neural regulation and since exercise-induced growth and adaptation of skeletal muscle are dependent upon protein concentrations, *neurotrophism* may be an important channel through which chronic exercise mediates its anabolic effects on skeletal muscle. Some of the variables known to affect muscle growth are shown in Figure 13–15.

A COMMON EFFECTOR SYSTEM FOR REGULATING MUSCLE GROWTH

All of the various stimuli that can induce muscle growth or hypertrophy must eventually act through the same cellular mechanisms. The rate of muscle-protein synthesis depends, in part, on the amount and activity of RNA, RNA polymerase, and/or DNA. In other words, regardless of the stimuli (whether exercise, testosterone, or muscle stretch), the ultimate mechanism is induced through DNA stimulation. The total protein that a fiber can maintain is DNA dependent. This means that fiber hypertrophy may be determined by the number of nuclei produced in the early phases of muscle growth or by the capacity of satellite cells to provide additional nuclear material for the muscle fiber to maintain additional protein synthesis.

Growth of Cardiac Muscle

Endurance exercise induces cardiac hypertrophy. The cardiac muscle fibers (predominantly those fibers in the left side of the heart) enlarge in diameter while changes in muscle-fiber length are minimal. As is true of skeletal muscle, cardiac muscle nuclei do not divide and form daughter nuclei. However, connective tissue nuclei do have this capacity; as in the case of skeletal muscle, the greater number of nuclei in the overloaded heart is due to cells other than cardiac fibers. The possibility that the heart has a cell analogous to the satellite cells of skeletal muscle fibers may also be important in the maintenance of cardiac fibers, particularly if chronically overloaded (endurance trained).

Increases in cardiac-fiber diameter with chronic endurance training have been discussed in Chapter 9. To review briefly, a higher workload on this muscular pump causes fiber enlargement due to the synthesis and maintenance of more myofibrils per fiber. The net increase of myosin protein following overload is due to enhanced synthesis rather than reduced degradation.

The same compensatory cellular events involved in skeletal muscle hypertrophy are also involved in cardiac hypertrophy. There is greater transcription activity of the DNA genes, resulting in an increase in RNA synthesis. At first there seems to be a greater synthesis of metabolically related proteins; subsequently the contractile proteins are increased.

Cardiac compensation to overload, as with skeletal muscle, is, in part, hormonally dependent; we know that testosterone enhances myosin synthesis. Diet is another important factor: protein anabolism of the heart can be stimulated if more amino acids are made available to the heart.

Growth of Tendons and Ligaments

General Properties of Tendons

In general the response of tendons to exercise and training has received little attention from research biologists—probably because we ascribe a passive role to tendons. Their metabolism is relatively inert, and they are usually free of disease. That they are free of disease, however, does not mean that they are free of trauma, particularly in athletics. The tendon serves to transmit tension from muscle to the bones. Teleologically, we can see other useful functions of tendons. For example they serve as a means of optimizing muscle-bone connections in relatively small attachment surfaces to provide the best mechanical advantage.

The tendon consists largely of parallel protein collagen fibers, which are flexible, strong, and relatively inextensible. Collagen bundles are about 1–2 microns in diameter. We know that from the fifth intrauterine month to 92 years of age there is only about a 7–10% rise of tendon collagen concentration. Complex sugars form an amorphous intracellular substance, which is thought to play an important role in the tendon's tensile strength—although they make up only about 1% of the tendon.

Effect of Muscular Force on Tendons

Although a tendon's maximum tensile strength is probably four times greater than the natural tension to which it is likely to be subjected, there is good evidence that endurance training can cause tendon cross-sectional area to increase significantly. The number of nuclei also seems to increase, as does the complex sugar content. It has been claimed that tendons of young animals hypertrophy in response to training whereas in the adult, internal qualitative modifications are induced that strengthen the tendons.

The specific factor that induces tendon adaptation to training is unclear. There appears to be no close relationship between muscle and tendon cross-sectional area. It has been suggested, however, that slow-twitch muscles have thicker tendons than any other muscles, which implies a direct relationship exists between tendon size and duration of work.

Ligaments and Joints

Trained rats have stronger ligament-to-bone attachments than control rats. Testosterone and estrogen supplementation also seem to strengthen the ligament-bone attachments. Testosterone has an anabolic effect on collagen metabolism which is consistent with its apparent effect on the strength of ligament-to-bone attachments.

On the other hand, only a few weeks of joint immobilization weakens ligamentous strength. These data clearly suggest that following joint immobilization and during rehabilitative procedure, the limb should be mobilized as soon as possible in order to maximize the rate and completeness of repair of the ligament.

It is a common experience that long-term immobilization of a joint leads to a loss in range of motion and a multitude of morphological alterations. Limited range of motion following immobilization is probably due to joint infiltration with fatty and fibrous tissues. Cartilage within the joint may be replaced by fibrous connective tissue. The bones lose their osteoblastic (bone-synthesizing) activity. In prolonged knee joint immobilization, the patella may actually adhere to the underlying bone. Immobilization lasting for a year almost completely destroys a joint. Manipulation of the immobilized joint reduces the magnitude of the stiffness that remains.

Neuronal Growth

At birth, most neurons are already formed although not fully developed. Axons and dendrites proliferate rapidly in the first months of life. These changes coincide with apparent improvements in muscular coordination. In cats and probably in humans, the dendrites of motoneurons controlling movements of the upper extremities are more developed at a given age than are those related to the lower extremities. It can be seen in Figure 13–16 that the growth curve of motor and sensory axons in cats is similar to the shape of the growth curves for whole-body weight and height curves, discussed earlier in this chapter. Note also that the motor nerves begin a rapid growth period earlier than the sensory nerves. How these neural maturation patterns relate to coordinated movements in young boys and girls is yet to be determined.

Although the neurons do not enlarge in the adult, they remain in a very dynamic state; new axoplasm is being formed continuously, and protein is being synthesized and transported down the axon from the cell body. Neurotubules and neurofilaments, which contain actomyosin-like chemicals (*neurostenin*) assist in the axonal transport of particles. Some of the components of the axonal transport are enhanced in growing neurons.

Axonal growth is subject to modification within and outside of the CNS; however, it is more limited within the CNS. The stimulus for the sprouting of axons seems to be some kind of lipid, perhaps similar to that found in *myelin*. The functional utility of such a substance is not presently known.

A functionally overloaded axon can hypertrophy if the treatment is sufficiently severe, but evidence that a chronic physical exercise program will induce such an effect is not convincing. In response to endurance training, axonal sprouting, particularly of the terminal branches that form the synaptic contact with a muscle fiber, does not seem to occur even though lengthening of the terminal axon does.

Myelin thickness and internodal length are directly proportional to axonal diameter. Both nerve diameter and internodal distance are proportional to conductive velocity. Internodal length of the myelin is increased with increasing stretch on the nerve. Although the stretch experienced during exercise could theoretically lengthen internodal distances, this parameter has not been studied sufficiently. Myelin of some axons is responsive to use: for example mice reared in the dark have less myelin in the optic nerve than those reared in the light. The importance of adequate myelinization of axons is also suggested by the observation that some mental retardation and some muscular maladies are myelin-defective diseases. The effects of postnatal physical activity on myelination of motor axons are unknown.

In addition to myelin changes induced by the environment, neuronal changes occur also. Neurons of the visual cortex of mice reared in darkness have fewer synaptic spines on their dendrites; this observation shows the effect of low afferent input to a neuron. The effect is particularly noticeable within the first ten postnatal days.

In the formative process of axonal connection to skeletal muscles, more neurons are present before than after maturation; in fact, there is an oversupply of axons. Since each adult muscle fiber can only maintain an axonal branch from one motoneuron, the neurons that do not make a permanent synaptic contact degenerate. Again, it is during this stage of development that normal motor activity seems to be the most critical.

The human infant's nervous and muscular systems are immature. The cerebellum of a monkey at birth is only 13% of its adult weight. At least as much postnatal synthesis of DNA occurs in the human cerebellum and areas of the forebrain as occurs in Rhesus monkeys. Although more mature at birth than humans, Rhesus monkeys and rodents have been used as models in neuronal developmental studies. The lack of neuronal maturity at birth makes the human nervous system highly susceptible to postnatal environmental factors that could affect the normal increase in cell number and connectivity. This CNS susceptibility to change can be used to advantage if we discover the optimal postnatal environmental conditions in terms of nutrition and activity to stimulate adaptation in the CNS.

Hormonal Control of Neuronal Growth

Testosterone is as effective as an anabolic hormone for neural tissue as it is for skeletal muscle. It stimulates neuronal and glial cell growth by stimulating RNA polymerase. Brain weight and cell enlargement during development are greater in males than in females; this relationship has been shown to be dependent on testosterone. The anabolic effect of testosterone on skeletal muscle is dependent on the presence of an intact neural supply to the muscle; this observation further demonstrates the neural control of cells in general.

Other hormones can also affect neuronal growth. *Thyroxine* enhances the rate of neuronal differentiation and can cause neurons

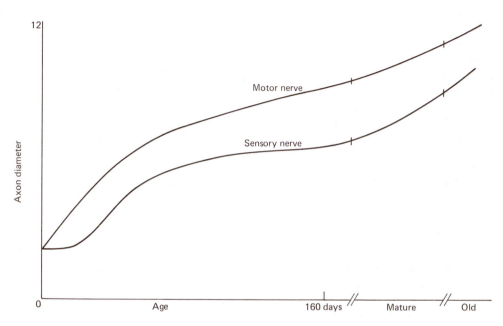

Figure 13-16. Relationship of axon diameter with age in motor and sensory nerves.

From S. Skoglund and C. Romero. Postnatal growth of spinal nerves and rods: A morphological study in the cat with physiological correlations. *Acta Physiol. Scand.* 66 (Suppl. 260):1–50, 1965.

to enlarge. An absence of the hormone in neonatal rodents causes decreased brain growth, less dendritic and synaptic development, defects in learning and behavior, and electroencephalogram changes. *Cortisone* is catabolic to neurons as it is to skeletal muscle: large dosages can decrease brain weight and DNA in rodents. *Growth hormone* is neurally anabolic, like thyroxine but its influence is more long-lasting. It causes an increase in brain weight, DNA, and number of dendrites in postpuberty rats. Rats given growth hormone are heavier; motor activity patterns occur earlier and more frequently than in nontreated rats; and dendrites longer than normal develop in the cerebral cortex.

Nerve Growth Factor

A neuronal growth-stimulating factor (NGF) has been found in mouse tumors (*sarcoma*) and salivary glands. Its effect is specific, limited to spinal, sensory, and sympathetic ganglionic neurons. Although it is extremely potent in some animals and in tissue cultures, its importance in normal development is unclear. It may be important in the regenerative function of a neuron. The effect of NGF is anabolic, enhancing protein synthesis and glucose oxidation. Exercise effects on NGF and on other neurally related growth mechanisms have been largely ignored by scientists.

Exercise and Tumor Growth

At least three published studies have reported that transplanted tumors were smaller or had completely regressed in exercised compared with control rats. An extract of electrically stimulated muscle, when injected into rats with transplanted tumors, inhibited tumor growth; while extracts of nonfatigued muscles failed to have this effect. On the other hand a postmortem examination of two hundred and twenty former athletes showed that the incidence of malignant tumors was significantly higher in this group than in a control population. Obviously no conclusions can be drawn with such limited data, and the topic needs to be studied more thoroughly.

Summary

Body growth is the result of proliferation of cellular components (*cellular hypertrophy*), which is sometimes followed by proliferation of cell number (*cellular hyperplasia*). Both of these effects are realized during growth—although hyperplasia of cellular organelles becomes a relatively more important mechanism of adaptation in the adult.

Exercise, or lack of it, can alter these normal growth processes in practically all body tissues. These multifaceted alterations result in an al-

tered tissue state that enables the tissue to tolerate more effectively subsequent stresses of a similar nature.

Well-defined height and weight growth-curves have been developed so that the adult stature can be predicted with reasonable accuracy within the first few years of life. But it is also apparent that these growth parameters are extremely susceptible to environmental influences such as hormones, diet, and physical activity—or lack of them.

Inactivation of the body or a portion of it causes rapid demineralization of the inactive bones, which makes them more susceptible to fractures and deformations. On the other hand excessive stress to bones will damage them, particularly during the prepuberty years when the bones are growing rapidly and consist of many injury-prone cartilagenous cells. There is evidence that there is an optimal amount of activity for proper bone growth and maintenance. Some evidence exists that children exposed to strenuous but not excessive physical labor grow taller and heavier than more sedentary children.

Skeletal muscle adaptation to overload depends on the nature of the stimulus. High-resistant low-repetitive exercise stimulates muscle hypertrophy. Mitosis of adult muscle nuclei does not occur with overload; however, the muscle does seem to have a reserve of nuclear material available in the form of the muscle satellite cells. The responsiveness of muscle to training is indirectly proportional to postpubertal age and is less in females than males.

Muscle growth is affected by diet, exercise, hormones, muscle stretch, and motoneurons, all of which are extremely important in assuring optimal performance capacity of muscular tissues.

Like muscle growth, neuronal growth is particularly responsive to environmental conditions such as diet, exercise, and hormones. Although neural tissue is adaptable during the maturational periods, this tissue becomes more resistant to environmental changes during adulthood. Synapses, dendrites, axons, and the cell bodies are all capable of adaptations to the environment during the developmental period and to some degree during the adult period.

In general, evidence indicates that exercise can be beneficial during the growth and developmental stages of bone, muscle, and neural tissue.

Study Questions

1. What is growth, hyperplasia, and hypertrophy?
2. What is the nature of the growth adaptation to exercise?
3. How does exercise influence normal body growth?
4. Describe the height and weight growth-curves.
5. How can exercise alter the growth of bones?
6. What injuries can be harmful to growth?
7. How does exercise influence the growth of skeletal muscle?
8. What sex-related differences are there in skeletal muscle?
9. What hormonal influences are important for growth?
10. How does testosterone alter glycogen supercompensation?
11. How do muscle stretch and exercise induce muscle hypertrophy?
12. What basic factors must be altered to achieve muscle growth?
13. How does exercise affect cardiac muscle?
14. How does exercise affect tendons and ligaments?
15. What effect does immobilization have on joints and ligaments?
16. How does hyperactivity affect neuronal growth?
17. Do hormones affect neuronal growth?

Review References

Adolph, E.F. Physiological adaptations: hypertrophies and superfunctions. *Amer. Scient.* 60: 608–617, 1972.

Bloor, G.M.; S. Pasyk; and A.S. Leon. Interaction of age and exercise on organ and cellular development. *Amer. J. Pathol.* 58:185–199, 1970.

Booth, F.W. and C.M. Tipton. Ligamentous strength measurements in pre-pubescent and pubescent rats. *Growth* 34:177–185, 1970.

Cheek, D.B., et al. Skeletal muscle cell mass and growth: the concept of the deoxyribonucleic acid unit. *Pediat. Res.* 5:312–328, 1971.

Golding, L.A.; H.E. Freydinger; and S.S. Fishel. Weight, size and strength unchanged with steroids. *The Physician and Sports Medicine* 5:39–43, 1974.

Johnson, L.C. and J.P. O'Shea. Anabolic steroid: effects on strength development. *Science* 164: 957–959, 1969.

Lamb, D.R. Androgens and exercise. *Med. Sci. Sports* 7:1–5, 1975.

Malina, R.M. Exercise as an influence upon growth. *Clin. Pediat.* 8:16–26, 1969.

Moss, F.P. and C.P. Leblond. Satellite cells as the source of nuclei in muscles of growing rats. *Anat. Rec.* 170:421–436, 1971.

Rarick, L., ed. *Human Growth and Development.* New York: Academic Press, 1973.

Rodahl, K.; J.T. Nicholson; and E.M. Brown, eds. *Bone as a Tissue.* New York: McGraw-Hill, 1960.

Key Concepts

• The fuel needed to supply energy to the body primarily comes from two basic sources, carbohydrates and fats.
• Vitamins and minerals are essential for proper metabolic functions.
• A sedentary way of life desensitizes the mechanisms that maintain a balance between caloric intake and output.
• Nutritional demands of a training regimen are specific to the nature of the training.

Introduction

The human body is composed of millions of cells, each of which is continuously producing energy for its own survival, maintenance, and reproduction in addition to the energy needed to perform the cell's special function. It is recommended that humans ingest 1500–3000 calories each day in order to produce enough energy to keep the cellular machinery operating. We can acquire 2500 calories by eating 25 bananas, 2000 radishes, 2.5 pounds of french fries, or 160 prunes. Any of the foods mentioned would be suitable diets if our only concern were a caloric source of energy.

In addition to energy considerations, other materials are also necessary to fabricate the cellular components needed to maintain cells. We might define nutrition, then, as the science of what and how much to eat in order to optimize efficiency of bodily functions. In striving to get enough of all the required types of chemical compounds, we must be wary not to overindulge—overindulgence causes the kidney and liver to work overtime, rejecting the excess or converting it to fat. It has been shown repeatedly in laboratory animals that longevity can be increased more than 2.5 times simply by restricting the daily number of calories consumed. The effect is greatest if food is

CHAPTER FOURTEEN

Nutrition and physical activity

restricted throughout an animal's lifetime, but beneficial effects can be seen even if restriction is not begun until the animal is an adult.

Many more far-reaching effects of malnutrition are becoming increasingly evident; some are more subtle than others. Malnutrition has been clearly indicated in children of urban American cities: it has been most obviously manifested in limited growth and mental development. Malnutrition during the stage of rapid brain development in laboratory animals results in less brain growth: the brain's weight and number of cells are less than those of well-nourished animals. Also malnutrition in early life has been associated with emotional disturbances in later life. The emotional disturbances are very similar to those characteristic of animals placed in social isolation. In other words the behavioral ramifications of malnutrition seem to involve the social development of the individual—ability to get along with others.

The fuel needed to supply energy to the body can come from three basic groups of compounds: carbohydrates, fats, or proteins. In simplistic terms carbohydrates are used as a short-term energy supply, fats as a long-term energy supply, and proteins used for the structural and enzymatic matrix of cells. This simplification suffers from the fact that, to a certain extent, carbohydrates and fats are also used as structural components and that proteins can be used, to a limited extent, as a source of energy.

Foundations of Nutrition

Carbohydrates

Carbohydrates are the sugars, starches, and cellulose of the plant kingdom. *Monosaccharides*, or single sugar compounds, are formed directly from CO_2 and water by the plant's photosynthetic process. These monosaccharides are fre-

quently joined together in chains to form *polysaccharides*. When glucose, the most common monosaccharide or single sugar, is used as the subunit of a long chain polysaccharide, the result is starch such as in potatoes, beans, and grains. When *glucose* is stored as a polysaccharide chain in animals, it is branched like starch in plants and is called *glycogen*—a ready reserve of stored energy in animal cells.

Common table sugar is sucrose, a paired combination (*disaccharide*) of glucose and fructose. Similarly lactose (milk sugar) is a glucose-galactose combination. The presence of carbohydrates in the mouth stimulates secretion of saliva, which lubricates the food and contains *amylase*, an enzyme that begins to break polysaccharides into shorter chains (Figure 14–1). We can note this reaction by chewing bread (starch, not sweet) until it becomes sweet (mono- and disaccharides). Heat helps to break these polysaccharides into subunits, explaining the mildly sweet flavor of toast and fried onions. The digestion of carbohydrates is aided by mastication, the mixing of food with saliva to increase the surface area upon which the amylase enzyme acts. About half of all starch digestion occurs when food is in the mouth or the stomach. Digestion of short polysaccharides continues in the small intestine in the presence of amylase from the pancreas. The monosaccharides (glucose, fructose, and so forth) are then transported to the liver, muscle, and other tissues where glycogen is synthesized as storage for glucose until it is needed.

Lipids

Lipids are greasy substances such as oils, fats, and waxes, which are not soluble in water; common examples are the oils on the surface of beef soup and the cream on raw milk. Lipids, important in the diet as a source of fuel and fat-soluble vitamins, are used in the production of certain hormones and are stored as a future energy source. The storage of fats, however, serves another purpose: the fat acts as thermal insulation and protects some organs from mechanical shock. The storage form of lipid is *triglyceride*—three fatty acids attached to a three-carbon compound, *glycerol*. Fatty acids attached to other compounds such as phosphoric acid (*phospholipids*), carbohydrates (*glycolipids*), and proteins (*lipoproteins*) and cholesterol are important in the membrane structure of all cells and many organelles.

The human organism is an intricate organization of structures within structures, all of which are composed in part of lipid compounds. The lack of water solubility of lipids is perhaps a major reason for its use in membrane structure. For instance, if the skin were not made of "waterproof" lipids, it would soak up water like a sponge or dissolve with the first spring rain. All lipids can be produced from a carbohydrate-protein diet if sufficient quantities of the one essential fatty acid (*linoleic acid*) are included. Linoleic acid must be ingested from the diet because the body cannot synthesize it.

Digestion of ingested fats, mostly triglycerides, phospholipids, and cholesterol begins in the duodenum where bile salts emulsify large lipid droplets into smaller ones, thus increasing the surface area accessible to the enzymatic degradation of fats into free fatty acids and small amounts of glycerol (Figure 14–2). These are then transferred to the liver by the blood or lymphatic system for distribution or storage.

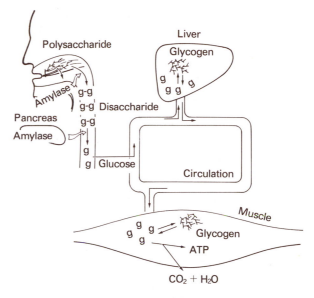

Figure 14–1. Carbohydrate digestion, absorption, and utilization (g = glucose).

Proteins

Proteins are structural components of cells, hormones, enzymes, and the molecules of muscular contraction. *Proteins* are long chains of nitrogenous subunits called *amino acids*; 23 varieties of amino acids occur naturally. The

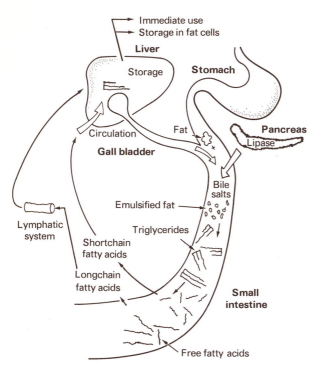

Figure 14-2. Digestion, absorption, and use of fat.

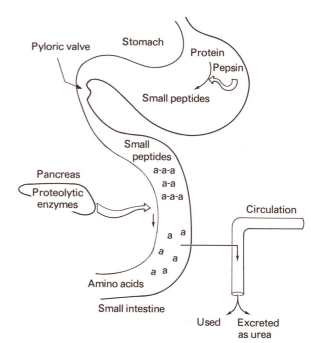

Figure 14-3. Digestion, absorption, and the fate of ingested proteins.

most dramatic type of protein is the special class called *enzymes*. An enzyme is the structure upon which chemical reactions take place.

The blood contains a variety of proteins such as *antibodies* and *plasma albumin*. Protein concentration varies from tissue to tissue: for example muscle has 75% water, 20% protein; collagen tendon has 67% water, 33% protein; and nerve tissue has 78% water, 10% protein.

In order to manufacture the variety of proteins needed, the body must have a source of amino acids. Most of them can be made by the existing enzymatic pathways within the cell by the utilization of available sugars or

lipids. However, eight amino acids are classified as *essential amino acids*; they cannot be synthesized in the body but are needed for the synthesis of a variety of human proteins. The eight essential amino acids that must be ingested in the diet are: *lysine, leucine, isoleucine, methionine, phenylalanine, threonine, tryptophan* and *valine*.

The most common source of amino acids in the United States is meat. Soybeans and meat are the only common foods that alone have a full complement of essential amino acids; other important sources of amino acids are milk, cereals, and beans. Since cereals have all the essential amino acids except lysine and since beans have all except methionine, the combination serves as a source of all of the essential amino acids and is an adequate protein substitute for meat.

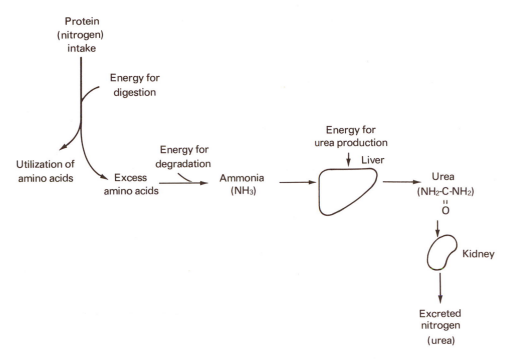

Figure 14–4. Protein catabolism.

In the stomach protein digestion involves the formation of shorter protein fragments by the action of the enzyme *pepsin* (Figure 14–3). Enzymes from the pancreas act on protein fragments in the duodenum to make smaller chains of amino acids, which in turn are further digested to free amino acids in the intestinal wall. Roughly 50–60% of ingested protein is digested and absorbed by the time the food passes through the duodenum; another 20% is ingested by the end of the next section of the intestine (jejunum). Since amino acids cannot be stored, excess amounts are broken down to urea and excreted by the kidney (Figure 14–4).

Vitamins and Minerals

Vitamins are needed in only small amounts, but they are essential for the body to function properly. They can be conveniently classified into two categories: water-soluble (B-complex and C) and fat-soluble (A,D,E,K) vitamins.

WATER-SOLUBLE VITAMINS

Vitamin C. Perhaps the most well-known of the water-soluble vitamins is vitamin C, the antiscurvy factor found in citrus fruit. Vitamin C, or *ascorbic acid*, plays a part in the maintenance of bone, dentin of teeth, and cartilage (collagen). Defects in these tissues are caused by the vitamin-deficient state. It probably plays a further role in the respiration of mitochondria and in the maintenance of the mechanical strength of blood vessels. Vitamin C also increases absorption of iron.

The taking of large doses (2–10 grams a day) of vitamin C has become popular as a result of the claims that it prevents respiratory ailments. The warning should be given, however, that a large dose of vitamin C (or of any other vitamin) may prove to have detrimental effects on other functions of tissues. Since vitamin C is not stored, it is lost in the normal process of metabolism. Therefore it must be replenished daily from common sources such as citrus fruits, green leafy vegetables, berries, or tomatoes. Oxygen destroys vitamin C, and water extracts it from food during cooking. If an excess of vitamin C is ingested, it is excreted and does not seem to be toxic even in the large doses of "megavitamin therapy."

Vitamin C and Exercise. A lack of vitamin C affects muscle metabolism in such ways that we might expect sub-par performances from an individual on a low vitamin C diet. Vitamin C deprivation causes a high plasma (blood) enzyme level, probably the result of elevated muscle-fiber membrane permeability, which allows enzymes to leak from muscles. Muscle glycogen is also reduced by vitamin C depletion. Furthermore animals deficient in vitamin C

are less tolerant of exercise; their muscular contractions become increasingly weaker. Each of these effects of reduced vitamin C suggests a major influence on muscle metabolism that would logically affect most physical performances.

Vitamin C has a protective effect against a number of stress agents. It is well known that a variety of stresses, including exercise, causes a reduction in the vitamin C content in adrenal glands. Intake of vitamin C during stress situations may be beneficial to the individual. However, we strongly caution anyone that no evidence suggests that a high vitamin C diet enhances muscle metabolism and physical performance above normal levels.

Vitamin B-Complex. The other component of the water-soluble vitamins is composed of the B-complex of compounds, each of which has a distinct function: *thiamine* (B_1), *riboflavin* (B_2), *niacin* (B_5), *pyridoxine* (B_6), *biotin* (B_7), *folic acid* (B_9), *cyanocobalamine* (B_{12}), *pantothenic acid*, and the psuedovitamin *choline*. Table 14–1 describes some of the properties, requirements, and food sources of the B vitamins. In general each of the B-complex vitamins is essential for the metabolism of carbohydrates and amino acids, the formation of active acetate, the synthesis of fatty acids, and the formation of nucleic acids for RNA and DNA.

FAT-SOLUBLE VITAMINS

Vitamin A. Among the fat-soluble vitamins, the role of vitamin A has been shown to be an integral factor in the production of *rhodopsin*, the substance in the rods of the retina that senses light. The rhodopsin, which is broken down by light to vitamin A and a protein, is resynthesized for subsequent use; the speed of this resynthesis is directly related to the amount of vitamin A available. Thus concen-

Table 14–1. Water Soluble B Vitamins

Vitamin	Physiological Functions	Requirement	Food Sources
Thiamine (B_1)	Coenzyme in carbohydrate metabolism	0.5 mg/1000 cal	Pork, beef, liver, whole or enriched grains, legumes
Riboflavin (B_2)	Component of hydrogen carrier compounds (flavin adenine dinucleotide: FAD)	0.6 mg/1000 cal	Milk, liver, enriched cereal
Niacin (nicotinic acid, B_5)	Component of hydrogen carrier compounds (nicotinamide adenine dinucleotide: NAD)	14–19 mg (niacin equivalent)	Meat, peanuts, enriched grains
Pyridoxine (B_6)	Coenzyme in amino acid metabolism	2 mg	Wheat, corn, meat, liver
Pantothenic acid	Coenzyme in formation of active acetate (Acetyl-CoA)		Liver, egg, skim milk
Lipoic acid (sulfur-containing fatty acid)	Coenzyme (with thiamine) in carbohydrate metabolism to reduce pyruvate to active acetate		Liver, yeast
Biotin (B_7)	Coenzyme in synthesis of fatty acids, amino acids		Egg yolk, liver
Folic acid (B_9)	Coenzyme for single carbon transfer	0.4 mg	Liver, green leafy vegetables, asparagus
Cobalamin (B_{12})	Coenzyme in formation of nucleic acid and cell proteins	5 μg	Liver, meats, milk, egg, cheese
Choline	Component of neural transmitter; acetylcholine		Meat, cereals, egg yolk

Adapted from Sue Rodwell Williams. *Nutrition and Diet Therapy*, 2d ed. St. Louis: C.V. Mosby, 1973, pp. 106–107.

trations lower than normal yield temporary night blindness when a bright light interrupts vision of a dim field. The speed with which one can recover vision of the dim light (re-synthesis of rhodopsin) is a good indicator of vitamin A deficiency. Vitamin A is known also for its role in maintaining the integrity of cell membranes. Sources of this vitamin include orange and green vegetables, like yams and parsley, meat and liver. Excess vitamin A is stored in animal liver until needed. In excessively large doses this vitamin, like vitamin D, is toxic.

Vitamin K. The action of vitamin K is involved at the ribosomal level in the biosynthesis of *prothrombin*, a precursor of a substance essential for the formation of the blood clot. Since it is produced by intestinal bacteria, a regular dietary source is unnecessary. Like vitamin E, it is found in green leafy vegetables.

Vitamin E. Vitamin E or *alpha-tocopherol*, an antioxidant, is found in such a wide variety of meats, vegetables, and cereals that a nutritional deficiency is highly unlikely. But laboratory rats without vitamin E exhibit atrophic spermatogenic tissue, sterility and, in females, resorption of the fetus.

One useful chemical property of tocopherol is its antioxidant capability: it is used in foods and pharmaceuticals to protect other vitamins from oxidative destruction by light. Vitamin E seems to play some role in lipid metabolism or fat storage. The claims—that vitamin E helps to remove scar tissue when applied to wounds, enhances sexual potency, and improves circulation—remain scientifically questionable.

Although vitamin E has been used in attempts to improve performance of athletes and racehorses, its beneficial effects as an ergogenic aid are questionable. It has been claimed that vitamin E helps to prevent atherosclerosis by preventing the abnormal buildup of cholesterol in the blood vessels. Theoretically a high vitamin E concentration could lower fatty acid utilization, an important source of energy in endurance-type exercises; however, any effect of vitamin E on the performance of athletes has not been found.

Vitamin D. The sunshine vitamin, vitamin D, is necessary for solid bones and teeth. It can be produced in the body by irradiation (exposure to sunlight) of a precursor compound. The precursor, being fat soluble, is carried to the surface of the skin during exposure to sun-light and is subsequently reabsorbed. The reabsorbed active form then facilitates the absorption of calcium and phosphorus in the gut. A parathyroid hormone has recently been implicated in this activation process.

Vitamin D is one of the most stable fat-soluble vitamins and is frequently used to fortify milk. Since vitamin D along with the other fat-soluble vitamins can be stored, it is found in high concentrations in animal livers. Because of this storage phenomenon and the increasing interest in "megavitamin therapy," a warning must be made here about massive doses of fat-soluble vitamins: they can be dangerous. Evidence indicates that large excesses of vitamin D can calcify or harden normally soft tissues.

MINERALS

Miscellaneous Trace Minerals. The fat-soluble vitamins and minerals are stored in the body so that a daily dietary intake is not essential; however, an adequate supply over a long period is required. Among the minerals required for normal bodily function are: iron, calcium, phosphorus, iodine, fluorine, sodium, potassium, magnesium, copper, sulfur, trace amounts of zinc, manganese, molybdenum, cobalt, selenium, chlorine, and bromine. These are widely distributed in plants and animals so that special foods need not be eaten to insure adequate intake of each; however, iron deserves special attention.

Iron Deficiency Anemia. Two-thirds of the body's iron, a protein and a heme molecule form *hemoglobin*, the oxygen-carrying component of blood. Iron deficiency anemia is manifested in low hemoglobin levels and can usually be alleviated by ingestion of certain iron compounds. The fully oxidized form of iron, such as in iron metal or raisins, is not absorbed and therefore is an unsatisfactory dietary source. Beef, kidney, liver, and beans contain the reduced form of iron and are excellent natural sources. Certain substances enhance iron absorption; others inhibit it. For example the vitamin C in a glass of orange juice markedly increases the proportion of iron absorbed from the diet, while eggs reduce the amount.

The prevalence of anemia is as high as 80% in young children in some rural areas of the United States. Iron deficiency is also found in a surprisingly large percentage of the people in tropical areas where hookworm is prevalent. In spite of the low oxygen-carrying capacity of those with anemia, adaptive mechanisms help to maintain an apparently normal functional capacity.

Routine iron-supplement therapy for children from birth to two years of age in a normal population in southern California revealed no beneficial effects in regard to their general health or freedom from diseases or infections. In iron-deficient subjects without anemia (who have low iron stores but normal hemoglobin), physical performance in most cases does not appear to be seriously affected. Iron deficiency anemia has been related to decrements in performance capacity and maximum oxygen consumption in humans—and in voluntary physical activity patterns of laboratory rats. However, if a pint of one's own blood is removed and stored for several weeks, the body replenishes this loss, given an adequate iron intake. The subsequent reinfusion of the pint of blood results in a greater oxygen uptake capacity. This technique is not recommended for the purpose of improving performances primarily because of the potential dangers involved in transfusions.

Iron imbalance may have more subtle effects than those immediately related to oxygen transport. For example wound healing is slower in iron-deficient rats; this may be related to the fact that iron is required in the synthesis of collagen in connective tissue. The potential effect of iron deficiency on catecholamines in the brain was discussed in Chapter 5. On the other hand, a very excessive dietary iron intake can result in iron poisoning, which can be fatal in some individuals with specific hemotological abnormalities. Also very high dietary iron intakes can depress the rate of growth in maturing rats. It seems, however, that barring these extremes, the body has an enormous capacity to prevent toxic iron levels from accumulating as well as a very good capacity to adapt to low levels of iron.

Metabolic Adaptations to Dietary Changes

The effectiveness of high carbohydrate diets in improving performance in certain types of exhaustive exercises has been demonstrated repeatedly. It is now well known that such a diet will elevate glycogen storage levels in skeletal muscle. Furthermore repletion of glycogen in exhaustively worked muscles occurs much faster and is much greater if the worked individual is placed on a high carbohydrate diet as opposed to a high fat-protein diet (Figure 14–5). The mechanism for these glycogen-synthesizing effects is not well understood.

High carbohydrate diets induce alterations in other types of metabolism since they indirectly affect fatty acid and cholesterol metabolism. As a consequence, carbohydrate diets have come under great scrutiny; evidence has been found linking high carbohydrate diets to high incidences of cardiovascular diseases.

A high-fat diet leads to a decrease in work time when the work rate is moderate (Figure 14–6). This decrement is related to lowered resting-muscle glycogen. Performances of a more intense nature are probably not affected adversely by a high-fat diet. This is probably because the point at which glycogen is depleted from muscles does not correlate well with the

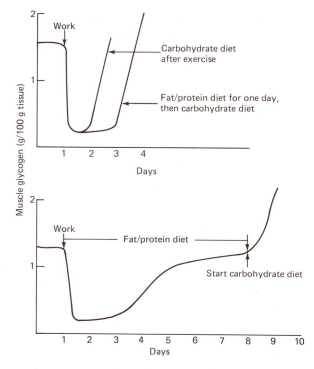

Figure 14–5. Effect of diet on resynthesis of muscle glycogen after exhaustive work.

Adapted from E. Hultman, Physiological role of muscle glycogen in man, with special reference to exercise. In *Physiology of Muscular Exercise* by C.B. Chapman. Monograph no. 15, 1967. By permission of The American Heart Association, Inc.

point of exhaustion when the exercise is either very intense or much more moderate and prolonged.

The most important point to recognize about a high-fat diet, particularly one of saturated fats, is the greater risk of heart disease. An example of special interest here is the Alaskan Arctic Eskimo, whose fat intake is up to twice that of an average American. However, the Eskimo's serum (blood) cholesterol does not differ from that of the general U.S. population; and high values (over 250 mg/100 ml) are

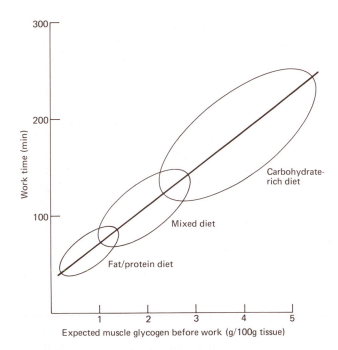

**Figure 14–6. Effect of muscle glycogen on
work performance capacity.**

Adapted from E. Hultman, Physiological role of muscle
glycogen in man, with special reference to exercise. In
Physiology of Muscular Exercise by C.B. Chapman.
Monograph no. 15, 1967. By permission of The
American Heart Association, Inc.

found in only 3% of the Eskimo population.
But heart disease is about ten times more prevalent in the general U.S. population than in Arctic Eskimos. Seasonal variations in abundance
of food and consequently hypo- and hypercholesteremia are among several factors that apparently prevent cardiac ischemic diseases in
Eskimos.

The effect of the caloric distribution ratio
between carbohydrates and fats does not affect
protein utilization providing that the diet does
not contain extremely high or low dietary protein percentages. A high-protein diet, however,
stimulates the glucose-synthesizing capacity of
the liver, thus assuring a source of glucose.

Nutritional Demands During Exercise and Training

Nutritional demands vary according to age, sex, metabolic properties, hormonal status, as well as physical activity—all of which are in some way interrelated.

Proteins: Demands of Exercise

Protein metabolism is affected by exercise as is the metabolism of carbohydrates and fats. Even with enhanced dietary protein utilization for energy during exercise, there is also a greater total body amino acid retention and protein synthesis. A characteristic adaptation to training appears to be that the lean body mass increases and that body fat decreases—with no net change in body weight. Exercise in hypercaloric subjects significantly raises excretion of nitrogen and sulfur. Similar changes in amino acid metabolism can be induced by low insulin secretion and elevated plasma corticosteroids, both of which can occur during exercise. Low insulin curtails glucose availability to some tissues, therefore producing a greater dependence on proteins; corticosteroids promote muscle protein degradation (muscle wasting).

It is claimed by many investigators that exercise does not increase the body's demand for protein. This conclusion probably results from lack of assessment of all sources of nitrogen loss (protein loss), such as sweat and epithelial cells lost from the epidermis. Nitrogen loss through skin may amount to about 14% of the total nitrogen loss during heavy manual work. Mean nitrogen balance is positive in competitive weight lifters during training; that is, more nitrogen is retained than is excreted. In weight lifters, the protein requirement is increased in proportion to body weight and work intensity. Daily protein requirement is about 2 g/kg of body weight when about 8000 kg of weight are lifted in a 24-hour period.

The effect of exercise on protein metabolism is made more apparent when marked proteinuria, hemoglobinuria, or myoglobinuria (loss of these compounds in the urine) occurs with extended, strenuous exercise. The catabolic effect of exercise, as manifested by hemoglobinuria, may be responsible for the occasional occurrence (2% in males and 2.5% in females) of anemia and iron deficiency in well-trained athletes. Iron deficiency without anemia was found in 15% of the tested female athletes who trained for the 1968 Olympics—a somewhat surprising finding in light of the generally accepted erythropoietic effect of training (synthesis of erythrocytes or new red blood cells). But the increase in degradation of red blood cells with intense training may be greater than the increase in synthesis; thereby the body will require additional protein for hemoglobin synthesis. If the rigorous training is continued without adequate iron intake, the hemoglobin concentration will eventually decrease.

Changes in metabolic rate are regulated largely by enzymatic proteins, and the exercising situation activates an enormous number of these enzymes in order to assure an adequate source of energy. Enzyme activity is enhanced by conformational and other minor modifications in the enzyme molecule itself and/or by increasing the quantity of the enzyme produced. The latter occurs particularly in response to training and, indeed, represents a positive nitrogen balance. Enzyme synthesis is, in fact, one of the major avenues of adaptability to training. Myoglobin and hemoglobin are other proteins that are increased by endurance training.

Carbohydrates and Fats: Demands of Exercise

More work can be performed at a high oxygen consumption rate when a person is on a high-carbohydrate rather than on a high-fat diet.

The intensity of the work load is an important factor in determining the proportion of fats and carbohydrates that are utilized for the energy required by the exercise stress. Light to moderate work loads (less than 70% of max $\dot{V}_{O_2}$) utilize a relatively greater amount of fat than heavier work loads. Glycogen has been shown to be a very important factor in determining exhaustive exercise times when continuous and constant work is performed at about 75% max $\dot{V}_{O_2}$. The relationship between work performance and muscle glycogen after three days on a high-fat, high-protein diet, a mixed diet, and a high-carbohydrate diet is shown in Figure 14-6.

Fat is also an important energy source during exercise. During sustained exercise at a light to moderate work load, the high plasma free fatty acid level promotes a greater role for its uptake and oxidation in muscle. During endurance training, the subject increases the capacity of the trained muscle cell to utilize free fatty acids—and at a greater rate. Training also elevates the intramuscular resting pool of glycogen, which may be used initially during exercise until plasma free fatty acids become available.

The Pregame Meal

The pregame meal—no matter what its relative concentration of carbohydrates, fats and proteins—is not likely to greatly improve athletic performance. Carefully controlled studies have demonstrated that a regular meal, regardless of the relative proportions of carbohydrates, fats, and proteins (but isocaloric), does not significantly affect performance. However, the meal can be detrimental if the food ingested precipitates a nauseous response, which sometimes occurs because of pregame nervousness.

In athletic events demanding a sustained effort for an hour or more (e.g., in marathon running) additional factors should be considered in determining the content of the pregame meal. First, the meal must be psychologically, as well as physiologically, satisfying. For psychological reasons, the large steak meal prior to an event may be beneficial. Nutritionally, however, it is more advantageous to ingest a meal consisting predominantly of *carbohydrates* for the following reasons: (1) carbohydrates during extended physical exertion become practically exhausted in the muscle, and they are also lowered in the liver so that a high-carbohydrate meal a few hours prior to the initiation of the exercise could slightly augment the carbohydrate pool; (2) carbohydrates are digested faster than proteins or fats—a factor that is particularly relevant in light of the potential inhibitory effect of pregame excitement or emotional stress on the rate of digestion and absorption; (3) carbohydrates can be made available as an energy source for muscular work with the least amount of energy output; (4) a high-carbohydrate diet helps to maintain normal plasma glucose levels; and (5) muscle glycogen is replenished faster after an exercise if a high-carbohydrate postgame meal is ingested.

Muscle glycogen can be elevated by almost 25% following four hours of glucose or fructose infusion. However, infusion of glucose during work does not affect the rate of glycogen loss in muscles, probably because glucose, although used as a supplement to glycogen, suppresses consumption of fats. Consequently performance is not greatly affected by glucose infusion during the work; there is, however, an alteration in the type of substrate utilized.

Liver glycogen is extremely responsive to alterations in carbohydrate ingestion and to exercise. Liver glycogen can be reduced almost 90% after 24 hours of fasting; it can be increased more than 180% after 24 hours of carbohydrate refeeding. These radical fluctuations in liver glycogen demonstrate the rapidity with which liver glycogen can be utilized.

Regulation of Food Intake

A Balance of Intake and Output

Regulation of food intake is such that energy input and output are relatively constant over extended periods of time; yet daily fluctuations in this balance occur repeatedly. When we consider the amount of food eaten over a ten-year period by some adults without a significant change in body weight, we can better appreciate the control mechanisms involved. Even a slight imbalance in the food intake over the energy output could lead to a serious overweight condition within a few years. For example an extra half slice of bread added to one's daily caloric intake would lead to an accumulation of 20 additional kg (44 pounds) of weight in 10 years, assuming no metabolic adaptations to the added caloric intake.

How much of the daily caloric intake is utilized in external work? Muscular activity utilizes less than half of the energy input of most individuals in our modern society; the remainder is used by the heart, smooth muscles, glandular secretions, liver metabolism, and, in general, is used for the maintenance of the cells' basic functions, such as the continuous discharge of electrical potentials in neurons.

Mechanism of Exercise-Induced Appetite Depression

Vigorous exercise is followed by a period of *hypophagia* (reduced eating) for up to several hours. In part, this lack of appetite is caused by lactate, a by-product of exercise, which inhibits appetite. However, lactate returns to normal fairly rapidly so that high lactate or some other independent agent produced during exercise must induce a more long-lasting effect on the food-intake regulator, the *medial hypothalamus*. Exercise as a tool for weight reduction has a twofold advantage: it not only requires a greater expenditure of energy but tends to reduce food intake.

Pharmacological Control of Appetite

Appetite can be stimulated pharmacologically. The agent called *cyproheptadine* stimulates appetite, weight gain, and linear growth in small children. Although its mechanism of action is not known, it has been used in underweight, asthmatic children. Amphetamines and weight-reducing pills, on the other hand, speed up metabolism and, to a certain extent, decrease appetite. Most of these drugs are not only dangerous but also offend all aesthetic sensibilities when imposed indiscriminately upon a natural organism as delicately balanced as the human body.

Humoral Control of Food Intake

As a possible regulator of food intake, a *humoral factor* has been suggested following laboratory experiments in which normally fed animals were transfused with blood from food-deprived animals. Immediately following transfusion, the food intake of the normally fed animals was considerably above normal.

There is some evidence that the amount of energy dissipated by the body is related to food intake; this might mean that the daily energy cost of a standardized activity will be greater if it follows absorption of excess caloric intake. If this is the case, we should not gauge weight loss or gain by only observing total food intake and physical activity since the metabolism may be reset at a lower or higher rate depending upon the time difference between exercise and eating and the normal hormonal balance for each individual.

Psychological and Sociological Aspects of Regulation of Food Intake

The effect of environmental factors in controlling food intake is often underestimated. In obese subjects, there is a higher degree of "external" environmental influence as compared to normal subjects whose regulation of food intake is more physiologically motivated. Stated simply, obese people are more motivated to eat in response to environmental stimuli (e.g., seeing delicious food) than by biological forces that serve as physiological indicators of when the body needs food. Thus it appears that the hypothalamus is reset in obese people to be less sensitive to the *satiety factor*. Palatability of food is an example of a factor that demonstrates the greater dependence of obese individuals on an external stimulus: if a food such as ice cream is given to an obese and a non-obese subject, the obese subject will eat the most. However, if quinine is added to reduce the palatability of the ice cream, the obese individual eats the least; this study suggests an enhanced sensitivity to taste factors in obese subjects.

Body weight is nearly equally influenced by diet and by genetics; body composition is influenced largely by dietary habits. Laboratory experiments have shown that the frequency of feeding affects the way foods are distributed and utilized in the body. Low meal frequency is characterized by higher body weight, greater skinfold thickness, hypercholesteremia, and diminished glucose tolerance compared to the physical state when the same amount of calories are ingested at a high meal frequency.

One meal a day also affects the normal diurnal rhythm of plasma free fatty acid levels and liver glycogen. Eating one large meal in the morning delays by about 12 hours the time for which liver glycogen will be the highest and plasma free fatty acids the lowest. Plasma free fatty acids and liver glycogen usually fluctuate inversely. Since glycogen levels in the muscle are directly related to exercise exhaustion times and since liver glycogen is considerably reduced during an endurance-type exercise, the time of the day an individual eats a large meal should be considered in relation to the time of exercise. Regardless of the availability of glycogen, the mobilization of free fatty acids during any part of the diurnal cycle is adequate for moderate exercise.

Social factors can also be related to obesity. Although obesity is more prevalent in the lower socio-economic areas, malnutrition is too. The diets among these groups are adequate calorically (high in carbohydrates), but the type of food is not adequate nutritionally. Control of obesity can only be achieved by understanding all of the underlying factors that influence weight gain or loss.

Adaptations to Obesity

A person can be "overweight" without being obese. For example "well-built" weight lifters can be matched by age and weight to less developed individuals, and obviously they differ in body shape, composition, and physiology. Metabolic differences between overweight obese and overweight nonobese individuals are due to the differences in relative amounts of fat and protein content in the body. When obese overweight men are compared to muscular overweight men, we find that the obese are hyposensitive to insulin: the obese individual must secrete more insulin in order to lower blood glucose levels than the nonobese subject. This lack of insulin sensitivity can be in muscles or fat, but the point is that the beta cells in the Islets of Langerhans (in the pancreas) have to produce so much insulin to overcome the muscle's impermeability to glucose that the pancreatic beta cells may become exhausted; the result may be the possible formation of diabetes. Theoretically the tendency of obese people to become diabetic may be alleviated by exercise since exercise makes the muscle cells more permeable to glucose, thus lowering blood glucose levels and sparing the beta cells from overwork. It is encouraging to know that correction of the obesity in nondiabetic subjects is accompanied by a return to normal glucose tolerance, insulin resistance, and insulin sensitivity to glucose.

Obese subjects generally have high cholesterol, triglyceride, free fatty acid, and phospholipid levels in the blood. These high values are reversed during the process of weight reduction. In the obese person weight reduction by fasting markedly reduces liver lipids because the lipids become a major source of energy. Up to 95% of the energy expenditure in fasting obese subjects can come from fats. Under these circumstances, the brain adapts: it increases its capacity to use ketones, a metabolic product of excess lipid oxidation. The decrease in lipid levels in the blood must be long lasting if it is to be effective in reducing the relative risk of heart disease. A summary of endocrine and metabolic changes in human obesity and of some of the countereffects of exercise is given in Table 14–2.

Determination of Obesity

It has been a practice in the past to predict a person's weight by his or her height. The obvious disadvantage is the inaccuracy introduced if an individual's basic body structure is not considered. For example most professional football players are "overweight" according to national height-weight norms, but they are not usually obese.

A more accurate way to determine the level of obesity, or lack of it, is to measure the percent of the body that is fat tissue. This measurement can be easily obtained because fat tissue is lighter than water, muscle, blood, or bone (fat has a low specific gravity). To determine body density and subsequently to calculate the percent of the body that is fat tissue, we simply weigh the subject in the air and then again in the water when the subject is

Table 14–2. Hormonal and Metabolic Adaptations to Obesity

Compared to normal, obese individuals demonstrate:

1. Greater insulin secretions
2. A greater capacity to synthesize lipids
3. A greater storage of lipids in the blood in the form of cholesterol, triglycerides, free fatty acids, and phospholipids
4. Depressed sensitivity of growth hormone (GH) secretion to hypoglycemia.*

As a countereffect, exercise tends to:
1. Reduce the need for insulin by duplicating its function
2. Reduce serum lipids (cholesterol, triglycerides, free fatty acids, and phospholipids) by increasing degradation
3. Elevate GH secretion (sixfold), which would lessen the effect of a reduced sensitivity of GH secretion to hypoglycemia.

*GH elevates blood glucose during hypoglycemia, and also raises free fatty acids in the blood. Obese individuals have to secrete a greater than normal amount of GH to overcome the hypoglycemic effect.

completely submerged. Body density can then be determined by dividing body weight in the air by the weight of the water displaced (determined by subtracting weight underwater (kg) from weight in air (kg) and multiplying by 1000, since 1 cc of water weighs 1 g).

$$\% \text{ fat} = \frac{4.57}{\substack{\text{specific} \\ \text{gravity}}} - (4.142)$$

Since this technique of underwater weighing requires special skills and equipment, a more practical way to determine obesity is by the use of anthropometric measurements, such as skinfold thickness. Skin calipers are used to measure skinfold thickness at representative sites on the body, for example the skin overlying the triceps. Some practice is essential in order to be able to repeat these findings reliably. The technique is reasonably accurate as a measurement of obesity since more fat (50%) is localized at an immediately subcutaneous level than anywhere else in the body. Lower limits of the triceps skinfold thickness in Caucasian American females and males, from 4 to 30 years old, are shown in Figure 14–7. Note that, at the onset of puberty, skinfold thickness decreases in males but steadily increases in females during this period.

Cellularity of Obesity

A survey of major U.S. cities shows that obesity is present in 10–40% of the population. Obesity in white males is about twice as prevalent as in white females; the opposite tendency seems to be true among blacks.

What is the nature of this "disease"? The etiology of obesity should be considered when one devises a rational dietary and exercise program, the purpose of which is a general reduction in body weight. The cause of obesity is frequently overeating, but the situation is much more complex than this. *Adiposity* is partially determined by adipocyte cell size and number—with an increase or decrease in *size* being most variable as an adult and difference in *number* being more variable at preadult stages of maturation (Figure 14–8). For example, when an adult loses weight, there is a decrease in adipocyte size—but no change in number. It has also been shown that adipocyte number apparently is determined by eating habits developed as early as the first six postnatal weeks of life. Adipocyte numbers appear to stop increasing during the 12–25 year period. With slight underfeeding, the number of adipocytes that have developed by adulthood is lower than would be expected in an individual without any dietary restrictions. Babies who gain weight more rapidly than average during the first year of postnatal life are more apt to become obese as adults. At least 50% of obese adults were obese babies; about 80% of obese juveniles will become obese adults. Parents themselves are closely linked with the prevalence of obesity in their children: if one or both parents are obese, the child is very likely to be obese.

We must be cautious in overextending the results of restricted feeding postnatally, since we do not know the level of caloric restriction that would be most effective in properly regulating adipocyte number without retarding growth or inducing any adverse malnutritional effects on other tissues, especially the brain. Adipocyte number is a critical parameter related to obesity: it appears that it is easier to reduce and to maintain body weight if the fat cell number is low than if a large number of adipose cells are present. For example, in formally treated obese patients, practically all will lose weight while hospitalized, but only 15% will maintain that loss. And only about 1% will reach an ideal weight. The individual

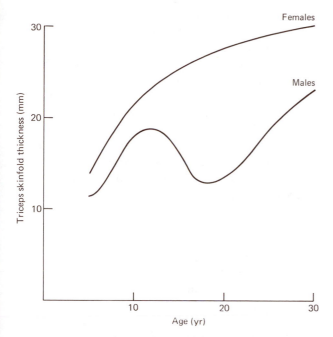

Figure 14-7. Lower limits of obesity in Caucasian Americans.

From Jean Mayer, *Overweight Causes, Cost and Control*, © 1968. Reprinted by permission of Prentice-Hall, Inc., Englewood Cliffs, N.J.

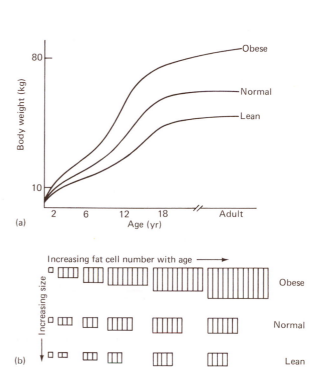

Figure 14-8. (a) Growth curve and (b) concomitant changes in fat-cell size and number with age.

From Jean Mayer, *Overweight Causes, Cost and Control*, © 1968. Reprinted by permission of Prentice-Hall, Inc., Englewood Cliffs, N.J.

with the larger number of fat cells appears to have greater difficulty in maintaining a constant weight than one with fewer fat cells.

Physical Activity and Weight Control

In spite of the usual, irresponsible, and naive citing of the amount of physical work that must be done in order to lose a given number of kilograms, exercise is a very useful tool in preventing weight gain or in inducing weight loss. Exercise with diet restrictions is more desirable than diet restrictions alone, particularly for slightly obese subjects who develop the con-

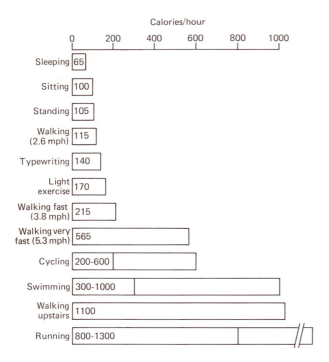

Figure 14-9. Energy expenditure.

dition during adulthood. However, exercise alone as a means of treatment for some obese subjects seems very unrewarding. An intelligent approach to losing weight does not involve the loss of several pounds a day, or even one pound a day, especially within a single exercise session. However, the small additional exercise-linked energy requirements each day total a significant weight loss in a matter of months, if we assume energy intake does not change. Figure 14-9 illustrates some typical caloric requirements of various physical activity levels.

The interrelationship of food intake to caloric output is extremely sensitive during hyperactivity, normal activity, and hypoactivity since appetite is altered by physical activity (Figure 14-10a). This graph illustrates that a sedentary individual's caloric intake is similar to that of an individual who participates in heavy work, while light work tends to minimize the caloric input. If we consider the combination of caloric intake and energy expenditure, the average body weight for the individuals in Figure 14-10a is given in Figure 14-10b. These results simply say that the complex mechanisms for controlling the balance of caloric intake and expenditure are desensitized at a relatively critical point of inactivity. Above this point, a direct relationship exists between caloric intake and output, resulting in a steady body weight. Similar results have been found repeatedly in animal experimentation. For example, rats that exercise about an hour per day eat less than sedentary rats; rats that exercise more than an hour per day eat more.

An additional problem of adiposity is its self-perpetuating properties. If one restricts caloric intake, there is a tendency to become less physically active; as a result, the restricted intake may be counteracted by a reduction in the normal daily caloric expenditure. This phenomenon is of particular significance in children when activity patterns are probably being formulated. Obese children have been shown to be less active than nonobese boys and girls. The relative inactivity of obese adolescents, while playing sports, has been verified by photographic analysis of playing patterns. Again we have a self-perpetuating condition whereby a slightly obese child becomes even less active as obesity develops—because to keep up with his peers the child has to expend a relatively greater amount of energy. Thus it would be only logical for this child to develop adverse attitudes towards physical activity during the early developmental years.

Perhaps another reason for the generally erroneous opinion that physical activity is an ineffective method of weight control is the usual overestimation of our actual energy expenditure. Studies have shown that women are usually less active than they see themselves to be. And the difference between how the male sees his energy output and his actual expenditure is even *greater* than in women. Neither men nor women like to view themselves as sedentary, we all like to think of ourselves as energetic and vivacious. Similar problems occur during pregnancy when there is a normal reduction of physical activity and consequently of caloric needs—in spite of the slightly greater demands of the developing fetus.

Exercise is probably more useful in the prevention of rather than the cure for obesity. An exercise program in addition to weight loss by food-intake modification go together well. In animals, loss of weight by food restriction causes loss of heart weight; but if food restriction is accompanied by an endurance exercise program, the heart is maintained at its predicted weight. A similar response is found in skeletal muscles: diet restriction results in a loss of muscle protein mass as well as fat tissue. An appropriate exercise program can counteract the protein catabolic effect while enhancing the fat and carbohydrate degradation effect.

Obesity and Cardiovascular Diseases

Obesity has been clearly associated with a relatively high incidence of heart disease. However, obesity itself makes no independent contribution to the risk of coronary heart disease: it is the factors that accompany obesity that contribute to the increased risk. About 60% of the extremely obese people tend to be hypertensive; when obese people lose weight, their blood pressure is reduced from its previously high level. This finding suggests that the more important relationship is between hypertension and heart disease rather than between obesity and heart disease. Also the relationship of the incidence of heart attacks, obesity, or

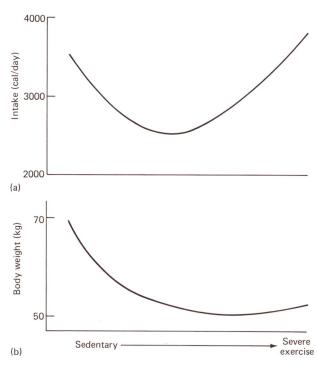

Figure 14-10. Interrelationship between daily exercise level, (a) normal caloric intake, and (b) average body weight for that level of exercise and caloric intake.

From Jean Mayer, *Overweight Causes, Cost and Control*, © 1968. Reprinted by permission of Prentice-Hall, Inc., Englewood Cliffs, N.J.

relative weight in Europe or the United States is not impressive if the factor of high cholesterol is eliminated.

In a sense the heart of an obese individual is "trained" because it is chronically overloaded. The degree of cardiac hypertrophy is directly related to the degree of obesity. The work overload is due to a number of cardiovascular adjustments to obesity. Although the obese subjects tend to be less active than nonobese subjects, it takes much less external movement in the obese to reach a given cardiovascular overload. The unusually high cardiovascular overload results in part from: (1) more tissues (fat) that need to be supplied with blood; (2) greater blood-flow demands to the skin for heat dissipation, basically because of the insulating effect of fat and the decreased sweat gland density; (3) the greater energy necessary to move a larger body mass; and (4) the added energy needed to support hyperventilation, characteristic of many obese subjects.

Blood Cholesterol and Lipids in Response to Exercise

Elevated blood cholesterol level has been designated as one culprit that contributes to the increased risk of cardiovascular diseases. Often it is not realized that cholesterol is an essential ingredient of every cell membrane as well as a chemical backbone for the synthesis of steroid hormones. Even if no cholesterol is ingested in the diet, the body will synthesize it from fatty acid precursors or metabolites. Cholesterol, like triglycerides, phospholipids, and lipoproteins, is transported in the plasma and is usually bound to some form of protein.

Dietary factors are probably the most important ones in relation to regulation of safe serum lipid concentrations. Saturated fats are twice as effective in elevating serum cholesterol as unsaturated fats. Concentrations of trigly-

ceride, cholesterol, and phospholipids can be reduced within a week by utilizing sunflower seed oil (45% of calories) and proteins (10% of calories). This cholesterol-lowering effect can be even greater if the carbohydrates ingested are glucose rather than fructose. The concentrations of fats, carbohydrates, and proteins are similar to those normally consumed. These findings demonstrate that the qualitative control of one's diet is not only extremely important in the prevention of *lipemia*, but also that lipid control is within the range of the normal diet. Serum cholesterol can also be lowered by reducing the amount of cholesterol consumed in the diet.

Lowering plasma cholesterol by dietary means decreases the combined incidence of myocardial infarction, cerebral infarction, and sudden death related to cardiovascular problems. Incidence of myocardial infarction or angina pectoris (chest pain) is better predicted by serum cholesterol level or by high diastolic blood pressure than by either systolic blood pressure or by body weight. Myocardial infarcts occur four times the normal rate and angina two times the normal rate when cholesterol levels are high (more than 290 mg/100 ml blood). A 15% reduction in cholesterol reduces the risk factor of coronary heart disease by 35%. A change of 15% in serum cholesterol —induced by dietary means—is about all that can be expected; however, this magnitude of change is certainly significant. Combining the effects of diet and exercise on the level of serum cholesterol should prove to be extremely profitable as a preventive approach to heart disease. By careful management of the diet, the incidence of coronary diseases can be reduced by 60% in some populations the first year.

The mechanisms involved in inducing *hypocholesteremia* are only partially known. Exercise elevates the oxidation of the side chain of cholesterol in rats, monkeys, and man.

Thyroid hormone is also involved in the control of serum cholesterol. In man other hormones, no doubt, are also involved. Each of these findings provides some evidence that may be helpful in eventually designing the optimal exercise, dietary and chemotherapy prevention, and rehabilitation of individuals with cardiovascular diseases.

Regulation of Cholesterol Metabolism

The single most important organ in the regulation of cholesterol metabolism is the liver. It is a source of cholesterol for other tissues and is involved in the conversion of cholesterol to bile acids. The liver contains about 3–5 grams of cholesterol; the blood has a total of approximately 10–12 grams. About 0.3 g/day is ingested in the normal daily diet, of which about 0.1 g is absorbed.

To maintain a homeostatic cholesterol level, the amount synthesized must be coordinated with the cholesterol and bile acids that are absorbed from the diet. Consequently efforts to reduce serum cholesterol are counteracted by the feedback loop, which enhances cholesterol synthesis when a reduction by diet or other means is induced. Theoretically a method to control serum cholesterol must involve simultaneous control of synthesis and degradation.

Summary

Carbohydrates and fats provide most of the fuels for energy production while proteins are utilized primarily to build new proteins within our bodies. Vitamins and minerals are essential for proper nutrition and are essential cofactors in many of the biochemical pathways.

Work time to exhaustion can be influenced by the type of diet. High-carbohydrate diets lead to increased glycogen levels and increased work times; high-fat diets result in shorter work times to exhaustion. By proper diet manipulation, it is possible to vary the amount of glycogen found in muscle; this technique may prove to be very useful in athletic competition.

The popular pregame meal is not likely to augment athletic performance since earlier preparation will have already established the levels of muscular energy sources. In addition the time span between the pregame meal and the athletic contest is probably too short to alter muscular metabolism.

Obesity usually results when a person's diet gets out of control. This condition also makes an individual a high-risk candidate for coronary heart disease. Nutrition early in life has been shown to be critical in the formation of new fat cells. In general we understand that the total number of fat cells can be controlled by proper nutrition during the preadult stages and that the size of the fat cells are more critical during the adult stages.

It is well recognized that physical activity can play a very significant role in weight control patterns. The effects of exercise are additive; even a mild amount of exercise, when carried out over a year, will result in a significant weight-controlling effect. The weight-controlling factor of exercise is of extreme importance due to the relationships between cholesteremia, obesity, and heart disease.

Study Questions

1. What are the purposes of each of the three basic groups of foods?
2. What are the important benefits of vitamins?
3. How can high-carbohydrate diets influence metabolism?
4. What effects do diets have on work time to exhaustion?
5. What special effects occur during exercise to alter the protein demand?
6. What is the effect of the "pregame meal"?
7. Show how caloric intake or output is additive?
8. How does exercise alter food intake?
9. How does meal frequency influence nutrition?
10. What are obesity and overweight?
11. What is the relationship between fat cell size and number and obesity?
12. How can exercise be used as a means of weight control?
13. How are obesity and hypertension related?
14. What is the effect of exercise on blood cholesterol and lipids?
15. Can exercise help control blood cholesterol and lipids?

Review References

Abrahams, A. The nutrition of athletics. *British Journal of Nutrition* 2:266–269, 1948.

Bergstrom L., et al. Diet, muscle glycogen and physical performance. *Acta Physiol. Scand.* 71:140, 1967.

Blix G., ed. *Nutrition and Physical Activity.* Uppsala, Sweden: Almquist and Wiksell, 1967.

Bogert, L.F. *Nutrition and Physical Fitness.* Philadelphia: W.B. Saunders, 1957.

Gollnick, P.D. and A.W. Taylor. Effect of exercise on hepatic cholesterol of rats fed diets high in saturated or unsaturated fats. *Int. Z. Angew. Physiol.* 27:144–153, 1969.

Keys, A., et al. Coronary heart disease: Overweight and obesity as risk factors. *Ann. Int. Med.* 77:15–27, 1972.

Mayer, J. *Overweight Causes, Cost and Control.* Englewood Cliffs, N.J.: Prentice-Hall, 1968.

Oscai, L.B.; P.A. Mole and J.O. Holloszy. Effects of exercise on cardiac weight and mitochondria in male and female rats. *Amer. J. Physiol.* 220:1944–1948, 1971.

Pauling, L., "Vitamin C and the Common Cold." San Francisco: W.H. Freeman, 1970.

Ribisl, P.M. When wrestlers shed pounds quickly. *The Physician and Sports Medicine* 2:30–35, 1974.

Key Concepts

· Training is the net summation of the adaptations induced by regular exercise.
· The specific training effects depend upon the individual, the type of exercise, and the previous level of training.
· An athletic event can be classified according to the specific demands of the event.
· There are specific guidelines that must be followed for optimal training adaptations.
· Training adaptation potential may be limited by genetic characteristics.
· Interval training techniques allow for maximum flexibility in planning training regimens.
· Specific training methods can be designed to develop strength, power, or endurance characteristics optimally.
· To obtain the optimal training adaptations, the training program must be planned.

Introduction

If we have succeeded in our attempts to make this book meaningful and understandable, you should be able to write this chapter. A chapter on the training of athletes—from a *physiological* viewpoint—must use the information contained within the first four units of this book.

The goal of an athlete is to produce the best possible performance; therefore, the athlete must mobilize the body to optimal limits. These limits are determined by the physical state of the body prior to and during the actual performance, that is, the preparation for and the execution of the activity. *Preparation*, another term for "training," is the subject of this chapter. *Execution* is another term for the exercising condition.

To consider a person "trained," we must first know the type of exercises or events in which that person is attempting to perform. A person trained for one event may or may not be

CHAPTER FIFTEEN

Physical training for athletes

trained for another event. Depending upon the type of activity, it is possible to train the *muscular system* for increased utilization of oxygen, for greater capacity to utilize stored glycogen, for increased capacity for the generation of contractile force—or for any other criteria that are specific to the performance of the event. The *nervous system* can be trained to increase its ability to stimulate the primary movers of a specific joint action and to decrease stimulation of nonefficient muscle contractions. The *cardiovascular system* can be adapted to increase its oxygen transport capability and to increase its ability to deliver oxygen to specific muscle fibers. The *ventilatory system* can be adapted to increase its ability to extract oxygen from the atmosphere and to facilitate gas exchange. The *endocrine system* can be trained to increase its ability to communicate the stress response and to increase substrate mobilization.

Training Theory

Our training theory is based on the concept that the training adaptations are specific to the training methods. We know that the training adaptations are a summation of the daily exercise bouts; thus it should not be surprising that our training theory is very similar to the "specificity of exercise" concept (see Chapter 1). We shall refer to this training theory as *specificity of training*.

The basis of our training theory is the assumption that it is only through regular exercise sessions that training adaptations can occur. For example daily strength activities (exercises) contribute to what we would call the *strength training* of an individual. Without the daily exercise sessions, there would be no training effect. If an individual exercises today, exercises tomorrow, the next day, etc., at some future time he or she will become a trained person. Although some athletes may

look for a more expedient way to become trained, there is no known short cut. Based upon our definition of the word "trained," a short-cut method is not possible.

Our performance on any given day is related to our state of training; our state of training is related to the number and type of recent exercise sessions. Thus the daily exercise sessions contribute to our state of training, and our state of training determines the quality of performance that we are capable of on any given day. This argument is identical for any strength, power, or endurance activity.

The "specificity of training" concept suggests that an individual who trains for only one specific event will be superior—in that event—to another individual, who trains simultaneously for a series of events. This prediction is especially true at the world-record class performances but becomes less true as the quality of the performances decreases. At the lower levels (below Olympic class) of performance, there are usually outstanding athletes who are superior to their competitors regardless of the competitors' level of training. This superiority points to the importance of genetic characteristics, which we will discuss later.

All exercises can be put on a time scale according to the total time of the event. If we assume that the goal of the athlete is to perform at a near maximum rate for the entire length of the activity, we then can predict the intracellular energy sources necessary for that activity. By identifying the specific energy source for an event, it should be possible to train that pathway optimally so as to provide the maximum energy for the performance.

Our theory would predict that there are at least six time classifications at the world-class level of competition, for which an individual cannot hope to excel in events encompassing three of the time intervals. Those time classifications are 0 to 1 second (strength), 1 to 10 seconds (high power), 10 to 30 seconds (power), 30 seconds to 2 minutes (power-endurance events), 2 minutes to 5 minutes (endurance), and 5 minutes and longer (high endurance) (Figure 15–1).

We recognize that application of the term *power* to events of up to two minutes duration violates the absolute definition of power. However, the concept of a powerful man is so prevalent that we have chosen only to modify that concept rather than to attempt to change it. We also should realize that, as soon as a movement is repeated, we are involved with an endurance component. The term *endurance* also has a popular connotation of representing something more than the mere repetition of a muscular contraction. The usual interpretation of the term *endurance* is applied to the longer lasting events. The above discussion forms the rationale for our use of the classifications of *strength, high power, power, power-endurance, endurance,* and *high endurance* to represent the types of events in athletic competition.

The above time classifications cannot be documentated in their entirety at the present time. The rationale is based upon the ability of one person to train optimally only one energy supply system. As soon as a person attempts to train more than one of the energy systems, he/she must divide his/her training time. Another person, concentrating on the training of only one system, will have a greater amount of time to train the one system—and thus is more likely to achieve a more successful training result. Therefore at a world class performance, where genetic differences are relatively minimal, the specialist in each event will normally be the victor.

A thorough classification of athletic events would be much more complicated than the time classification we have presented because of *skill factors, motivation, event complexity, team competition* and so forth—and the corresponding interrelationships of all these factors.

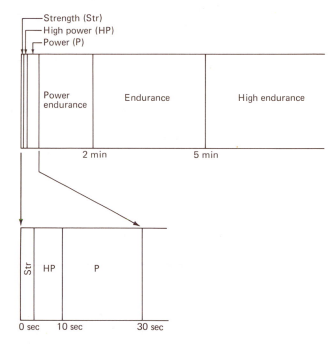

Figure 15-1. Time classification of athletic events.

The specificity of the event will also be a factor; that is, even within the same time classification, it is highly unlikely that an athlete will win, in world class competition, in an activity that requires primarily arm activity and also win in an activity that requires primarily leg activity.

Guidelines for Training

Simply put, the aim of training for athletic competition is to bring about those adaptations that allow the body to improve its ability to perform a specific task. Athletic events that allow for the measurement of individual effort (e.g., track and swimming) are the easiest for which to design training schedules. Team sports must utilize not only individual physical training methods but also team training methods. Team sports are also affected by strong psychological and sociological influences that must be recognized. Since this is a physiology book, we have concentrated on the physiological considerations of physical activity. We hope, however, that the reader is also aware of these other considerations and will be able to construct the appropriate trade-offs.

The following guidelines illustrate some of the physiological considerations that contribute to training-induced improved performance.

Overload is required for improved performance. This guideline implies that to achieve a training effect it is necessary to exercise at an intensity level greater than one's already existing capacity. A person cannot expect to improve his or her performance capability without daily exercise (overload) sessions. Furthermore a person cannot expect to improve if the exercise is at a level less than that at which he/she hopes to perform.

Rate of improvement is directly related to the intensity of the exercise sessions. The adaptation mechanisms within the body act in proportion to the stress rate. If the daily exercise sessions are very intense and at a high level in relation to one's training level, the body will adapt at a greater rate than if the daily exercise sessions are at a lower intensity than one's training level. However, there is a maximal tolerance level beyond which the body cannot sustain itself—and which leads to a state of exhaustion.

The effects of training are specific to the overload used. This principle is essentially "specificity of exercise" or "specificity of training." The adaptations a person can expect can be predicted from the daily exercise sessions. Not only is the amount of improvement related to the level of exercise but also the adaptations within the specific muscles and systems are dependent upon the nature of the daily exercise sessions.

Repetition is essential for muscular efficiency. We can find a partial justification for this guideline in the higher centers of the nervous system. Increased muscular coordination results in a more efficient exercise performance. As a muscle is continuously exercised, it begins to adapt to the exercise stress and eventually increases its efficiency. For example, as an activity is repeated, it becomes more automatic, which may be the result of a more automatic response from the nervous system (e.g., resistance of the antagonists is decreased).

Repetition also serves as the stimulus to bring about the adaptations characteristic of the trained state.

Response to training is individual and unique. Individuals respond in different ways to the same stimuli; in fact the same individual may respond differently to the same stimuli on different days. For one individual the exercise session may be an optimal session for improvement in that specific event; for another individual there may have been a different workout that would have elicited the optimal adaptation. "Treating everyone the same" does not require that everyone do the identical workout; this is a great fallacy that we see with many coaches. Under the guise of "treating everyone equal," the coach may insist that every individual on the team perform the identical workout at the identical rate. However, it is unlikely that each individual will develop to his optimal potential by doing group workouts.

Motivation is essential for effective training. Motivation is one of the key concepts for learning, whether it be increased physical performance or increased academic performance. Without the proper motivation, the individual will not proceed with the vigor, incentive, dedication, pride, and so forth, necessary to carry out an optimum training schedule. Athletic training, in the physical sense, is one of the most demanding of all disciplines, if the expected level of achievement is excellence.

Training for Specific Time Classifications

The basis for our time classification of exercise is the specific energy sources and the nature of the contractile activities. Therefore we maintain that in establishing training methods a person must emphasize those training techniques that optimally adapt the specific factors involved in the activity. *The exercise task must be specific to the training goal.*

We know that practically all activities have endurance, strength, and speed components. However, we do not have available, as of yet, information relating to the absolute contributions of each of these components to any given exercise. Most training programs recognize the contribution of these individual components and allow for "over distance" or "under distance" and quality and quantity during each practice session.

Interval Training

There appears to be no magic formula by which all athletes improve and prosper. Each activity is different, and each individual is unique. Knowledge of each athlete's goals, background, and physical abilities plus the correct blend of "under" and "over" distance within the framework of the "specificity of training" concept, will allow for optimal development.

Any training program may be defined as *interval training* if the exercise sessions are considered the *work intervals* and if the *daily rest intervals* (approximately 22 hours) are considered time between workouts. However, the further organization of the daily exercise sessions into interval-training workouts is the normal procedure in our exercising technique. The *interval-training workout* (or exercise session) can be expressed by specifying the *work distance* (or load), the *work speed* (or rate), the *rest periods between the working periods* (work-rest ratio), and the *total length* of the workout. A typical workout might be three sets of six 100-m swims at 60 seconds with a work-rest ratio of 2:1 (30 seconds of complete rest)—and 20 minutes between sets. In this example the athlete would perform each 100-m sprint at 60 seconds, with a rest interval of 30 seconds between each 100. At the end of six sprints the individual would rest for 20 minutes before proceeding to the second set.

The number of sets and the number of repeats within each set control the total amount of work the athlete will do. Multiple sets present an opportunity to vary the work rate and/or work-rest ratio between sets. Varying the number of sets and the rest interval between sets allows the athlete to repeat the sets at varying degrees of total recovery. The number of work intervals (*repeats*) and the distance of each work interval provide flexibility to allow for individualized workouts: either four repeats of 100 m or eight repeats of 50 m total 400 m, which may be the actual race distance. For motivational purposes it is possible occasionally to total the repeat times to get the 400 m time. This "broken" 400 can be compared to the best race time of that individual.

The number of work intervals and the distance of the work interval do not provide complete information about the workout; we also need to know the speed and work-rest ratio of the repeats. Four repeats of 100 m at 60 seconds (with a work-rest ratio of 1:1) is less strenuous than four repeats of 100 m at 60 seconds (with a work-rest ratio of 4:1). The former workout allows for 60 seconds of rest between 100s while in the latter workout the rest interval is 15 seconds.

An alternative method to assess the length of the rest interval is to rely upon the athlete's heart rate returning to some predetermined level before setting off on the next work interval. After a work interval the heart rate may be expected to be 160–180, or above, which is approaching maximum. The succeeding work interval should not be initiated until the heart rate ranges from 140–160. When the athlete achieves this lowered heart rate, it is used as the signal to begin the next interval. However, the correct heart rate for each individual must be determined by experience, considering the goals of the athlete, coach, and team. It should be remembered that the average heart rate that can be maintained during exercise is about 180—higher for events of less than two minutes in duration. The use of the 140–160 mark for the beginning of the next work interval appears to be a reasonable estimate as to when the body is capable of another repeat. The degree of "physical fitness" will influence the amount of time required for the heart rate to return to a level of 140–160.

In general, a work-rest ratio of 4:1 indicates an endurance workout, and a work-rest ratio of 1:4 indicates a power workout. The 4:1 ratio provides very little rest time; thus the speed at which the work interval is performed should be less than the work speed when the work-rest ratio is 1:4. Sprinters usually need more of 1:4 type of work; distance and middle-distance athletes need more of the 4:1 workouts.

It is possible to vary the type of activity during the rest periods—although complete rest appears to allow for the quickest and most complete recovery. Light activity during the rest intervals can be used to vary the workout.

The distance of the work interval (repeats) can be varied according to the needs of the athlete, but very seldom should the distance exceed one-half of the race distance. Once the longer distances are reached it is hard to maintain a *speed overload*; that is, at the longer distances the speed of the repeat will be nearer to, the same as, or slower than the race speed. The value of the workout becomes questionable when the athlete is performing the workout at a slower speed than that speed at which he or she wants to race. Obviously speed is not the only overload criteria; in an interval workout, the total work distance is also overloaded.

It appears that it would be more desirable to refrain from repeat distances greater than one-half of the race distance. Our recommendation would be to spend most of the workout periods at repeat intervals of one-fourth, or less, of the race distance. The overload is provided by an increased *speed*, increased *distance* (by an increase in the number of repeats), and increased *load* (by an increase in the work-rest ratio and in the number of sets).

Power, Strength, and Endurance Training

Motor ability tests most often identify strength and power criteria as important for high levels of performance and for a wide range of physical activities.

Strength can be defined and/or measured as static strength or dynamic strength. Static strength and isometric strength are often used interchangeably. Technically definitions of isometric strength allow no change in the muscle length, while static-strength measurements allow some movement, such as the compression of a spring in the hand dynometer.

Dynamic strength is that maximum load that can be moved throughout the total range of motion. In strength activities the contractile force is applied to a resistance and maintained for a relatively short period of time.

Power can be defined as the performance of work per unit of time. Strictly speaking, it should be determined by the speed at which a maximum load can be moved. Decreasing the load and increasing the speed of movement may increase the effective power. For our discussion of physical activity, we have found it convenient to discuss power performances in terms of high power, power, or power-endurance activities. Therefore we have extended the actual definition of power to include the relative maximal load that can be moved at a maximal speed for either a low, medium, or high number of repetitions.

Endurance can be defined as the ability to repeat a given movement. We can evaluate endurance performance by measuring the ability to repeat the movement for a *longer time* than previously was possible or by measuring capacity to repeat the movement for a set time period with an *increased load.* The latter situation is the most critical for success in athletic competition of endurance events because of the nature of the definition of competition; that is, the person who completes a predetermined race distance in the fastest possible time is declared the winner. In athletic competition the magnitude of the work load increases as the time of the events decreases and as the competitor approaches "better" times.

Strength

A physiologist would define the strength of a muscle as the maximum force a muscle can exert along its longitudinal axis. An athlete is more interested in defining strength as that force that a muscle can exert on an external object. A practical aspect of strength includes a consideration of the lever system on which the muscle must operate. To the physiologist, the strength of the muscle is directly related to the cross-sectional area of the muscle. The maximal force that the muscle can exert is approximately 3–4 kg per cm^2 of cross-sectional area. *In vivo* the applied force will vary according to the mechanical advantage of the muscle, which is determined by the lever system across the joint.

Another consideration, when assessing muscle strength, is the initial length of the muscle (compared to its resting length). For practical considerations, this factor can be expressed as a function of the joint angle; that is, the length-tension relationship can be expressed as the tension developed at a specific joint angle. If a muscle is in a position where it has a poor angle of pull, the minimum force will be exerted across the joint. Usually a poor angle of pull coincides with a stretched or a nearly totally contracted muscle. The optimal length for force application is where there is the optimal combination of the number of cross bridge attachments and the joint angle for maximum mechanical advantage.

The results of animal studies have demonstrated that increased strength and increased cross-sectional area are related to increases in myofibrillar protein (actin and myosin). Muscle hypertrophy (increased size) results in increased numbers of myofibrils per fiber and increased contractile filaments per myofibril. Without a doubt, the increased strength that accompanies muscle hypertrophy is related to increased contractile proteins. (See also Figure 13–14.)

Two other ways to increase strength are to increase the *mobilization* of motor units activated per contraction and/or to increase the *frequency* of impulses to the activated motor units. These adaptations may be "learned" responses and could account for the rapid increase in strength during the initial practice sessions. Mobilization or disinhibition of motor units may account for the untold number of instances of acts of superhuman strength in times of emergencies.

Isometric and isotonic techniques have been used to increase strength of a muscle group. The increased strength resulting from isometric techniques is limited to the *specific joint angle* that is used during the isometric exercise. The strength increases diminish as the joint angle deviates from the specifically trained angle. The "specificity of exercise" concept can be applied to this level of response since the *involved* motor units are specific to the joint angle or stretch of the muscle. In a similar way isotonic techniques are limited to the range of the specific movement through which the training takes place.

Increases in strength may occur at a rate of improvement of 5–12% per week. The percent of increase depends upon the relative strength of the muscle at the beginning of the training period. The muscle that is close to its genetic limit of strength does not have the same potential for improvement as the muscle that is far from its genetic capacity. At the present time we do not have the methods to assess genetic capacity, but we do know that certain body types are more likely to have higher strength capacities than others.

Strength-training methods have been found to produce the best results if the number of repetitions during any one set is limited to less than ten. The total number of sets should be three or less; and the resistance should be that weight that a person can lift no more than 10 times (10 repetitions maximum). On certain days it may be an advantage to work with five repetitions maximum and to do three sets of five. The gains in strength will be related to the relative strength at the beginning of the exercise and will be specific to the range of motion of the exercising condition.

Intuitively it appears that high strength is very critical for high performance levels in most sports. However, it is now known that many events do not require a high level of strength during the actual activity. For example swimming has been estimated to require from 20–50% of maximum strength during the 100-meter sprint, one of the swimming events that requires the most power. Although the required strength, as a percentage of maximum, is not extremely high, the ability to maintain this output during a stroke cadence of 60 contractions per minute is very critical.

The ability to maintain a work rate is greater if the repetitions demand a lower percent of the maximum strength. Although high levels of absolute strength are not necessary for the actual activity, if the absolute strength can be increased, the functional strength requirement of the activity will decrease (as a percentage of the absolute maximum strength). This concept of relative strength is depicted in Figure 15–2.

Power

Our functional definition of power suggests that a person can apply a great force to an object at a great rate of speed. The number of times he or she must do this is related to the type of power event. High power events that include high strength components, such as the shot put, require a minimum endurance component (if it exists). Power-endurance events,

such as the 200-meter dash, require a greater amount of endurance since more muscular contractions are required.

It is possible to train for power events by means of a weight-lifting program in which the weights are of a magnitude that can be lifted a total of 50 to 100 times at a predetermined speed; the selected speed should be at least as fast as the race pace. This type of training is very exhausting and probably cannot be maintained for any prolonged periods. As a general rule a power-training method that simulates the actual activity will result in the most beneficial results. Power training has become increasingly popular since the development of isokinetic apparatuses capable of being set at very high speeds.

Endurance

Endurance events require a lesser percentage of an individual's absolute strength capacity than power events, but the movement must be repeated more times. As the endurance component of an activity increases, the power and strength components decrease and the nature of the motor-unit activity changes. Instead of recruiting additional motor units and/or increasing the firing frequency for increased contractile force, the motor units are patterned to allow for sequential firing. Sequential firing (also called asynchronous firing) produces greater rest intervals between the repeated firings of any specific motor unit.

Animal and human studies have shown that, in response to endurance-type training, the muscle responds with increased levels of activity of enzymes involved with energy production, especially those enzyme systems within the mitochondrion.

Studies from the 1940s measured oxygen uptake as high as 80 ml/min/kg, which is the same as that of champion athletes of the present day. Since that time, the world record for the mile run has improved from 4:07 to 3:49.4. Apparently the modern endurance-training methods are not training the *maximal aerobic capacity* of athletes any better than did the training programs of the 1940s. Thus the endurance-training effect must be something other than an increased capability for oxygen uptake.

High endurance training is best achieved by working with interval-training regimens that have a high work-rest ratio. The interval distance should be less than one-half of the race distance; the speed of the interval should be as fast as can be maintained with the high work-rest ratio. For endurance athletes the total workout may be divided into two or three practice sessions per day. Endurance or power-endurance athletes should not cover the same total distance as the high endurance athletes, as the former athletes must perform more speed work (higher quality and less quantity). very high speeds.

Training for a Season

Individual Goals

The establishment of a personal achievement *goal* is a very complicated process and requires information input from many different sources. The individual must be able to assess his or her *abilities*, *motivation*, and specific *willingness* (dedication) to train in a manner compatible with these goals. A total training program that is compatible with the achievement of minimum goals may not be compatible with the development of those training adaptations necessary to obtain the maximum improvement in performance.

The first item that must be considered when establishing goals is an assessment of your performance potential. If you decide that your goal is within your estimated potential, you

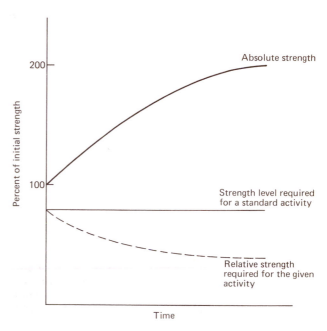

Figure 15-2. Effect of a strength gain in the performance of a standard activity.

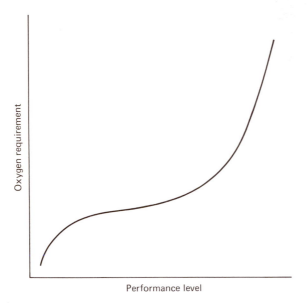

Figure 15-3. Schematic representation of the relationship between level of performance and the oxygen requirement for that performance.

then must determine what physiological adaptations are necessary to achieve that goal. You also must make an assessment of the differences between your present capabilities and those capabilities that are required for the desired performance. Finally you must determine the time course necessary to allow for the specified degree of improvement and consider what type of exercise sessions are necessary to stimulate the desired adaptations.

For example Figure 15–3 gives an idea of the relationship between oxygen uptake and endurance performance levels. As performance level increases, the oxygen requirement increases. At the higher performance levels, a slight increase in performance requires a great increase in oxygen consumption. World records have continued to fall—even though present-day champions do not exceed their 1940 counterparts in relation to maximal oxygen consumption. Thus we may rightly assume that there must be other considerations for championship performance. However, the gross relationship between oxygen uptake and performance does provide us with a means of identifying adaptations that must be achieved to reach new performance goals.

Time Course of the Training Program

You will remember from our guidelines for training that improvement is related to the degree of stress imposed; the greater the stress, the faster the rate of improvement. We also know that some evidence shows that the rate of improvement is inversely related to the eventual magnitude of the improvement. Furthermore the rate of improvement is inversely related to the length of time that a person can hold the maximum peak conditioning level. These three points are schematically illustrated in Figure 15–4.

The physiological evidence to document the above relationships is not as yet complete. However, we can speculate that these time and work-rate dependent adaptations are related to the stress concept. A high rate of improvement taxes the body to such an extent that the increased physiological load cannot be maintained because of failure of some critical system. It is possible that the adrenals become "exhausted" at the higher stress rates. Slower rates of improvement do not require the constant daily maximum stress, and thus the body can control the stress within its adaptative capability. The improvement rate associated with lower daily stress rates is maintained for an increased length of time, and the absolute potential is increased. This concept is very important in planning a training season. If the calculated goals are great and the required amount of adaptation is great, then the workouts must be initiated sufficiently in advance of the competitive season.

The rate of improvement is quite critical to

the competition schedule. If a critical competition day falls early in the season, there will be great pressure on an individual to attempt to improve too rapidly to be in keeping with the overall goals. This situation is very common and must be resolved prior to the season; it is another reason why the athlete must identify the goals for the season in relation to his or her training potential. Intermediate goals should be established to monitor the success of the training program as the athletic season progresses.

Peaking Techniques

Peaking is a term that describes the special preparation of an athlete to achieve a highly trained state for a maximum performance. Obviously if very tired, he or she is less likely to achieve a peak performance. Similarly, if his or her nutrition is out of balance or if under unnatural stress, performance will not be at a peak level.

The peaking technique that is adopted may be nearly as important as the long-term training used throughout the entire season—especially if the performance is under highly competitive conditions. The peaking schedule is closely related to the achieved degree of ''competitive fitness,'' and is unique to the individual. In general the better condition the athlete is in, the more time that can be devoted to the peaking process.

Peaking requires some rest from practice—or at least an easing up on the daily stress of the practice sessions. Each individual must experiment with the length of time required for the peaking process as well as with the rate of work load reduction. In general the peaking concept involves reducing the work load over a period of time and thus letting the body build up to peak strength, power, or endurance potential. The absolute peak should correspond to the actual day of competition.

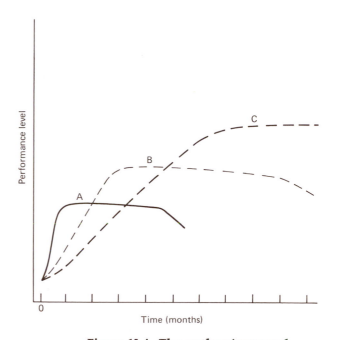

Figure 15-4. The authors' *proposed* relationship between the rate of improvement and the absolute level of improvement (during one competitive season). Workout schedule A is very intense and the corresponding rate of improvement is very rapid as compared to workout schedule C, which is less intense. Workout schedule B falls between them. The rate of improvement is inversely related to the final level of performance and to the length of time an athlete can maintain maximum performance.

We have previously discussed the relationship of glycogen content of muscles to work time, and we have discussed the ability of the muscle to "supercompensate" its glycogen stores in response to an exhaustive exercise. We can use this information in planning peaking techniques. The performance of an exhaustive exercise five days before the day of the major competition, followed by two days of a low carbohydrate and high fat-protein diet and three days of a high carbohydrate diet, is known to increase muscle glycogen levels to values much above normal resting levels. This higher glycogen level should be an aid in performance efforts which require about 70% of maximal $\dot{V}_{O_2}$.

Trade-offs

The discussion thus far has been related mainly to the individual, whose performance can be individually monitored. Training for team competition, where each member contributes to the team success in a cooperative effort, is more difficult. The main reason is that peak physical performances in team competition are not the only criteria for success.

The athlete or coach involved with team sports cannot spend the total practice session working on the "competitive conditioning" of the individual team members; a certain amount of time must be allotted to actual team practice. The innovative coach can identify specific team drills that emphasize team conditioning as well as individual conditioning. Furthermore, if it is possible to assess each team member's play responsibility, it should be possible to identify the varying physical needs of each position; given his/her particular responsibility, the individual can train specifically and also can stay within the time constraints of the team practice.

In team competition morale is a concern; in fact team morale may be more critical than the competitive conditioning of individual members. Therefore the athlete or coach has to make a *trade-off*: a decision must be made as to what are the most critical needs—in view of the goals—of the team. In this case, the individual's goals are probably secondary to the team goals. For example situations will arise when it is deemed more important to drill for teamwork than to drill for individual physical fitness.

Other Considerations for Peak Performance

Warming Up

Warming up is a technique to prepare the body for exercise at a competitive rate. Warming up is associated with increasing muscle temperature, activating energy sources within the muscle, activating hormonal resources, alerting the nervous system, and/or increasing core body temperature. All of these as well as other factors have been shown to be affected by sufficiently intense warm-ups. In fact *passive methods*, such as hot showers or massages, have been shown to affect one or more of these factors favorably, with a resulting increase in performance. The research done so far in this area does not give conclusive proof that warming up is a necessity, but we feel that the available evidence supports the need for warm-ups.

Figure 15–5 shows the relationship between muscle temperature and performance in a power event. We can see that the better performances are associated with the higher muscle temperatures.

Overloading a muscle group prior to a power activity increases the performance. The probable explanation for this is an increased level of excitation of motor units required to handle the increased load. The mobilization of the additional motor units, it is hypothesized, is carried over to the actual performance.

The value of the warm-up is less obvious in high-endurance activities. As a muscle begins to work, it takes a given amount of time to mobilize the oxidative energy sources. If this mobilization time takes place during warm-up, the oxidative energy sources are operating even during the initial part of the event.

Pacing

Most races or competitions are such that it is impossible to perform at the maximum rate of energy expenditure throughout the entire race. Therefore *pacing* is employed as a technique to distribute more evenly the exercise stress over the total race time. An even pace throughout the entire race results in the most economical energy expenditure; with this race plan the athlete sets a pace that he or she can maintain throughout the entire race. This method allows the body to arrive at a "steady-state" energy expenditure with rhythmic muscular contractions.

The even pace may satisfy the physiologist, but there are competitive advantages in a "non-even" race plan. Making several "moves" during a race may worry and upset the opponent; it may help the athlete gain confidence and perhaps bring about the unexpected bit of energy that everyone seems to acquire at one time or another.

Especially in endurance performances, it is important not to expend an excessive amount of effort early in the performance. A rate of energy expenditure equal to that of the power-endurance classification will severely tax the ability of the muscles to provide energy. The increased substrate demands early in the race will not be replenished, and the extra energy substrates will not be available for the final push to the finish. If a significant portion of

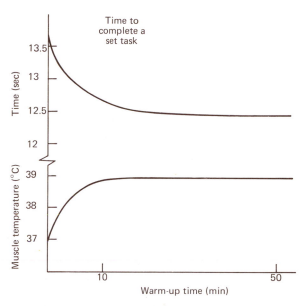

Figure 15-5. Muscle temperature and time to complete a task as a function of warm-up time.

Adapted from E. Asmussen and O. Boje. Body temperature and capacity for work. *Acta Physiol. Scand.* 10:1–22, 1945.

the energy stores are used early in the race, they not only become depleted, but their by-products remain in the muscle and/or blood stream. And these have been postulated to contribute to fatigue.

Quick changes in race pace or speed can initiate the use of reserve energy supplies. If it is necessary to increase the pace, the change should be as gradual as possible to allow steady-state mechanisms to adjust to the higher rate of energy expenditure. However, the most experienced athlete will know his or her capability and be able to plan and execute a race plan that calls for a rate of energy expenditure equal to the maximum rate of aerobic supply; that is, the rate of oxygen and substrate delivery to the active muscles will match the rate of oxygen and substrate utilization by the muscles. This pacing method allows for a complete oxidative energy supply throughout the race, until the final finishing spurt. The length of the finishing "kick" must be timed so that the finish line is reached at the precise instance that the nonoxidative energy substrates are in short supply.

Muscle Soreness

Everyone has experienced muscle soreness that arises in response to extra work performance of muscles—beyond that for which they have been trained. This soreness is thought to arise from the incomplete removal of muscle metabolites. Either the production was so great or blood flow was so reduced (or a combination of both) that the result was an accumulation of substances that were toxic to the muscle and the nerve endings. Excessive fluid accumulation might account for a swelling of the exercised muscle to such an extent that the extra fluid pressure would sensitize the nerve endings. The increased pressure would severely hamper blood flow and thus prolong the pain or soreness.

Fatigue

Fatigue exists in many forms. General body fatigue is the body's inability to meet the stress of the forthcoming event. Local muscle fatigue is the inability of the muscle group to contract. There are also degrees of fatigue.

Athletes are most accustomed, in general, to local muscle fatigue where one muscle or group of muscles is unable to work at the rate of energy expenditure that is necessary for the task.

The possible sites of fatigue are numerous and range in scope from the central mechanisms controlling movement to the actual mechanical generator and energy support mechanisms. It appears reasonable to propose that the actual site of fatigue is specific to the individual and that the site of fatigue is related to the specific type of exercise and the state of training of the individual.

For its practical application to athletics, it would appear to be beneficial to locate the proposed site of fatigue during competition and then to attempt to train that specific site. The location of the general fatigue site is not difficult: it is common for the athlete to be aware of when and how he or she begins to lose critical aspects of performance. After identification of these weaknesses and after the proper training, these specific limitations to performance should be diminished.

The above technique should help the athlete avoid staleness and excessive plateauing. In summary, our recommendation is to evaluate every performance in terms of the weakest component of the performance and to devote the succeeding practice sessions to training that proposed limiting factor. This process is then repeated during each performance of the task. With this technique there is a constant evaluation of performance in light of possible specific ways to improve the performance. This technique avoids the "just more of the same" workouts. Nevertheless there will be instances when "more of the same" is the correct solution to the problem.

Genetic Selection

Genetic Traits

The physiological principles discussed in this book are common to all humans. However, there are genetic factors that modify the response of the energy systems and the energy-support systems to exercise and training. The most obvious differences between humans are body build and sex: these are only two of the factors that may influence the response of the body to exercise and training. Other factors, such as intelligence, race, and anthropometric data, may or may not play an important part. The following discussion represents a hypothesis concerning the role of genetics in athletic selection.

Everyone is born with characteristic genetic traits, the overwhelming majority of which are not obvious to the eye. We can assume that there is an identifiable genetic coding for every trait that can be described. Furthermore we can assume that each of these traits can be trained; that is, if we can identify it, we can devise a system to adapt (train) it to an imposed stress. Most investigations have shown that there is a limit to the ability of the body to adapt any specific trait. For our purposes we will assume that the trainability of any specific trait is approximately 30%. Thus if a person does not use or train the specific trait, he or she may lose 30% of his or her capability, whereas if he or she trains this talent, he or she may increase by 30% his or her "normal" potential.

When comparing individuals in relation to a specific genetic trait, we will assume that there is a "normal statistical distribution" of this trait. Some people have a very high "natural" ability in this specific trait whereas other people will have a relatively low "natural" ability. Depending on his or her degree of training or detraining, an individual may move approximately 30% up or down the "normal distribution" scale. Thus with training, a person with a lower natural ability may be able to "outperform" a person with a higher natural ability who has not trained (and has perhaps even detrained up to 30% of his or her original ability). However, it is our opinion that there are circumstances in which the person with the lower specific genetic talent may improve 30% in the genetic trait but not be able to outperform the individual endowed with a sufficiently higher natural talent in that trait.

We hypothesize that athletic ability is the summation of many such genetic traits. And it is reasonable to speculate that athletic ability may be determined by either a few "key" genetic traits, or it may be determined by an "average" level of several traits. This "genetic talent" hypothesis must stand the test of scientific investigation. However, it gives us a basis to sympathize with and understand the athlete who, no matter how hard he or she works, cannot defeat a given opponent. This situation remains one of the most difficult situations for a coach to explain in terms of justification of hard work and motivation.

Somatotype

Fortunately there are some genetic traits that are plainly evident, and among them is the body type of each individual. Almost every individual can find a sport or activity in which his or her particular body type is beneficial—or at least is not a hindrance—to the ability to perform that activity.

Included in a discussion of body types are such characteristics as various body dimensions, percent of body fat, bone length, thigh to foreleg ratio, and lean body weight. It should be obvious that these anthropometric characteristics will influence a person's ability to perform at a high level in certain activities. However, in most cases, these factors are not critical for the same person's enjoyment of the activity; it is our opinion that body-type variations should not be a factor in a person avoiding participation in a given activity. We do suggest, however, that body-type assessments be made to establish a level of expectation, within a given probability of success, which will form the basis for the individual's mental approach to the activity.

Racial Considerations

To our knowledge the physiological concepts discussed in this book pertain to all races. The black has been the subject of much research over the past decade; very little evidence for any organic differences between whites and blacks have been described.

Anthropometric data are available to suggest differences in limb-to-trunk ratios as well as in absolute limb lengths. These differences may be attributed to *selective* genetic development brought about by the ancestral origin of blacks and whites. The whites generally came from the cooler northern climates and the blacks from the warmer equatorial regions. Shorter limbs, which appear to be a characteristic of whites, allow for less surface-to-volume ratio and thus aid in maintaining body temperature during exposure to cold.

It is our opinion that the physiological concepts in this book apply equally to whites and blacks. The difference within groups far exceeds any difference in athletic ability between races.

Summary

Relatively uncomplicated competitive events (such as running and swimming) require maximum efforts in relation to the total distance (or time) of the competition. In other words, energy is expended such that at the end of the race, regardless of the distance, the specific limitation to the performance should have been expressed. We can define at least six time classifications of athletic events based upon energy production and energy-support systems: (1) strength, (2) high power, (3) power, (4) power-endurance, (5) endurance, and (6) high endurance. We suggest that, as the competition approaches world-class levels of performance, it becomes increasingly difficult to achieve success in more than one time classification. This prediction is based upon the *specificity of training* concept and upon what we have concluded in our guidelines for training.

At the higher levels of competition, genetic differences are minimal; specialization allows competitors to train one energy system or energy-support system optimally. Competing in dissimilar events—that cross time intervals—necessitates the training of more than one energy system. Attempting to train more than one energy system results in a loss of the optimal training capability of any specific system.

Active planning of the training pattern, peaking technique, pacing pattern, and warm-up procedure is necessary for success in athletic events. These factors can be manipulated and "trade-offs" made to achieve both individual and team goals.

Study Questions

1. How would we define a "trained" person?
2. Describe our training theory.
3. What are the time classifications for exercise?
4. Why is overload required for a training program?
5. What is the basis for a "specificity of training" model?
6. How can interval training be manipulated to satisfy a number of individual training requirements?

7. What are the strength adaptations that take place during training?

8. What are some of the considerations that should affect the designing of a power-training program?

9. What are the endurance-training adaptations that take place in muscle?

10. What considerations are important in the planning of a training program for a whole season?

11. How does an athlete plan a "peaking" program?

12. What benefits are achieved by a warm-up?

13. How important is pacing?

14. What are some possible sites of muscle soreness?

15. How do genetic factors influence athletic ability?

16. Are body types important considerations for athletic participation?

17. What racial considerations are important for athletic ability?

Review References

de Garay, A.L.; L. Levine; and J.E.L. Carter, eds. *Genetic and Anthropological Studies of Olympic Athletes.* New York: Academic Press, 1974.

Faulkner, J.A. New perspectives in training for maximum performance. *J. Amer. Med. Assoc.* 205:741–746, 1968.

Fox, Edward L. and Donald K. Matthews. *Interval Training.* Philadelphia: W.B. Saunders, 1974.

Hettinger, T. *Physiology of Strength.* Springfield, Ill: Charles C Thomas, 1961.

J. Amer. Med. Assoc. Olympic Issue, vol. 221, September 1972.

Jensen, C.L. and A.G. Fisher. *Scientific Basis of Athletic Conditioning.* Philadelphia: Lea and Febiger, 1972.

Keul, J., ed. *Limiting Factors of Physical Performance.* Stuttgart, Germany: Georg Thieme Verlag, 1973.

McCafferty, W.B. and D.W. Edington. Subcellular basis of competitive swimming. *Swimming Technique* 10:109–111, 1974.

Muller, E.A. Influence of training and of inactivity on muscle strength. *Archives of Physical Exercise and Rehabilitation* 51:450–462, 1970.

O'Shea, John P. *Scientific Principles and Methods of Strength Fitness.* Reading, Mass.: Addison-Wesley, 1975.

Key Concepts

• Absolute work capacity of males exceeds that of females by approximately 20%.
• There are specific performance-related tasks in which some women exceed some men.
• For work-related tasks, the upper performance distribution for women often exceeds that of the mean performance for men.
• Special considerations should be given to exercise during pregnancy.

Introduction

Work tolerance of females has become of increasing interest as we have developed social attitudes against discrimination on the basis of sex. Discriminatory actions are obvious in many occupations as well as in sport activities that have been reserved for men heretofore. Are these discriminatory actions justified? Can women attain equal levels of performance or social benefits from such sports activities?

Work Tolerance of Females

Actual Work Performance Capacity

In general the absolute work performance capacity of males exceeds that of females by 20% or more. The differences in performance are due to body size; however, when workload is expressed per kilogram of body weight, the performance differences remain up to 15% higher in males. In these comparisons, the nature of the work test should be noted; some women can exceed some men (and vice versa), in selected types of performances. This point is usually overlooked but can be clearly demonstrated.

A basic point regarding sex discrimination in job selection is that personnel for some professions are selected on the basis of physical-performance demands unrelated to the job description. Although, in general, men exceed

CHAPTER SIXTEEN

The female in athletics

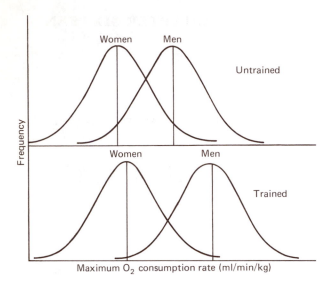

Figure axis labels: Frequency (vertical), Maximum O$_2$ consumption rate (ml/min/kg) (horizontal)

Labels in figure: Women, Men, Untrained, Women, Men, Trained

Figure 16-1. Maximum O$_2$ consumption rate (ml/min/kg) in normal men and women and in trained men and women.

women in physical performance, particularly in tasks in which strength and size are an advantage, it is also clear that the physical performances of some women exceed some men. This awareness logically leads us to conclude that personnel selection should be based on the demands of the job and not sex. The specific work-capacity tests should be designed to determine the physical qualification needed for the specific job; needless to say, the tests should be relevant to the job demands.

In Figure 16-1 we can see the normal population distribution of young adult men and women on the basis of maximal oxygen uptake rate. If one were to use the average O$_2$ consumption rate we can see that it would select against 76% of the women and against 47% of the men. This is an example of performance discrimination (on the basis of a single work-

related parameter) rather than of sex discrimination. For trained individuals the sex difference is more marked.

There are some physically demanding performances in which women, as a discrete population, exceed men; that is, the relative position of the population distribution of men and women is reversed. Working off the coast of Korea and Japan a special group of women divers can tolerate extremely cold temperatures. More than 70% of these divers are located on a single island, Cheju Island. Numbering 30,000, these women, called *Ama*, dive repeatedly for plant and animal life at depths of 20 meters below the surface. The better divers called *Funado* remain at the greatest depths in cold water for the longest periods of time. The air temperature in the winter approaches 0°C, and the water approaches 10°C. Body temperature, as measured by an oral thermometer, commonly drops from 37°C to 33°C or less.

How do these women tolerate working at these low temperatures? Some acclimatization has been identified. Their basal metabolic rate is about 35% higher in winter than summer. Although these women have no greater subcutaneous fat than nondivers, they are better insulated in the winter—probably because of an undefined acclimatization of the vasculature. Regardless of any acclimatizing that tends to maintain core temperature, the fact remains that for most other people the core temperature drops to temperatures below the tolerable level. Therefore the main acclimatization seems to be one of tolerance of low temperatures more than maintenance of normal temperatures.

Another example of women exceeding men in performance capacity appears to be tea pluckers of Sri Lanka where the women carry baskets of the picked tea leaves, which weigh approximately 40% of their body weight. The work is done on steep mountainsides at elevations exceeding 1500 meters. Their husbands do other general maintenance labor work on the tea plantations, such as pruning.

Sex Differences in Physiological Systems

Skeletal Muscle

The ratio of muscle to whole-body mass is similar in males and females as indicated by creatinine excretion rates (1 g creatine excreted per 24 hours = 20.5 kg muscle mass) for men and women up to 17 years of age. Obviously mean body weight for females is considerably less than the mean body weight for males. Muscle mass in males, as determined from potassium counts, may exceed females by 50%.

There are qualitative differences and similarities within the skeletal muscle mass of males and females. Figure 16–2 shows that the age at which the total number of muscle nuclei in males exceeds the number in females is at or near the onset of puberty. The total protein-to-DNA-concentration ratio remains about the same in males and females. Muscle strength per cross-sectional area also is similar in males and females (3–4 kg/cm^2). The proportion of fast- and slow-twitch muscle fibers in males and females seems to be similar in most muscles.

The *trainability* of muscle of males and females appears to differ. It appears that young adult males can improve their strength to a greater percent than young adult females. This trainability difference is also seen in the observation (referred to earlier) that there is a greater sex difference in maximum oxygen uptake rates in trained than in nontrained subjects. These differences appear to be related to, although they may not be caused by, higher levels of anabolic hormones in males.

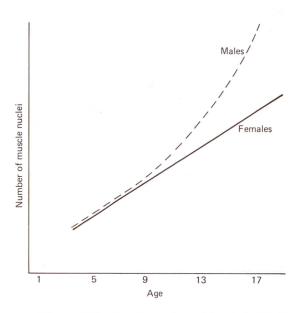

Figure 16–2. Number of nuclei of skeletal muscle at different ages in boys and girls.

Adapted from D.B. Cheek; A.B. Holt; D.E. Hill; and J.L. Talbert. Skeletal muscle cell mass and growth: The concept of the deoxyribonucleic acid unit. *Pediatric Research* 5:312–328, 1971.

Cardiopulmonary Function

Maximal aerobic capacity (or the ability to take up oxygen) is dependent, in part, on cardiopulmonary function. Maximum oxygen consumption rates are about 50% higher in young adult males than females. When maximal oxygen consumption is corrected for body size and lean body mass, the untrained male still exceeds the untrained female by 15–20%. In trained subjects the difference is about 20–25%. These figures suggest that if a male and female are working at the *same work load* the female will be working at a level proportionately higher in relation to her maximal aerobic capacity. Consequently it is unlikely that the female will be able to sustain that "submaximal" work rate for as long as the male. The proportionately higher work load means that blood lactate will rise more markedly and that heart rate will be closer to maximum in the female at that given work load. Also, expressed as per increment of work load, the arteriovenous oxygen difference will be greater in the female. We should remember that there is a distribution of males and females in all of these work tests and that the upper end of the female distribution probably exceeds the mean of the male distribution.

Several factors can explain the sex difference in aerobic capacity: women, in general, have less blood and a smaller heart per body mass, lower hemoglobin levels, and a smaller cardiac output than men. Stroke volume of males and females is the same if it is expressed in relation to the percent of maximum oxygen uptake. Lung volume is also less in females as manifested in vital capacity tests and in volume of blood within the alveolar capillaries. Maximum oxygen uptake is closely related to vital capacity values and to maximum ventilation rates. Maximum ventilation is about 50% higher in males than females in the 20- to 30-year-old age group, and is 35% higher in teenage

males than in teenage females. Peak heart rate is similar in males and females; however, due to the greater stroke volume, the cardiac output is greater, and thus the maximum oxygen uptake is greater in males. As always, a correction for body size reduces the magnitude of the sex differences but does not completely eliminate the differences.

An adult woman has to pump more blood to achieve a given oxygen uptake rate than a man. The relationship of heart rate to oxygen uptake in males and females is shown in Figure 16–3. Note that the female heart rate is about 10 beats greater per oxygen uptake rate. This difference becomes less obvious after the age of 40. Similarly a woman must ventilate more air to achieve a given rate of oxygen uptake.

The trainability of the cardiopulmonary system in women, although high, seems to be somewhat lower than in men. However, prepuberty and early puberty girls, after participating in an endurance program for a few months, can improve their maximal aerobic capacity by more than 25% (a rate very similar to that achieved by boys of the same age). The percent of improvement in this parameter, as in most work-related parameters, is determined in large part by the level of physical fitness that exists before the training begins.

Hormonal Influences

Sex hormones have specific effects on many functions known to be important factors in physical performance. *Testosterone* stimulates muscular growth (stimulates the protein-synthesizing apparatus); estrogen has a slightly atrophying effect. However, in animals it has been shown that muscle hypertrophy can occur without the presence of testosterone. During prepubertal years there are few sex differences in the human musculature; the more marked differences occur in young adults. As the testosterone secretion rate is gradually reduced during aging, so is the sex difference in amount and strength of the musculature. Some experimentation in laboratory animals suggests that the proportion of slow- and fast-twitch muscle fibers in some muscles can be altered by manipulating testosterone levels. However, no fiber-type differences between sexes in adult leg muscles have been found in humans.

Testosterone also affects muscle glycogen levels. For example normal male guinea pigs, when trained, have higher muscle glycogen levels than trained males that were previously castrated. This information suggests that the male sex hormone, testosterone, plays an important role in the well-known phenomenon of glycogen *supercompensation* in response to endurance training. The obvious question, which has not been investigated, is whether females can supercompensate their muscular glycogen levels to the same degree as males.

The maximum oxygen consumption rate is also influenced by hormonal levels. For example anabolic steroids taken over a period of weeks seem to augment the maximum oxygen uptake capacity: this suggests that one possible explanation for the greater aerobic capacity of males is their normally greater testosterone levels.

Sex-related hormones may be a deciding factor in the difference between the longevity of males and females. *Estrogen*, for example, has a protective role in relation to cardiovascular disease; one reason for this may be due to its ability to lower blood cholesterol levels. Estrogen, given to male cardiac patients, significantly reduces the fatality rate below the expected level.

The temperature-regulating mechanisms of males and females also differ. The female sex hormone, estrogen, has an inhibitory effect on sweating. Although it is not known if the effect is caused by hormones, women begin to sweat at lower temperatures during the earlier

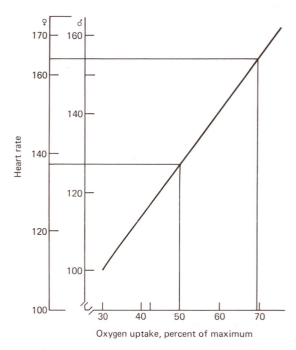

Figure 16–3. Heart response to exercise as related to percent of maximum O$_2$ uptake rate of men and women.

From *Textbook of Work Physiology* by P.-O. Astrand and K. Rodahl. Copyright © 1970 by McGraw-Hill, Inc. Used with permission of McGraw-Hill Book Company.

phase of menstruation. The difference in subcutaneous fat content also affects temperature regulation—because of the insulative effect of fat. Work at high temperature is more stressful if it is more difficult to dissipate the body heat; on the other hand at low temperatures heat conservation may be advantageous.

Training itself may affect hormonal secretion rates. Females participating in track and field events reach puberty about a year later than more sedentary females. The general hormonal response to chronic exercise is not as well known for females as it is for males.

Sex-related hormones also affect social behavior. For example, when androgens are given to an infant female rhesus monkey, her aggressive behavior increases to the point where she may actually replace males at the top position of the social hierarchy without modifying her feminine sexual behavior. This behavior lasts for months after the last injection of androgen and suggests a somewhat permanent effect of the hormone on the central nervous system—since the long-lasting effects could not be due to continued high levels of circulating androgens. Male rhesus monkeys show masculine behavioral development despite prepubertal castration; in fact they are more aggressive than the normal prepubertal monkeys—even though testosterone levels are low at this developmental stage. These tentative findings require further clarification.

ORAL CONTRACEPTIVES

The oral contraceptive Enovid reduces daily voluntary running activity of female rats. However, after a month of administration of this contraceptive, the depressing effect on voluntary activity is lost. Progesterone also reduces running activity if it is administered in combination with estrogen. To our knowledge, studies on humans related to these findings have not been done.

LONGEVITY AND HEART DISEASE

The incidence of heart disease in white males is five times higher than in white females. It is only twofold higher in black males than black females (we might also note that the incidence rate of white females exceeds that of black males). This sex difference is due in part to the depressing effect that estrogen has on blood lipids, which are known to be one of the key risk factors in heart disease. Male sex hormones are directly related to increased lipid levels in the blood. This relationship may be due to the stimulatory effect that dietary lipids have on gonadal secretion rates, which seems to be mediated via the pituitary gland and the male accessory sex structure, the seminal vesicles.

The higher incidence rate of heart disease in males may also be directly related to hemoglobin levels. Mean hemoglobin values in males are greater than females. Heart disease is significantly higher in males who show a hemoglobin concentration at the upper range of normality (hematocrit of more than 46%). However, it should be recognized that the mortality rate of males and females in the anemic range is also higher than normal (hematocrit of less than 36%). It has been shown in Sweden that most heart disease in women occurs after the menopause; again the implication is a sex-related hormonal factor. In summary the typical risk factors associated with myocardial ischemia in males seem to be of less imporance in young and middle-aged females.

Exercise and Pregnancy

For many years pregnancy was regarded as a period of illness; the pregnant woman was in a delicate state and needed pampering and overprotection. However, the many examples of women participating in strenuous exercise during pregnancy discourage the outmoded idea that exercise during pregnancy is harmful.

During the 1952 Olympics, a diver who was three and one-half months pregnant, placed third.

With the greater participation of women in sports and in all occupations and with the increased emphasis on physical fitness, there is intense interest in what happens to the mother and child if regular exercise is continued during pregnancy. Many women want to know just how much physical activity they can engage in during pregnancy. The usual advice stresses the need for caution and for common sense. Pregnant women frequently are told to avoid all violent exercise, such as tennis, riding, cycling, or swimming, especially during the last six months. The justification for this arbitrary recommendation is minimal in most circumstances. "Common sense" would not seem to be the best recommendation considering the wide range of common sense.

Natural childbirth programs often include exercise routines from the very beginning of pregnancy. Some obstetricians are convinced of the benefits of a physically active expectant mother. There is rather limited information concerning the actual effects of exercise during pregnancy on the mother; information is even more limited in relation to the effects on the fetus. Recently increased efforts have been made to determine the specific training adaptations that occur during pregnancy—in an effort to understand both the beneficial and detrimental effects that exercise might have on the mother and child. Exercise imposes additional requirements on the cardiovascular, respiratory, muscular, and metabolic systems above and beyond those changes already established because of the pregnancy.

Cardiovascular and Respiratory Effects

Cardiac output increases and is redistributed during exercise. After five months of pregnancy, resting cardiac output is increased by approximately 30–50% over nonpregnant resting levels. The reason for this increase is not as obvious as it might seem since the greatest increment in cardiac output occurs early in pregnancy when uterine and placental blood flow is only slightly elevated. Also the maximum blood volume occurs at about the eighth or ninth month of pregnancy.

Cardiac output, as expressed per unit increase in oxygen uptake, at different work levels is the same throughout pregnancy and for three months after birth. However, the demands of any specific workload on the heart are higher during pregnancy; this means that the pregnant woman reaches her maximum cardiac output at a lower level of work. In response to moderate exercise, the corresponding increase in cardiac output is progressively smaller as pregnancy nears term. This progressive decline in cardiac reserve may be due to the peripheral pooling of blood which is due to the fetal obstruction of the inferior vena cava.

Minute ventilation during a standard exercise increases during the latter phases of pregnancy, but the rate of increase remains proportional to work load on a bicycle ergometer. These results, in general, mean that during pregnancy the metabolic efficiency remains the same, but the energy cost at a given work load is increased. Exercise during pregnancy requires more oxygen primarily because of the increase in body weight.

In general the cardiovascular and respiratory systems do not exert any great limitation on exercise, except perhaps in late pregnancy. This is particularly true when the exercise is done in a non-weight-bearing posture.

Some interesting research has been done in determining the adequacy of oxygen delivery to the fetus during maternal exercise of sheep. After exercise, pregnant ewes hyperventilate; a respiratory alkalosis occurs, which results in a decrease in fetal P_{CO_2} and an increase in pH. Also a decrease in P_{CO_2} occurs after exercise is initiated. Generally the responses of the fetus to maternal exercise tend to follow those of the mother, and recovery from all the above changes occurs rapidly. In fetuses of sheep that have some defect in the blood supply, moderately severe exercise may be detrimental to the fetus; however, the normal fetus is able to endure maternal exercise. These results suggest that women experiencing some difficulties with the pregnancy should be cautious about participating in heavy, prolonged exercise.

Clearly the question of exercise during pregnancy deserves more study. There have been reports of human fetuses exhibiting changes in preexercise and postexercise heart rates, suggesting uteroplacental insufficiency.

Muscular and Structural Effects

Exercise before, during, and after pregnancy theoretically might have a number of effects on labor, delivery, and postpartum conditions. Labor and delivery demand heavy muscular effort; we conclude from this that those women with stronger muscles are in an advantageous position. It has been shown that women who are physically fit have shorter durations of labor during childbirth. One of the main functions of the natural childbirth courses, such as that of Lamaze, is to increase the tone and control of back, abdominal, and perineal muscles in order to aid in labor. However, when "Lamaze" mothers were compared with "non-Lamaze" mothers, length of labor was similar in the two groups.

Zaharieva, in studying former Olympic athletes, found that the average duration of the first stage of delivery was about the same for Olympians as for nonathletes. But the length of the second or expulsive stage was one and one-half times shorter for the sportswomen, perhaps owing to the stronger abdominal muscles of women athletes.

Some women suffer from postpartum backaches frequently because they fail to do the exercises prescribed by their doctors. These women may reduce their activities to such a point that the slightest exercise bothers them. It has been found that nearly all of the women who have postpartum back problems have a history of extremely sedentary behavior throughout their school years and marriage.

Some disadvantages for the pregnant woman in terms of participation in athletic competition, comes about from the structural changes that take place: she is handicapped by changes in balance and center of gravity. Simple awkwardness might limit her physical performance. During pregnancy a shift in the center of gravity can produce *lordosis* or abnormal anterior-posterior curvature of the spine; however, this condition in itself does not create a physical hazard to the mother. The awkwardness of movement in pregnant women is probably one of the main reasons for the limited capability with which most pregnant women are believed to perform everyday tasks. Obviously, pregnant women are heavier than before and will have a greater work load to deal with in any weight-bearing task.

Nutritional Demands of the Female

The metabolic characteristics of females are essentially the same as for men, at least in a quantitative way. For example, if the rate of oxygen consumption at rest is expressed per mass of active tissue, then the metabolic rates in women and men are equal. Absolute basal metabolic rate per kilogram of body weight is lower in women (Figure 16–4). Body composition in terms of fat to lean body mass is higher in women than in men.

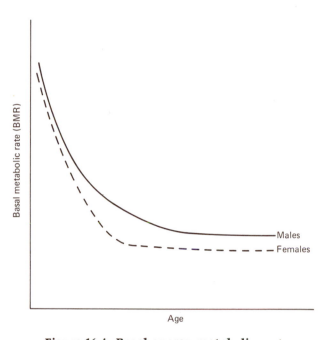

Figure 16–4. Basal energy metabolism at various ages.

Data such as the above have been interpreted by some to mean that women are at an advantage in marathon races because they have a larger store of fat from which to derive their energy. Although their additional fat can be advantageous from one respect, under some circumstances it is detrimental to performance in that it means more work must be done because of the greater body weight.

The need for specialized foods such as vitamins and minerals is the same for men and women. Some women may need to supplement their dietary iron intake if menstrual flow is high for several consecutive months. It has been shown that women who have high menstrual flows tend to have lower hemoglobin levels; they are also much more likely to have reduced or no iron stores and are thus more susceptible to developing anemia.

Proper nutrition during pregnancy is very crucial. Although more American than European doctors have stressed the need for minimal weight gain during pregnancy, it is now known that the weight of the child at birth was higher in those mothers that gained slightly more weight than was customary. As pregnancy proceeds, there are some obvious supplemental nutritive needs. The amount of dietary food need not be increased markedly since metabolic rate can be expected to increase less than seven percent. There are demands for more minerals such as Ca^{2+} to support new bone growth and lactation and for more iron for hemoglobin synthesis. The demand for most of the other minerals and vitamins is similar in the nonpregnant and pregnant state.

Pregnancy requires additional amino acids, as do skeletal muscles in adapting to training. During pregnancy amino acids are mobilized from skeletal muscle to support fetal growth. It has been shown in laboratory animals that amino acid absorption by the gut is depressed by single bouts of exercise in nontrained, but not in trained, pregnant animals.

The Menstrual Cycle and Physical Performance

A general conclusion on the effects of menstruation on physical performance or on the effects of training on the menstrual cycle cannot be drawn from existing information. Most studies thus far have concluded that menstruation does not adversely affect peak performance; in a large population this apparently is the case. Although some women menstruate with minimal or no discomfort, others are affected emotionally or experience painful cramps or both. Seventeen percent of the female athletes studied in Japan felt that menstruation adversely affected their performance. Other reports have shown that as high as 50% of women athletes perform at subpar levels during menstruation. It would be difficult for a coach to convince a female athlete who is in discomfort during menstruation that her peak physical performance is independent of the menstrual process. The only apparent physiological reason for a poor performance due to menstruation is the blood loss, but this amount is insignificant during any one cycle. However, over a period of time, the blood loss volumes could reduce iron stores and hemoglobin levels and thereby lower maximum oxygen consumption. It is also worth noting in this respect that birth control pills tend to reduce dietary iron absorption.

Even less is known about how training and competition affect the menstrual cycle. Of the female Olympic athletes in Japan mentioned above, 41% stated that training or competition affected or disturbed their normal menstrual cycle in some way. If the menstruation becomes more regular and if less discomfort is experienced, the effects of training could be beneficial.

Precise measures of blood flow and careful monitoring of behavioral characteristics during successive menstrual cycles before, during, and after training have not been conducted.

Summary

The topic of women in sports has aroused a great deal of interest in the 1970s. In light of the present evidence, we see no reason why women should not be allowed to and encouraged to participate in all sports. The question of participation with and against men should depend upon the nature of the specific activity. The choice between separate and integrated sports programs must be examined in terms of each sport and of the philosophical perspective of the participants involved. Obviously, in some activities where absolute strength, power, and body size are advantageous to an athlete, the woman is normally at a distinct disadvantage.

Except for the physical characteristics and the differential action of specific steroid hormones (estrogen and testosterone), we find it difficult to identify other specific physiological limitations of women in sports. Many investigators have demonstrated differences between men and women in such physiological parameters as hemoglobin count, oxygen uptake, basal metabolic rate, and so forth. It is very difficult to assess this data knowing that the women in these studies probably had not engaged in relatively intense lifelong physical activity—in contrast to their male counterparts.

Activity during the menstrual cycle may be impaired during the two or three days preceding menstruation; however, there is no concensus of agreement on this point. Physical activity during pregnancy should be limited to the noncontact sports. Considering the added burden of fetal circulation and nourishment that the female has to bear during pregnancy, we recommend very limited competitive activity during the later stages of pregnancy. However, we are not discouraging physical activity for the benefit of health and fitness, which we feel should be continued and perhaps even increased during a normal pregnancy.

One of the problems of examining the role of the female in athletics is the relatively small population of female athletes. Female athletes in the contact sports are extremely few in number. Swimming, gymnastics, and track and field attract the greatest number of qualified, highly trained female athletes. Even in these sports, the percentage of female participants is markedly less than the percentage of male participants.

Study Questions

1. How does body weight influence the relative work capacities of women and men?
2. What differences are there in skeletal muscles between women and men?
3. How does trainability differ between the sexes?
4. What are the sex differences in aerobic capacity?
5. What are the influences of the sex-related hormones on trainability?
6. What are the influences of the sex-related hormones on longevity?
7. How can exercise aid and hinder pregnancy?
8. What are the special nutritional demands of the female?
9. What effect does the menstrual cycle have on performance?
10. What are the significant sex differences in determining athletic performance?

Review References

Astrand, P.O., et al. Girl Swimmers. *Acta Pediat.* Suppl. 147, 1963.

Drinkwater, B. Physiological responses of women to exercise. In *Exercise and Sport Science Reviews*, Vol. 1, edited by J.H. Wilmore. New York: Academic Press, 1973.

Kilbom, Asa. Physical training in women. *Scand. J. Clin. and Lab. Invest.*, vol. 28, Suppl. 119, 1971.

Klafs, Carl E. and M. Joan Lyon. *The Female Athlete*. St. Louis: C.V. Mosby Co., 1973.

Pomerance, J.J.; L. Gluck; and V.A. Lynch. Physical fitness in pregnancy: It's effect on pregnancy outcome. *Amer. J. Obstet. Gyn.* 119: 867–876, 1974.

Research studies on the female athlete. *Journal of Health, Physical Education and Recreation* 47:32–45, 1975.

Seitchik, J. Body composition and energy expenditure during rest and work in pregnancy. *Amer. J. Obstet. Gyn.* 97:701–713, 1967.

Thomas, C.L. Effect of vigorous physical activity on women. In *American Academy of Orthopedic Surgeons Symposium on Sports Medicine*. St. Louis: C.V. Mosby Co., 1969.

Ulrich, C. Women and sports. In *Science and Medicine of Exercise and Sports*, W.R. Johnson, ed. New York: Harper and Row, 1960.

Zaharieva, E. Olympic participation by women—effects on pregnancy and childbirth. *J. Amer. Med. Assoc.* 221:992–995, 1972.

Key Concepts

• Temperature regulation in the body is involved with maintaining a balance between heat gain and heat loss.
• Temperature regulation mechanisms are responsive to local as well as to central temperature receptors.
• There is little evidence that the inhalation of pure oxygen aids performance.
• Acclimatization to altitude includes physiological adaptations conducive to increased performance.
• Exposure to smoking and smog may result in decreased performance.

Temperature Regulation

Introduction

The human body normally functions within a relatively narrow temperature range, between 23.0°C and 37.5°C. The lower temperature is the temperature of the exposed skin; the higher is the central body core temperature. Although there is a temperature gradient from the central body core to the skin's surface, the largest drop occurs near the skin (Figure 17–1). Clothing can be viewed as adding another layer of insulation that decreases heat loss and decreases the temperature gradient near the surface of the skin.

To maintain a relatively constant body temperature, there must be a long-term balance between heat gain and heat loss. Although there are short-term imbalances due to extreme cold (excessive heat loss) and due to intense exercise (excessive heat generation), these temporary imbalances activate the appropriate temperature-regulating mechanisms.

Heat Gain

Heat gain within the body occurs primarily from the metabolism of foodstuffs and the

Exercise and environmental conditions

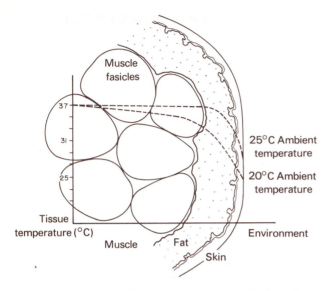

Figure 17-1. Temperatures at various points from core to skin. Note insulating effect of the fat layer.

Heat energy can be quantified in units called *calories*. Since the body is nearly 80% water it is not surprising that the specific heat of the body is approximately that of water; one calorie will raise the temperature of one gram of tissue 0.83°C.

Under normal resting conditions the metabolic rate (heat production) of the human is about one calorie per gram of body weight per hour or 80 kilocalories per hour for an 80-kg individual. Under these conditions, body temperature would increase 0.83°C per hour. During intense exercise the metabolic rate may increase as much as 10–20 times. The increased metabolism during exercise increases heat production to such an extent that the body temperature could rise at the rate of 8–16 degrees per hour unless controlled by the temperature-regulating mechanisms.

utilization of oxygen. At rest, heat production within the body is distributed among the various tissues as follows: the brain, 18%; heart, 11%; kidney, 7%; hepatic-portal, 20%; muscle, 20%; skin, 5%; and 19% from the remaining tissues. In proportion to tissue mass, muscle, which is 50% of body mass, produces a relatively small proportion of the heat. During exercise the working muscle can cause a tenfold increase in total body heat production. Work at 75% of max $\dot{V}_{O_2}$ elevates core temperature to 39°C (Figure 17-2). Core temperatures are known to increase during exercise to as high as 40°C. Working muscle temperatures rise even higher than rectal temperature (Figure 17-2). A reading as high as 42°C has been recorded in skeletal muscles of exercising animals. Skin temperature is reduced during exercise, primarily because of evaporative heat loss due to sweat gland activity. Unlike core temperature, skin temperature is relatively independent of work intensity.

Heat Loss

The transport of heat from the muscle to the skin surface is facilitated by the circulatory system. As the cooler blood perfusing the active muscles reaches thermal equilibrium, it "picks up" and carries the heat to the skin where the thermal gradient is in the opposite direction. The primary ways the body can dissipate the heat gained from metabolism are through water evaporation from the skin and by the movement of air or water over the surface of the skin.

Sweat glands, when stimulated by high temperatures or the autonomic nervous system, secrete water onto the surface of the skin. As the water evaporates, the heat necessary for the heat of vaporization (580 calories per gram of water) is taken from the elevated skin temperature. During exercise the amount of heat loss by water evaporation from the skin depends upon the amount of sweat and the environmental water vapor saturation. It is possible to perspire at the rate of 3 liters (3 kilograms)

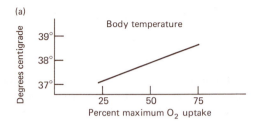

(a)

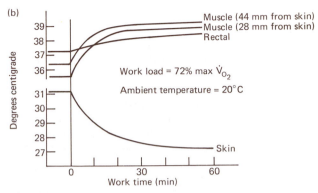

(b)

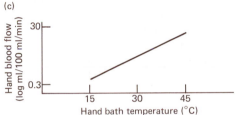

(c)

per hour during maximal exercise. If all of this water was evaporated from the skin, it could account for the removal of 1740 kilocalories per hour (580 × 3000). If perspiration proceeds without evaporation, as may be the case during exercise in increased humid environments, no heat loss will occur; and the body may become dangerously overheated.

The loss of water during excessive perspiration not only results in dehydration but also in salt imbalances within the body, which may lead to serious consequences such as tiredness, cramps, inefficiency, and nausea. Sweat glands are located over the entire surface of the body with a concentrated number in the palm of the hand, the feet, head and axilla.

A greater number of sweat glands are found in people that live in the tropics. To maximize heat acclimatization as an adult, there appears to be an advantage to having lived in the tropics as a child. In spite of the fact that acclimatization to heat involves increasing the number of functioning sweat glands, less sweat is lost in the form of water droplets because the sweat is more uniformly distributed over the body and thus the surface area for the collection of the heat of evaporation is maximized. Other sweating adaptations to heat include a quicker onset of sweating during exercise-related increases in temperature

Figure 17–2. Thermal compensations.

(a) and (b) Adapted from K.E. Olsson and B. Saltin. Diet and fluid in training and competition. *Scand. J. Rehab. Med.* 3:32, 1971. (c) Adapted from L.D. Carlson and A.C.L. Hsieh, "Cold." In *Physiology of Human Survival* (O.G. Edholm and A.L. Bacharach, eds.). London: Academic Press, Inc., Limited, 1965, pp. 15–52.

Table 17-1. Adaptive Changes to Exercise in the Heat

Mechanism	Adaptation
Sweating	Increased capacity*
	Quicker onset*
	Better distribution over body surface*
	Reduced salt content*
Cardiovascular	Greater skin blood flow*
	Quicker response*
	Blood flow closer to skin surface
	Better distribution over body surface
	Reduction in counter-current blood vessels
Metabolic	Lowered basal metabolism rate (BMR)
	Lowered energy cost for a given task
Respiratory	Hyperventilation (panting)*
Heat storage	Increased tolerance to higher body temperature
	A lower resting body temperature*
Anatomical	Change from short and stocky to long and thin*

*Adaptations for which there is evidence.

From O.G. Edholm and A.L. Bacharach. *The Physiology of Human Survival.* London: Academic Press, Inc., Limited, 1965, p. 73.

and a reduced salt content of the sweat. Other systemic changes also occur (Table 7-1).

Temperature-Regulating Receptors

The monitoring of body temperature takes place on at least two levels, central temperature control receptors and local receptors (Figure 17-3). The central control is monitored by temperature-sensitive cells within the hypothalamus. When the core temperature of the body rises, these cells discharge impulses to the autonomic nervous system, which then innervates sweat glands. Sweat glands are innervated primarily by the *cholinergic parasympathetic nerves;* however, a sympathetic effect is suggested by the profuse sweating rate during exer-

cise or by that sweat on the palms during emotional stress. When the sweat glands are stimulated, water is secreted onto the surface of the skin; the process of vaporization of the sweat acts to cool the skin. The decrease in skin temperature is gradually reflected in a decrease in core body temperature.

There exists, in addition, a local mechanism for temperature regulation: the local heating of a body segment will elicit sweat secretion in that specific area. Also blood flow is regulated locally and in direct proportion to the environmental temperature (Figure 17-3).

Abnormal Environments

Oxygen Inhalation

Inhalation of air containing elevated oxygen concentrations is a common practice in specific therapeutic conditions in hospitals as well as a common practice of many athletes. Theoretically only limited value can be derived from these procedures assuming pulmonary function is normal; however, respiratory abnormalities are greatly alleviated by this procedure.

Oxygen inhalation therapy tends to decrease the production of surfactant by the lung, particularly in the newborn. Because of the increased surface tension to the alveoli not supplied with sufficient surface-tension-reducing substance, *surfactant*, the alveolar-capillary gas exchangeable surface area is reduced; this leads to alveoli collapse (*atelectasis*).

In neonates there is good evidence of pulmonary dysfunction resulting from O_2 therapy, a common practice at one time for infants with respiratory stress syndrome. Vital capacity, respiratory rate, minute ventilation, compliance, pH of the blood, and arterial P_{O_2} are all adversely affected by prolonged O_2 therapy. Also the air-blood barrier increases in thickness, surfactant-producing cells hypertrophy, and edema develops.

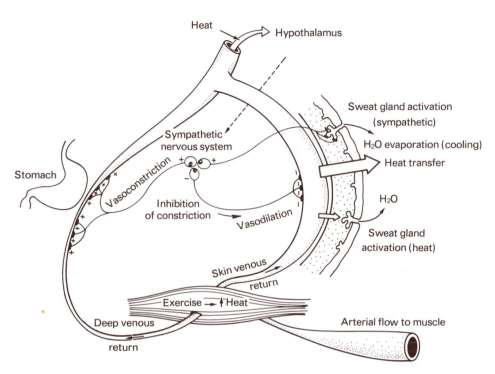

Figure 17-3. Response of central and local receptors to changes in blood temperature with exercise.

In spite of the toxic effects of oxygen, there is no known contraindication to the use of pure O_2 for brief periods. Effects resulting from repeated brief administrations of oxygen over long periods of time have not been noted. It is thought that pure oxygen can be breathed up to 24 hours before toxic effects will become evident. But in light of the frequency of oxygen inhalation therapy currently in use in sports, more research is needed.

Altitude

The problems encountered with respiration at high altitude stem from the reduced atmospheric pressures and not the relative concentrations of the respiratory gases. The atmospheric pressure decreases from the sea level pressure of 760 mm Hg to 226 mm Hg at 30,000 ft (9000 m). But the composition of the air is still primarily 79% N, 20.95% O_2, and 0.04% CO_2. At 10,000 ft the barometric pressure is reduced to approximately two-thirds of the sea-level pressure; the absolute amount of O_2 in the air is two-thirds of the amount of O_2 at sea level. To obtain oxygen uptake values that compare with those at sea level, the ventilation rate must be increased by at least one-third and perhaps higher—because of the increased oxygen cost of breathing, the decreased P_{O_2}, and the decreased saturation of the arterial blood. The lower limit of oxygen concentration that man can survive has not been determined. Natives living at an altitude of 19,000 ft have been found to have alveolar oxygen tension as low as 44 mm Hg (150 is normal), and 22 mm Hg (100 is normal) in arteries.

At any altitude the gases inside the body are exposed to the same temperature and watery environment; this means that the water vapor pressure remains at 47 mm Hg. As the altitude increases, the proportion of total alveolar pressure due to water vapor pressure changes from approximately 33% (at sea level) to 100% (at 30,000 ft) of the sum of the O_2 and CO_2 pressures.

At rest, the adequacy of ventilatory processes are noted by the fact that acute exposure at altitudes up to 10,000 ft does not result in an increase in the ventilatory rate. However, during exercise at these altitudes, the nonacclimatized man has a decreased work capacity of up to 40% of sea level capacity.

Altitude acclimatized rats (4 weeks at an oxygen pressure of about 4400 ft) have a greater work capacity and a greater oxygen uptake than nonacclimatized rats. The former are also more resistant to loss of body heat when swimming in cool water (22°C), indicating a positive cross adaptation between altitude and cold.

People who have become acclimatized to altitude or who have lived at high altitudes all their lives have several distinguishing characteristics of acclimatization: (1) increased pulmonary ventilation, (2) increased cardiac output, (3) increased diffusing capacity of the lungs (probably due to more blood in the pulmonary capillaries), (4) increased red blood cell count (also increased hemoglobin content), (5) increased vascularity of the tissues, (6) increased ability of the peripheral cells to utilize oxygen, (7) increased myoglobin content of the muscles, and (8) increased oxygen available to tissues because of elevated 2,3-diphosphoglycerate.

Anyone interested in improving exercise performance will note that these adaptations to altitude are directly related to the adaptations that one would like to see occur with endurance training at sea level. An immediate application of the altitude adaptations is to train athletes at a moderate altitude, for example 6000–8000 ft. Physical training at these moderate altitudes induces the normal training and altitude adaptations, whereas training at higher altitudes has been shown to be relatively ineffective. The return to sea level, theoretically, allows the individual to perform at a higher level, for a limited amount of time, than he or she would have prior to the altitude training. This technique has been shown to be useful for many athletes, but as one might expect, the success rate is not 100%.

Smoking

A substantial amount of evidence has been published linking smoking to cancer of the lungs and to coronary heart disease. The exact mechanism of the link between smoking and these diseases has not been described in complete detail; however, there is enough evidence to convince even the most skeptical person that smoking is related to both these diseases.

The effects of smoking on the respiratory tract are caused by at least three factors: (1) the inhalation of submicroscopic particles, (2) the inhalation of carbon monoxide, and (3) the inhalation of nicotine. The submicroscopic particles can cause irritation and blockage of the minute respiratory channels in spite of the action of cilia, which move the inhaled particles into the mouth. The constant irritating effect has been linked to the origin of cancer within the lungs. Blockage of respiratory channels would effectively close down the operative alveoli by preventing air from being ventilated. These particles have also been shown to increase the resistance of the airways.

The inhalation of carbon monoxide has the effect of noncompetitively binding with hemoglobin: carbon monoxide has an affinity for hemoglobin 200 times that of oxygen. Thus the binding of carbon monoxide excludes the oxygen binding and decreases the oxygen saturation. This effect occurs within a few seconds after smoking and lasts for several hours. Over 10% of a heavy smoker's hemoglobin is bound by carbon monoxide; consequently the oxygen-carrying capacity of his hemoglobin is reduced by that same amount.

The inhalation of nicotine causes constriction of the respiratory channels and the pulmonary capillaries; after being absorbed, this drug acts to stimulate the resting heart rate.

The inhalation of the hot, dry air from smoking can also irritate the throat passages. The incidence of throat cancer is higher in smokers than nonsmokers.

Figure 17–4 illustrates the effect of cigarette smoking on the incidence of myocardial infarction in active and nonactive men. As expected, the active nonsmoking men have less than one-third of the infarcts of the less active smokers and only one-half of the infarcts of the least active nonsmokers.

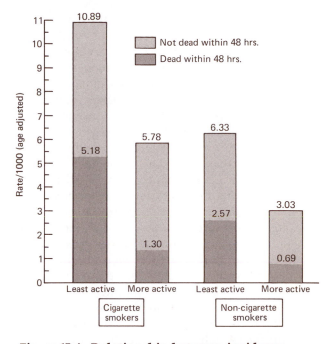

Figure 17–4. Relationship between incidence of first myocardial infarction, physical activity level, and cigarette smoking in men.

From S. Shapiro. The incidence of coronary heart disease in a population insured for medical care. *American Journal of Public Health*, vol. 59 (Suppl. 2), 1969.

Smog and Physical Performance

The components of smog are specific to the city and its industry. Some of the components that are more likely to affect physical performance immediately are the oxidants, which are primarily ozone (O_3), carbon monoxide (CO), nitrous oxide (NO), nitrogen dioxide (NO_2), sulfur dioxide (SO_2), and larger particulate matter such as asbestos. Of course any impurities inhaled over long periods of time that affect one's health affect one's physical performance capacity. The immediate effects of smog on performance are evident—even though the human organism is supplied with numerous back-up and compensatory mechanisms that could mask potential detrimental effects of smog on a single physiological parameter such as oxygen consumption.

Both young and old rats and mice that were lightly exercised for 15 minutes per hour for six hours per day and that were exposed to ozone for the whole six hours had a much higher mortality rate after 200 days than rats only exposed to ozone. These findings clearly suggest that exercise can potentiate the detrimental effects of oxidants on survival rate.

Voluntary activity level of animals has been markedly depressed by exposure to concentrations of oxidants equal to those of a first-stage smog alert in Los Angeles. The correlation between oxidant levels and low spontaneous voluntary activity is extremely high in mice; as the oxidant level in the air goes up, spontaneous activity decreases proportionately. Similarly exposure of mice to automobile exhaust fumes for six hours may decrease spontaneous activity by more than 50%.

Pollutants in general induce bronchiolar constriction, which increases air resistance; more mucus is formed; reduced surfactant and edema occurs. All of these conditions impair O_2 and CO_2 exchange. Long-term exposure to pollutants results in loss of *alveolar macrophages* (which defend the body against toxic particles) and *emphysema* (loss of functional alveoli). Acute exposure to ozone increases respiration rate and decreases tidal volume. Nitrous oxide probably hinders normal oxygen transport by binding to hemoglobin. It has been shown convincingly that carbon monoxide at concentrations that are common for the freeways of Los Angeles represses cardiopulmonary function since cardiac patients' tolerance to exercise is severely decreased for two hours after such exposure.

Studies related to the effects of atmospheric pollutants on maximal work capacity are less numerous, and evidence of a decrement in maximal performance is not as well documented as might be expected. However, performances over a five-year period of high school track athletes in the Los Angeles area were studied. The records were rather convincing that, on the days on which the smog levels were high, performances were low.

A number of physiological parameters have been monitored during a submaximal exercise: they do not seem to be dramatically affected by smog. Again a point to remember is that, when demands are submaximal, there are reserves or compensatory mechanisms available to the working body that may camouflage any direct effect.

At this time there is a glaring need for more research in the area of work performance and atmospheric pollutants. These studies are needed not only for those interested in athletics but also for those interested in the effects of industry on the atmosphere.

Summary

Body temperature is maintained within a narrow range—with skin temperature several degrees lower, depending on clothing and ambient conditions. Exercise, which elevates metabolism approximately 10–20 times, elevates body temperature; these two factors act to stimulate the sweat glands that assist in cooling the body. Water loss due to chronic, excessive perspiration can cause salt imbalances, which can lead to serious consequences. The development of more sweat glands, a quicker onset of sweating upon a rise in temperature, and reduced salt content of the sweat occur as a result of training adaptations to chronic work in hot environments.

Oxygen ventilation provides no known benefits to exercise other than a possible increased recovery rate. Acclimatization to altitude (6000 ft) results in increased training adaptations that could be shown to be beneficial to performance. Cigarette smoking and smog subject the body to conditions nonconducive to good physical performance.

Study Questions

1. What are the temperature extremes for human performance?
2. What accounts for heat gain?
3. What accounts for heat loss?
4. What receptors sense and regulate body temperature?
5. How does the body react to oxygen therapy?
6. What are the special considerations for exercise at altitude?
7. What effect does smoking have on performance?
8. How can smog influence performance?

Review References

Adams, T. and P.F. Iampietro. *Temperature Regulation in Exercise Physiology*, H.B. Falls, ed. New York: Academic Press, 1968.

Adaptation to the environment. In *Handbook of Physiology*, D.B. Dill, ed. Washington, D.C.: American Physiological Society, 1964.

Balke, B. Work capacity at altitude. In *Science and Medicine of Exercise and Sport*, W.P. Johnson, ed. New York: Harper & Row, 1960.

Dill, D.B. Physiological adjustments to altitude changes. *J. Amer. Med. Assoc.* 205:123–130, 1968.

Gisolfi, C.U. and J.R. Copping. Thermal effects of prolonged treadmill exercise in the heat. *Medicine and Science in Sports* 6:108–113, 1974.

Jokl, E. and P. Jokl, eds. *Exercise and Altitude.* New York: S. Karger, 1968.

Londeree, B.R.; W.F. Updyke; and J.J. Burt. Water replacement schedules in heat stress. *Research Quarterly* 40:725–732, 1969.

Wells, C.L. and S.M. Horvath. Metabolic and thermoregulatory responses of women to exercise in two thermal environments. *Medicine and Science in Sports* 6:8–13, 1974.

Wyndham, C.H. The physiology of exercise under heat stress. In *Annual Reviews of Physiology*, J.H. Comroe; I.S. Edelman; and R.R. Sonnenschein, eds. Palo Alto, Cal.: Annual Reviews, Inc., 1973, pp. 193–220.

Key Concepts

• Many drugs are designed to mimic the normal body functions.

• Pharmacological agents are used frequently in athletics to optimize performance without knowledge of the long-term side effects.

• Although a drug is taken for a specific effect, there are side effects due to the nonspecificity of the drug.

• The effects of acute or rapidly acting drugs are usually of short duration (minutes), whereas more slowly acting drugs induce effects of long duration (days).

Introduction

Several classifications of drugs have significant effects on physical activity. Certain drugs are directly applicable either to improving exercise performance, or to promoting training adaptations, or to increasing resistance to pain. Other drugs are involved with the treatment and/or prevention of metabolic diseases or with the alteration of some physiological function that may indirectly influence the capacity for physical activity.

Throughout this book we have attempted to discuss pharmacological considerations and how they applied to the specific topic under discussion. However, the role of drugs in contemporary society warrants a brief, special chapter devoted to the subject. What is important about pharmacology and drug usage for people concerned with exercise? We should be aware that a single dose or a series of doses of a drug can alter physical performance. Effects of stimulants occur within a few hours, whereas the effects of anabolic steroids can only be realized over a period of weeks. A general knowledge of the chemical means to alter the mechanisms involved in exercise is important in order to appreciate the potential exercise effects.

CHAPTER EIGHTEEN

Pharmacological considerations for physical activity

We can easily understand the competitive problems facing the athlete and appreciate his or her goals and attitudes. For the most highly motivated athletes, the concept of winning and being acclaimed the champion often overrides most of the other considerations in their lives at the time of the competition. This high level of motivation is not unique to athletics but exists in other fields of endeavor such as in the classroom, in business activities, and in medical research. Given this high motivational level, any agent that will aid in the quest for excellence will have a strong appeal.

In the late 1960s and early 1970s, several popular books and articles were written accusing the sports establishment of encouraging athletes, and especially professional athletes, to improve performance through the use of drugs. The issue of drugs in athletics is still very much alive—even for athletes in high school and perhaps in junior high school.

Drug Design

In many cases drugs are designed to supplant the natural biological activity of the body. In fact nearly half of the medically prescribed drugs are derived from naturally occurring materials. The hormones of the body are the most commonly mimicked biological substances, especially those drugs proposed to aid the actual performance or the training adaptation.

Stimulants

AMPHETAMINES

Amphetamines are a group of chemicals that have a strong, general excitatory effect on the central nervous system. The name *amphetamine* is derived from the first letters of the chemical name <u>a</u>lpha <u>meth</u> phenyl <u>e</u>thyl <u>amine</u>. Amphetamines are sympathomimetic: they mimic the effects of the sympathetic autonomic nervous system, which releases

norepinephrine (noradrenaline) and the primary adrenal medulla hormone *epinephrine* (adrenaline). Both of these hormones are normally elevated in the blood during exercise of a reasonable intensity or merely by stress alone. Two common trade names for commercially available amphetamines are Benzedrine and Dexedrine. The effects can generally be classified as related to the central nervous system, heart, blood pressure, peripheral circulation, and metabolism.

Amphetamines are very effective in excitation of the respiratory centers; animals show increased motor activity upon administration of amphetamines. Amphetamines are known to induce a state of wakefulness and to reduce the sensation of fatigue. Some side effects of amphetamines are dizziness, headaches, confusion, and eventually extreme fatigue. Appetite is reduced by amphetamines; consequently amphetamines serve as a questionable treatment for obesity. Acceptable dosages increase metabolic rate, which further aids in weight reduction. Although blood fatty acid levels are elevated by amphetamines, carbohydrate metabolism is affected minimally. A marked tolerance to amphetamines has been recorded in some "users"; addiction to amphetamines may occur.

Effect of Amphetamines on Performance. Performance of some simple motor tasks appears to improve with the intake of amphetamines. In general and under some conditions, the duration that work can be tolerated is lengthened. Performance time of rats doubles within 15 minutes after the administration of amphetamines at a dosage of 20 mg/kg. Therapeutic dosages for people usually range from 2 to 10 mg per individual. It is difficult to compare the dosages in rats and people because of the difference in metabolic rates. Available data on people, in contrast to rats, suggest that amphetamines do not enhance the duration of performance when the work is of low intensity; however, performance time is lengthened at higher work intensities.

The effect of amphetamines on performance must be considered in light of the physical and mental state of the recipient of the drug. Since amphetamines relieve some of the sensations of fatigue, it seems that they would be more effective when used by a subject in a fatigued state. But this assumption has been tested and was not supported. Also it is felt that taking the stimulant during a state of fatigue is potentially more dangerous than taking it when in a normal resting state.

Amphetamines exert their effects by: (1) increasing the release of catecholamines from the natural stores (sympathetic nerve terminals and adrenal medulla); (2) inhibiting re-uptake of released catecholamines into storage vesicles; (3) directly activating the brain receptor sites for catecholamines; and/or (4) any combination of these.

Dangers of Amphetamines. There are logically sound reasons for controlling the use of amphetamines as a means of enhancing performance. As stated earlier, an individual can become addicted—or at least emotionally dependent on them. A great tolerance, with respect to some of the specific effects, is built up with continual usage; thus larger dosages are required. The effect of long-term use of amphetamines is not fully known: they may induce cardiovascular pathology when taken in large doses. The potential enhancement of performance will not be necessarily realized because of the specific *nature* of the performance required and because of the *variation* of the drug effects in relation to the general physical and mental condition of the individual. In terms of maximal athletic performance, the fact that a dangerous and unknown element is involved when the body is stressed to a maximal effort makes it undesirable to administer amphetamines.

CAFFEINE AND RELATED DRUGS

A phenomenal amount of *caffeine* and related compounds is consumed daily by people of practically all nations. These compounds are found in plants throughout the world—the seeds of coffee being the most common. Tea contains *theophyline* as well as caffeine. Cola drinks also contain caffeine, which comes from the nuts of a cola tree. A single cup of coffee contains from 15–300 mg of caffeine.

Caffeine, like the amphetamines, stimulates the central nervous system. Caffeine is one of the substances that most rapidly passes across the blood-brain barrier, similar to alcohol and morphine.

Caffeine, theophyline, and threobromine exert their effects via the *cyclic adenosine monophosphate* (cAMP) system. For example theophyline is a drug used routinely in experiments that inhibit the enzyme *phosphodiesterase*, which normally degrades cAMP. Because cAMP is a metabolic stimulant of glycogenolysis and lipolysis, inhibition of phosphodiesterase potentiates the cAMP effect on metabolism, and metabolism may be elevated by 25%.

Effect on Performance. A threefold increase in running time to exhaustion has been demonstrated in mice after ten minutes, one hour, and two hours of the oral administration of caffeine. Caffeine usually eliminates drowsiness and fatigue, and facilitates clearer thoughts. Sensory stimuli are more noticeable, and sensory reflexes are more rapid than normally. Several cups of coffee might be enough to hinder performance of fine neuromuscular control; however, skills such as typing are improved with the intake of caffeine. Because of the calcium-activating effect of caffeine, as well as its stimulatory effect on the central nervous system, it enhances strength of muscular contraction and makes the muscle less fatigable.

The overall effect seems to be a much more practical means of stimulation than amphetamines: caffeine enhances strength and endurance, and its side effects are better known and less feared due to the traditional scope and frequency of its use. It is astonishing how little this drug has been tested in relation to its effect on physical performance.

Dangers of Caffeine. Caffeine markedly enhances urine formation. Gastric secretions likewise are augmented and can eventually induce ulcers. The reduction of blood-clotting time with high dosages suggests an increased susceptibility to intravascular clotting. There is some evidence that several cups of coffee ingested daily over a long period of time increase an individual's susceptibility to coronary heart disease.

Tranquilizers

In general tranquilizers act on the central nervous system to separate intellectual capabilities from emotional reactions. The potential for this class of drugs in physical activity is to relieve the acute and chronic state of anxiety without dulling the senses or reactions. Some tranquilizers, however, depress transmission in the spinal cord and thereby cause a reduction in muscular strength.

Given intravenously or intramuscularly, tranquilizers can be used as relaxants. Drugs such as Valium are used for this purpose. This drug seems to have no neural depressant effects; it acts solely on the muscular tissue.

All tranquilizers given in large enough quantities will produce drowsiness, sleep, and eventually coma. There is little chance for addiction; however, habituation can occur, necessitating increasingly larger doses.

Common tranquilizers are a derivative of phenothiazide; trade names include Comazine, Thorazine, Stelazine, and Mellaril. Although not a derivative of pheothiazines, Meprobamate is one of the most common drugs for the treatment of anxiety. Among the minor tranquilizers (meaning that they are not used only for hospitalized and/or psychotic patients) are Librium and Valium.

No evidence is available on the effect of these drugs on exercise performance. Perhaps a more important question is what are the tranquilizing effects of exercise instead of the effects of tranquilizers on exercise. It has been found that a brief walk at a pace sufficient to maintain a heart rate of 100 beats/min is a more effective muscle relaxant than a common tranquilizer, Meprobamate. The muscle relaxant effect lasts for at least an hour. Any side effects that we might incur from the exercise will probably be desirable, unlike the case with pharmacological agents.

Barbiturates

The site of action of the barbiturates is the central nervous system. These drugs are applied in graded degrees of potency and can be used for quick anesthesia in hospitals, prescribed for insomnia, and in the less potent form can be used as depressants.

Common trade names for these drugs are *Butisol, Nembutal,* and *Seconal.* These drugs act as depressants, and it is difficult to hypothesize any possible beneficial effect on physical activity. They have been used as a method to "get high," but their overall effect is to bring about a depressed effect, similar to alcohol. The "perceived" effect may mask the actual biological mode of action.

Pain Relievers

Aspirin (*salicylates*) and morphine are commonly known pain relievers. These two drugs are the extremes of possible pain relievers that can be used. The mechanism of their effect on the central nervous system is not understood. Aspirin stimulates respiration and increases O_2 consumption and CO_2 production. These effects may be related to the fact that salicylates cause uncoupling of oxidative phosphorylation in mitochondria. High salicylate dosages also increase epinephrine secretions, which in turn induce hyperglycemia by activating glycogenolysis. Similarly salicylates enhance fatty acid oxidation within skeletal muscle. Hyperventilation occurs when the plasma concentration of salicylate is about 35 mg%. Although ACTH secretion is enhanced by large dosages of salicylates, this is not the mechanism through which it induces its antiinflammatory effect. Interestingly, most aspirin contain caffeine also.

The *phenacetin* in aspirin also reduces fever. Chronic ingestion of phenacetin can lead to anemia because of the shortened life span of the red blood cells. Peptic ulcers are also caused by a chronic intake of aspirin.

To summarize, aspirin has proven to be a relatively effective and safe agent in relieving the pain associated with a variety of adverse conditions. Its usefulness in maximizing physical performance is probably limited to the role of a pain killer. There is no reason to suspect that it can improve performance in a normal individual free of injury.

Specific narcotic drugs used to relieve pain are Demerol, Codeine, and Percodan. Nonnarcotic pain relievers include Talwin, Darvon, Tylenol, and Norgesic. Specific muscle relaxants include Norgesic, Parafon-forte, and Equagesic.

Antiinflammatory Drugs

Antiinflammatory drugs are used to reduce the pain and inflammatory process resulting from sprains and strains. The pain from a muscle "pull" is usually thought of as a tearing of muscle fibers. This may lead to an intracellular "leak" into the surrounding interstitial space and activate the antiinflammatory action of the body's defense mechanism. Either swelling takes place and puts pressure on the nerve endings, or direct physical trauma to the sensory nerve endings causes the pain. *Adrenocortical steroids* and their synthetic analogues play a key role in the maintenance of injured muscles and joints in a condition adequate for continued physical performance.

As discussed in Chapter 11, *adrenocorticosteroids* are secreted by the adrenal cortex in response to release of *adrenocorticotrophic hormone* (ACTH) from the anterior pituitary. Exercise and any stress will initiate the release of ACTH. Although a major effect of adrenocorticosteroids is metabolic, its use in athletics is basically for its antiinflammatory action. In fact the metabolic effect is undesirable promoting muscular protein *catabolism*, whereas muscle growth (*anabolism*) is generally desired by athletes. Furthermore it has been shown that muscle atrophy induced by steroids is selective for fast-twitch muscle fibers, the ones which have the greatest tension-producing capacity.

Corticosterone and *cortisol* are two of the common steroids secreted by the adrenal cortex. They are synthesized from cholesterol and differ chemically from the anabolic steroids in that the corticosteroid molecules have 21 carbons as opposed to 19.

Commonly used synthetic steroids are *Prednisolone* and *Triamcinolone*. These steroids reduce local redness, swelling, and tenderness characteristic of the inflammatory process. Some of the nonendocrine-related diseased conditions for which these steroids are used therapeutically are progressive arthritis, rheumatic carditis, renal diseases, bronchial asthma, chronic colitis, skin diseases, and so forth. This multitude of maladies for which steroids are used is of interest to the exercise biologist since in the future it is reasonable to expect that exercise programs will be formulated that can counteract the proteolytic or muscle-atrophying effect of the steroids. This may be one way to minimize a critical side effect of a medicinal agent that is essential for the maintenance of life for some individuals. Peptic ulceration and osteoporosis (bone dissolution) are harmful side effects of steroids; proper exercise programs theoretically would have beneficial effects.

Anabolic Steroids

Anabolic steroids refer to those drugs that mimic the effects of testosterone minus the androgenic (masculinizing) effect. The purpose of these drugs is to promote additional muscular growth without affecting sexual characteristics. These drugs have been used to promote increased body weight by increasing muscle mass. In those sports where strength and high power are important components, this additional body weight is an aid since we know of the close relationship between body weight and strength.

The growth-promoting effects of these drugs are well established. However, the athlete subjects his body to potential risks including liver damage, abnormal sperm counts, and loss of libido. Furthermore the use of these drugs in immature individuals should be *avoided* without question because of the possibility of premature closure of the epiphyseal plates and of other harmful influences on growth. The use of these drugs in women is discouraged for similar reasons, and also to avoid the possible masculinizing effects.

Unlike the acute response of stimulatory drugs, the potential beneficial anabolic steroids effects are realized only after several weeks. Maximal muscular growth and strength occur when the anabolic steroids are ingested in concert with adequate availability of protein in the diet and with strength-demanding exercises.

Since the effects of anabolic steroids are long

term, it has been and continues to be a problem to detect the use of these drugs by athletes. For example the beneficial effects (and detrimental side effects) are realized long before the athlete arrives at the stadium. The athlete can abstain from taking a steroid for a few days prior to competition, and the drug will not be detected on the day of the contest. However, chemical analysis of the blood (newly developed in England) can determine whether anabolic steroids have been taken on days preceding the test day.

The use of these drugs in pathological conditions is well established, and the physical educator may encounter individuals under medication in the normal course of the physical education classes. However, the use of these drugs for athletic purposes, in order to gain advantages over "normal" physiological function, is questionable—to say the least. The possible gain of athletic achievement should be weighed against the possible detrimental effects on the body.

Not only is the taking of anabolic steroids of questionable use to all individuals, but it may upset the normal hormonal functional balance, discussed in Chapter 11. Furthermore, in their zeal for excellence, athletes have been known to take up to 50–100 mg of these steroids per day when the maximum daily recommendation of the manufacturer is 5–10 mg.

General Aspects of Drugs

Drugs and Nutrition

In general we like to think of drugs as having beneficial effects in improving nutrition or in limiting disease. Medical personnel constantly must be aware of the differences between individuals and be sensitive to these differences. Similarly, in the use of pharmacological agents for the improvement of ability in physical exercise or in training, we must be ready to expect different reactions in different people. Pharma-

cological agents have not proven to be substitutes for the normal well-balanced diet. It is not known whether this inability to find the superior food additive is due to our not utilizing the correct variables or dosage levels or whether, in fact, no dietary supplement will increase the functional efficiency of the body above that of the normal well-balanced diet. Nutritionists tell of us that the latter case is closest to the truth.

Summary

Drugs are designed to stimulate naturally occurring chemicals (such as hormones) so that normal functions of a particular system can be manipulated. In one respect, practically all drugs are used to enhance performance: when a person is ill and is taking medication, the medication is to improve performance—performance in this case being the usual daily routine. All drugs, or potential ergogenic aids, have some side effects—that is, effects that occur in addition to those for which the drug was taken.

Stimulants (*amphetamines*) are used by some athletes, professional and amateur, to improve performance. They mimic the effects of the sympathetic nervous system and adrenal medulla. There is evidence to support the idea that stimulants augment performance but that the dangers outweigh their potential minimal benefit.

Caffeine is also used as an ergogenic aid, more commonly by office workers than athletes. Chronic coffee drinkers as a group have been linked to those with a higher incidence of heart disease and ulcers.

Muscle relaxants would seem to have some therapeutic value if taken during recovery from muscular injuries. As far as aiding the voluntarily relaxation of skeletal muscles, mild exercise can be more effective than the commonly used pharmacological muscle relaxants and, furthermore, has no undesirable side effects in a healthy individual.

Pain relievers have been used to maximize physical performance, but their effects are variable. The risks associated with the potential benefits do not warrant their use.

Anabolic steroids have been shown to improve strength-related performance when taken with an adequate diet and a physical training program. Again their potentially adverse side effects do not justify their use by normal individuals.

Study Questions

1. What is important about pharmacology and drug usage for people concerned with exercise?
2. What is the basis for the design of most drugs?
3. What are the primary effects of amphetamines?
4. What are the possible dangers associated with the use of amphetamines?
5. How can caffeine possibly alter human performance?
6. What is the primary use of tranquilizers?
7. Can pain relievers aid performance?
8. What are the uses of antiinflammatory drugs?
9. Can anabolic steroids make a difference in strength development?
10. What are the dangers associated with anabolic steroids?
11. How do drugs alter dietary needs?

Review References

Anders, M.W. Enhancement and inhibition of drug metabolism. *Ann. Rev. Pharmacol.* 11:37–56, 1971.

Bhagat, B. and N. Wheeler. Effect of amphetamine on the swimming endurance of rats. *Neuropharmacology* 12:711–713, 1973.

Cooper, Donald L. Drugs and the athlete. *J. Amer. Med. Assn.* 221:1007–1011, 1972.

deVries, Herbert and G.M. Adams. Electromyographic comparison of single doses of exercise and meprobamate as to effects on muscular relaxation. *Amer. J. Phys. Med.* 51:130–141, 1972.

Halberstam, Michael. *The Pills in Your Life.* New
  York: Grosset and Dunlap, 1972.

Kouroounakic, P. Pharmacological conditioning for
  sporting events. *Am. J. Pharmacy* 144:151–158,
  1972.

Morgan, William P., ed. *Ergogenic Aids and
  Human Performance.* New York: Academic
  Press, 1972.

Riker, W.F. and M. Okamoto. Pharmacology of
  motor nerve terminals. *Ann. Rev. Pharmacol.*
  9:173–208, 1969.

Villa, R.F. and P. Panceri. Action of some drugs on
  performance time in mice. *Il Faraco* 1:43–48,
  1973.

Williams, M.H. *Drugs and Athletic Performance.*
  Springfield, Ill.: Charles C Thomas, 1974.

CHAPTER NINETEEN

The role of exercise in cardiovascular disease

Key Concepts

• A desirable physical fitness program stresses the attainment of cardiovascular fitness.

• Cardiovascular fitness can be assessed by electrocardiogram and by oxygen uptake monitoring during maximal and submaximal exercise stress tests.

• Cigarette smoking, diet, and lack of physical activity are indicators of a high risk factor for the occurrence of coronary heart disease.

• Exercise is a useful tool in the prevention and treatment of, and the rehabilitation from, cardiovascular diseases.

• Drug therapy may be used to help control cardiovascular disease.

Physical Fitness

The role of physical fitness during a person's lifetime has varied dimensions from early childhood to old age. One's type of activity, one's motivation, and one's ability to perform the movement skills are constantly changing. Therefore we do not propose to present a concise definition of the term *physical fitness.* We feel that physical fitness can be described as a state of "feeling good about yourself" and being relatively free from chronic diseases. Physical fitness as we will use it in this chapter is primarily concerned with cardiovascular fitness. In the process of training for cardiovascular fitness, other standard physical fitness criteria, such as flexibility, are often included in the training schedule.

A thorough search of bookstores will uncover a multitude of books concerned with physical fitness and health. These programs are based on the development of a high level of cardiovascular fitness and on the maintenance of weight control. The details of these various fitness programs are outside the scope of this book, but it is hoped that the reader will explore the possible fitness programs and evaluate them in light of fitness goals. Special fitness goals are unique to a given individual, and fitness criteria should be designed specifically for these special goals.

The need of a fitness program is very often masked by the gradual deterioration of fitness indicators. If the rate of change is sufficiently small, the difference in the widening gap between two points becomes less perceivable. Similarly, if we gradually lose physical fitness characteristics, we are less aware of the loss of fitness than if the loss took place over a short time span.

We generally consider preadolescents to be in reasonably good physical condition; they can run and jump with great vigor for long periods of time. As they proceed from the junior high school years, through the mid-twenties, and into the early forties, physical fitness can gradually deteriorate, if not maintained by regular exercise sessions. After completing the years of formal education, (at approximately 20 years of age) an individual is very often involved with becoming professionally established; this process consumes time until the mid-forties. The stress of becoming involved in work-related activities, coupled with possible marriage and family obligations, may occupy the "apparent" energy of these people. It is not until some later point in their lives that enough daily time becomes readily available for consideration of personal physical fitness. By this time many of the fitness components have deteriorated, and these individuals find that bodily functions are hampered by years of relative inactivity. Thus we believe it is important to convince young people of the need to invest in their bodies by reserving the time for daily exercises.

We feel that the most convenient way to inform people about the advantages of personal fitness is through physical education classes in the public schools. These classes provide the time to teach about the body and to develop skills that can be used in the attainment and maintenance of personal fitness.

Carryover Activities as Taught in the Schools

The skills necessary for participation in activities that have lifetime participation possibilities are often taught in physical education classes. These skills usually involve equipment that is readily available, and the activities are such that a great number of additional skilled participants are not required. The more highly organized athletic activities are very seldom participated in after the formal education years (e.g., football as played in the United States).

Swimming, running, and cycling are sports that do not require highly organized structures; the amount of attainable cardiovascular fitness in these sports is very high. Although other carryover sports (bowling, golfing, softball, and archery) provide personal entertainment, their contribution to cardiovascular fitness is much less. A third classification of sport activities that can contribute to cardiovascular fitness includes such activities as tennis and handball, but these require the use of an appropriately skilled partner and available facilities.

As a first choice, we would prefer to see taught those individual sports that stress cardiovascular fitness, do not require a partner or extensive facilities, and can be self-motivating. Second, we would recommend those activities that need a minimum of participants and few readily available facilities.

Motivation is one of the key factors in the promotion of physical fitness programs for either individuals or groups. So many times we have seen fitness programs begin with 20 to 30 participants, yet within the short period of two weeks the number has dwindled to less than ten. This dropout rate can very often be correlated to lack of good leadership, minimum variety of activities, and low motivating factors in the program. A group program of calisthenics, jogging, and other exercises is not enough if the program does not stimulate internal motivation on the part of each participant. Supplemental techniques, such as information about the mechanisms by which daily exercises can alter the human body, may be used to supply the critical internal motivation.

Cardiovascular Fitness

An exercise program designed for cardiovascular fitness is not a great time consumer if done every day. Basically, people only need to exercise enough to bring their heart rates to about 140 for 20 minutes, five days a week. This exercise intensity is only a guideline for fitness; it is possible to develop many variations of this training schedule. For example the exercise may be of such a nature that the heart rate may not rise above 120, but the total exercise time will be extended to over one hour. Perhaps a heart rate of 150 for 15 minutes is more suited to the time goal or constraints of a given individual. We highly recommend that anyone interested in cardiovascular programs evaluate any of the several fitness programs.

Measuring Cardiovascular Fitness

Directly or indirectly measuring the ability of the body to utilize oxygen in maximal or submaximal exercise tests is routinely used as a measure of cardiovascular fitness. In the laboratory the direct measurement of oxygen uptake is the preferred method since it is considered to be the most valid and reliable. Indirect methods have been devised to measure physical fitness; most often these techniques utilize submaximal tests. The indirect tests estimate the maximal capability through the measurement of performance criteria either related to the heart rate or work performance.

The common methods employed to measure fitness are the bicycle ergometer, the treadmill, the step test, and the running performance tests. The objective of these tests is to relate oxygen uptake to work rate. In general, if one person's oxygen uptake or heart rate increases less than another person's, in response to the same work task, we would conclude that the former is more physically fit.

As a person exercises, using the above exercise tests, the oxygen utilization rate is a function of oxygen extraction by the lungs, oxygen delivery by the cardiovascular system, and oxygen utilization, primarily in the leg muscles.

It is possible to increase heart rate and thus oxygen uptake by swimming (primarily arm exercise) or by running (primarily leg exercise). The work of the heart may be comparable whether the exercise is through swimming or running. However, since the leg muscles have a greater total muscle mass than the arms, the increase in the rate of oxygen uptake may be greater in a training program involving running. In this chapter, we are not critically interested in the training specificity difference of swimming versus running: the difference is not what we would consider critical for the purpose of determining fitness for prevention of cardiovascular diseases.

The specific methods to assess cardiovascular fitness were discussed in Chapter 10.

Cardiovascular Diseases and Physical Fitness

In the coming decades, the philosophy of the medical profession must be geared to the preventive aspects of medicine rather than to a concentration on identification and treatment of diseases. Prepaid medical care will make it economically beneficial for the physician to maintain the general population in a healthy state. This concept will drastically alter medical services in that medical income will not depend upon the number of sick, but the income will be relatively increased in proportion to the number of healthy patients; that is, a group of doctors with healthier patients will be in a position to enlist more persons in their prepaid plan.

Coronary heart disease is singled out in this chapter because of the established role of inactivity in the etiology of the disease. Cardiovascular diseases claim the lives of over 50% of all men over 40 in the U.S.A.; a white male's chances of having a clinically diagnosed coronary heart disease before the age of 60 is one in five. For any given year the trend has been that over one million people suffer heart attacks and over 600,000 deaths are attributed to heart disease. These facts are presented to give an indication of the prevalence of the disease. We say prevalence because the end point —heart attack—is only the culmination of the disease, which may begin as early as childhood.

It appears that the etiology of coronary artery disease (*atherosclerosis*) does not involve any single factor but a multitude of factors acting independently or in combination. The disease involves a gradual decrease in the size of the arterial lumen through which blood can flow (Figure 19–1). In general the stages of the development of this disease include: (1) a sensitizing of the artery, (2) lipid accumulation, (3) disruption of the ecology of the artery, (4) necrosis of the arterial walls, (5) calcification and ossification, (6) rupture of the arterial wall, (7) thrombus formation, and (8) cardiac infarction. These proposed stages are not discrete: one stage leads into another without any clear differentiation.

We know from epidemilogical studies that there are definite *risk factors* involved with the etiology of coronary heart disease. These risk factors do not operate separately but in a combination that increases the chances to develop coronary heart disease. The most prevalent risk factors are cigarette smoking, poorly con-

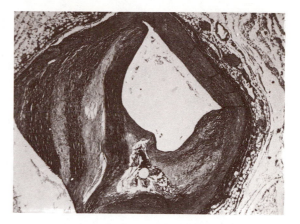

Figure 19–1. Plaque buildup in human coronary arteries.

From Meyer Friedman, M.D. The coronary thrombus: Its origin and fate. *Human Pathology* 2(1):81–128, 1971.

trolled diet, and lack of physical activity. Other factors include a family history of coronary heart disease, elevated blood pressure, elevated blood lipids, increased dietary intake of fat and cholesterol, increased body weight, diabetes or elevated levels of blood sugar, and increased socio-psychologic tensions.

In treating the disease, the practicing physician is most likely to discourage cigarette smoking, identify and control hypertension, prescribe dietary regimens that lower blood lipids and blood sugar, advise weight reduction for the overweight, and encourage regular habits of exercise. Figure 19–2 illustrates the complexity of the life of the modern citizen and the contributing factors to coronary heart disease.

Risk Factors

CIGARETTE SMOKING

Cigarette smoking is known to increase susceptibility to coronary heart disease by 300%. Smoking practically always reduces the air exchange in the lungs, and the smoking particles may cause clogging of the alveoli. Nicotine causes the heart to beat faster, causes constriction of the small arteries, and combines with

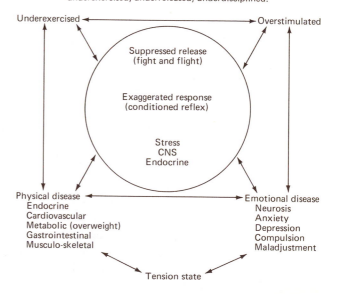

Figure 19–2. The mechanized, urbanized, unbalanced individual.

From H. Kraus, Preventive aspects of physical fitness. In *Prevention of Ischemic Heart Disease* (Wilhelm Raab, ed.). Springfield, Ill.: Charles C Thomas, 1966.

heparin, thereby inhibiting the anticlotting mechanisms in the blood. The inhaled carbon monoxide reacts with hemoglobin and displaces oxygen that is bound to hemoglobin, thereby reducing the oxygen-carrying capacity of the blood (see Chapter 10).

DIET

Being overweight and choosing a diet that contains a high percentage of fat are two prime risk factors in coronary heart disease. Being overweight alone is not thought to be a primary causative factor; but the loss of diet control, which often accompanies this condition, may be the prime factor. High levels of lipids in the diet and/or in the blood stream have been related to an increased risk factor of about three to one over those individuals with lower lipid levels. Blood cholesterol levels, triglyceride, and other lipid components combine to increase the risk factor. Cholesterol levels above 285 mg% result in an increased incidence of coronary artery disease of 60–70%. A very low percentage of people with blood cholesterol levels below 200 mg% exhibit the advanced stages of coronary heart disease.

A diet high in polyunsaturated fats in relation to saturated fats lowers serum cholesterol. In a nine-year study at the Los Angeles Veterans Administration Center, 424 men on an experimental diet, containing a normal and equal amount of fat except with a lower proportion of saturated fats, had a higher survival rate than 422 men on a conventional diet. Death, as a result of heart diseases, heart attacks or cerebral strokes, occurred in 84 subjects on the regular diet and in 52 on the special diet. An additional 11 on the regular diet and 4 on the special diet died as a result of artherosclerosis. The most dramatic results were in the "younger" subjects who were from 55 to 65 years when the study began.

PHYSICAL ACTIVITY

Physical activity has been negatively correlated with the incidence of coronary heart disease in all of the major comprehensive studies. Figure 19–3 illustrates that the daily level of physical activity is related to the occurrence of coronary heart disease. In this case the exercise was classified in terms of hours per day, and even the shortest time classification provided a protective effect.

Many, but not all, clinicians feel that lack of physical activity is an important factor in the development of heart disease. Those athletes who remain active have lower coronary heart disease (CHD) mortality than those athletes who become inactive. Also, it seems clear the participation in athletics only as a youth does not have any beneficial effects in terms of longevity. Physically inactive people have a mortality rate roughly seven times higher than even smokers of the same age who are active.

Many studies have been conducted on personnel in various occupations with varying requirements of physical activity. Most of these studies show that the jobs necessitating the greatest physical activity have half the mortality rates; a similar correlation appears in the relationship between physical activity and the severity of the heart attack when it occurs (Figure 19–4). Patients with the highest job-centered physical activity and also with high off-the-job physical activity had the lowest mortality rate four weeks after myocardial infarction; those with the least activity had the highest mortality rate.

HEREDITY, HYPERTENSION, AND STRESS

Heredity is one factor that we cannot change: we cannot choose our own parents. We must examine our hereditary traits and try to modify any condition that is unfavorable. Almost

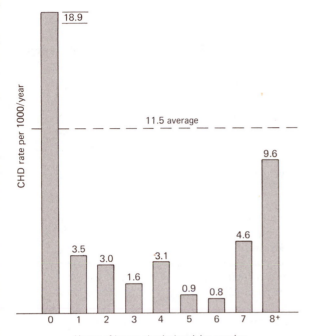

Figure 19-3. The relationship between hours of job-connected physical activity per day and the incidence of coronary heart disease. Study taken in a six-county area of North Dakota (1957), based on a 10% population sample of the males over 35 years of age.

From William L. Haskell. Physical activity and the prevention of coronary heart disease: What type exercise might be effective. J. S.C. Med. Assoc., vol. 65 (Suppl. 1), December 1969.

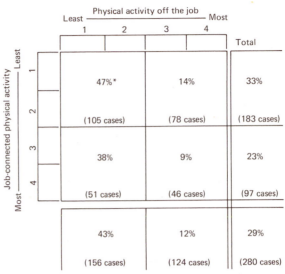

*Percentage of patients dead within 4 weeks

Figure 19-4. Distribution of CHD patients and mortality when patient population is expressed as a function of job and off-the-job physical activity.

From C.W. Frank; E. Weinblatt; S. Shapiro; and R.V. Sager. Physical inactivity as a lethal factor in myocardial infarction among men. *Circulation,* 34:1022–1033, December 1966. By permission of The American Heart Association, Inc.

twice as many people contract coronary heart disease who have a history of the disease in the family. Body build is another factor in heredity; mesomorphic individuals seem more prone to heart attacks. It has been advanced that higher incidences of CHD are found in those people who are excessive in their habits—working too hard, playing too hard, and given to excesses in eating, smoking, and drinking alcohol.

Table 19-1. Cardiac Risk Index

Age	10 to 20 (1)	21 to 30 (2)	31 to 40 (3)
Heredity	No known history of heart disease (1)	One relative with cardio-vascular disease over 60 (2)	Two relatives with cardio-vascular disease over 60 (3)
Weight	More than 5 lb below standard weight (0)	Standard weight (1)	5–20 lb overweight (2)
Tobacco smoking	Nonuser (0)	Cigar and/ or pipe (1)	10 cigarettes or less a day (2)
Exercise	Intensive occupational and recreational exertion (1)	Moderate occupational and recreational exertion (2)	Sedentary work and intense recreational exertion (3)
Cholesterol or % fat in diet	Cholesterol below 180 mg; diet contains no animal or solid fats (1)	Cholesterol 181–205 mg; diet contains 10% animal or solid fats (2)	Cholesterol 206–230 mg; diet contains 20% animal or solid fats (3)
Blood pressure	100 upper reading (1)	120 upper reading (2)	140 upper reading (3)
Sex	Female (1)	Female over 45 (2)	Male (3)

Hypertension is diagnosed when systolic blood pressure exceeds 160 mm Hg and/or diastolic pressure exceeds 90 mm Hg. A diastolic pressure of over 95 has been related to an increased risk factor of about six times; those with systolic pressures of over 160 have four times the risk of cardiovascular diseases.

Mental tension, brought about by the constant strain of pushing oneself throughout the day or week or year, has been identified as an increased risk factor. This mental tension can often be correlated to the other risk factors of hypertension, overweightness, cigarette smoking, and lack of physical exercise. Mental tension in combination with other factors can increase the risk factor by five to one.

SUMMARY

It is possible to grade ourselves on the above risk factors, and it is likely that all of us run some risk of coronary heart disease (Table 19-1). However, the risk greatly increases when combinations of the above factors exist. For

Table 19–1. (cont.)

41 to 50 (4)	51 to 60 (6)	61 to 70 and over (8)
One relative with cardio-vascular disease under 60 (4)	Two relatives with cardio-vascular disease under 60 (6)	Three relatives with cardio-vascular disease under 60 (8)
21–35 lb overweight (3)	36–50 lb overweight (5)	51–65 lb overweight (7)
20 cigarettes a day (3)	30 cigarettes a day (5)	40 cigarettes a day or more (8)
Sedentary occupational and moderate recreational exertion (5)	Sedentary work and light recreational exertion (6)	Complete lack of all exercise (8)
Cholesterol 231–255 mg; diet contains 30% animal or solid fats (4)	Cholesterol 256–280 mg; diet contains 40% animal or solid fats (5)	Cholesterol 281–330 mg; diet contains 50% animal or solid fats (7)
160 upper reading (4)	180 upper reading (6)	200 or over upper reading (8)
Bald male (4)	Bald, short, male (6)	Bald, short, stocky male (7)

Numbers in parentheses indicate scoring points. To obtain your total score, find a score for each of the 8 categories listed in the far-left column. For example, in the "age" category, if you are between the ages of 21 and 30, give yourself 2 points. In the "heredity" category, if you have two relatives under 60 with cardiovascular disease, give yourself 6 points. If you are 21–35 lbs overweight, give yourself 3 points, and so forth. After the total score is computed, the test should be graded as follows: Group I, 6–11 points, very low risk; Group II, 12–17 points, low risk; Group III, 18–25 points, average risk; Group IV, 26–32 points, high risk; Group V, 33–42 points, dangerous; Group VI, 43–60 points, extremely dangerous.

With permission of J.L. Boyer, M.D. Exercise Physiology Laboratory, San Diego State University, California.

example, the combination of a lack of physical exercise, cigarette smoking, and high-fat diet will increase the risk far above that of any of the three separately.

The Role of Physical Activity in Controlling Coronary Heart Disease

Most experts agree that coronary heart disease (CHD) represents the greatest health hazard known to exist at the present time. Physical

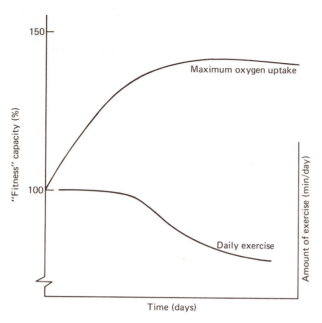

Figure 19-5. Relationship between "getting into shape" and maintaining physical condition.

activity can play a significant role in the prevention and detection of, and rehabilitation from, these diseases. A discussion of exercise in connection with coronary heart disease is related to our emphasis on long-term regular exercise. Although the initiation of an exercise program must be at a low level, a gradual progression in intensity should take place over a period of 6–12 weeks. To get "in shape," a person must gradually increase exercise to the point where he or she can participate in as much as five 60-minute exercise sessions per week. Figure 19-5 shows a possible improvement rate and the amount of daily exercise necessary to maintain the improved level of fitness. We can estimate that the minimum weekly exercise level should consist of three 20-minute exercise sessions. This should be considered as a *maintenance* level.

Prevention

The prevention of coronary heart disease is the area where we believe physical activity can play the most significant role. It is not enough to know that the lack of physical exercise can increase the risk factor by three times: a person must be motivated to perform the daily activity. High levels of regular physical activity can modify the repercussions of most of the other risk factors.

Table 19-2 lists the possible benefits gained from regular physical activity in relation to the etiology of coronary heart disease. Sufficiently intense exercise is known to decrease the resting levels of norepinephrine in the heart. This decreased content would allow for less norepinephrine to be released during times of great stress; the result would be a less immediate inotropic response to the stress.

The development of coronary collateral circulation is an important mechanism by which exercise can protect the body from crippling myocardial infarctions. Collateral circulation implies that each myocardial fiber is supplied by more capillaries; thus, if one of the capillaries is blocked, the muscle fiber will not be without a supply of oxygen and nutriments. In a situation where the only available supply of oxygen is shut off, tissue necrosis occurs, which may result in a myocardial infarction. Alternative capillary supply routes could prevent a single fiber from becoming necrotic.

It has been stated that an enlarged but normal heart, like the heart of many athletes, is not necessarily beneficial since the blood-diffusion capacity is reduced, i.e., the distance from a capillary to the center of a hypertrophied fiber is greater than in a normal-sized fiber. It seems possible that the increased vascularization (collateral circulation) more than compensates for the increase in actual muscle mass or fiber diameter.

Other beneficial effects of the "trained" heart probably result from an improvement in the efficiency of the heart. For example bradycardia, greater stroke volume and cardiac output, increased diastolic period (cardiac muscle is not perfused with blood during systole), decreased heart rate per work load, hypertrophy, occasionally reduced systolic blood pressure, and more efficient distribution of blood to the active muscles are typically found in trained individuals. This last factor reflects adaptation within the skeletal-muscle vasculature and perhaps other peripheral, sensory, and CNS adaptations involved in regulation of cardiac physiology during exercise.

It is apparent that whatever cardiovascular benefit that can be derived from exercise is best achieved by participating in exercise activities throughout a lifetime. We cannot store up the benefits. One of the beneficial side effects of regular exercise may be an individual's increased awareness of living more prudently in relation to smoking, weight control, nutritional considerations, physical activity, and methods of handling stress.

Detection

As in the case of most diseases, the earlier the detection of heart disease, the more likely the prescribed cure will alleviate the condition. By placing the heart under the controlled stress of exercise, it is possible to observe the performance of the heart under conditions of stress. For this purpose it is possible to use maximal or submaximal exercise stress tests. During the exercise testing period, the ECG is monitored and observed for abnormalities. The ECG, during the subject's recovery, can also lend information to the detection of cardiac malfunction. The most common ECG abnormality seen in coronary disease is a depressed ST segment (see Chapter 8).

Table 19-2. The Apparent Benefits of a Trained Heart

By Increasing:	By Decreasing:
Coronary collateral vascularization	Serum lipid levels, triglycerides, and cholesterol
Myocardial efficiency	Glucose intolerance
Efficiency of peripheral blood distribution and return	Arterial blood pressure
Fibrinolytic capability	Neurohormonal influences
Red blood-cell mass and blood volume	"Strain" associated with psychic "stress," resulting in less depression, better sleeping habits, and improved self-image
Tolerance to stress	
Prudent living habits	
Physical work capacity	
Stroke volume	
Diastolic period of the heart beat	Heart rate
Maximal cardiac output	Lactate levels following all levels of work
Maximal oxygen uptake	
Arterial-venous oxygen difference	
Restitution of heart rate post exercise	
Myocardial hypertrophy	
Pulmonary blood-flow distribution	
Vital capacity	
Maximal breathing capacity	

Adapted from S.M. Fox and W.L. Haskell, "Physical activity and the prevention of coronary heart disease," *Bull. N.Y. Acad. Med.* 44:950–967, 1968.

Another method used to assess cardiovascular function is oxygen uptake tests of the maximal or submaximal variety. If a subject has coronary heart disease, we would expect to see a reduced capacity for exercise and reduced oxygen uptake values.

The ECG test records the electrical characteristics of the heart; the oxygen uptake measurements indicate the functional capability of the heart (cardiac output). Decreased oxygen uptake may reflect a decreased cardiac output induced by lowered ventricular filling, an increased peripheral resistance, a regurgitating flow, or a weakened myocardial contraction. Changes in the ECG indicate abnormal electrical conductance in the heart, such as alterations due to necrotic tissue that cannot react to the electrical signals. We should be aware that the interpretation of the normality of an ECG varies widely from one observer to another. Fourteen cardiologists who assigned abnormal responses to ECG after exercise varied from 5% to 58% in their ratings. And this figure was even higher when diagnosis was made during, as opposed to after, the exercise test. The combination of the two methods (ECG analysis and oxygen uptake tests) gives the most complete evaluation of the functional capacity of the heart.

Rehabilitation

It is now commonly accepted that the most efficient way to recover from a myocardial infarction is to begin a controlled exercise program within a few days following the attack. The rationale for this approach is to retrain the heart for increased efficiency of operation.

The methods used in the prevention of cardiac infarctions are almost identical to those that can be used in the rehabilitation programs, although the initial stress should be less in the latter. The peripheral changes in the cardiovascular systems will be similar, if not identical, to those changes we normally see in well-trained individuals. All of these peripheral changes will decrease the load on the heart.

Daily exercise during rehabilitation from myocardial infarction will decrease the amount of catecholamines in the heart. This in turn will decrease the excessive oxygen utilization associated with increased sympathetic response to stressful situations. The adaptations to the regular exercise regimen allow for more oxygen to become available through collateral circulation and decrease the possibility of massive catecholamine release during stress.

The exercise program should be one that gradually evolves from a short walking exercise to a mild jogging program. Careful monitoring of the ECG and oxygen uptake values should be performed as often as possible. Most rehabilitation programs use a supervised, graded physical activity program. These programs have reported increased vital capacities, maximal breathing capacities, ECG improvements, and increased "overall fitness".

The training program can increase a cardiac patient's aerobic working capacity to a degree comparable to a healthy subject. Reductions in systolic and diastolic blood pressure—at rest and throughout exercise—may be experienced with training. After two months of training, increased exercise tolerance can be related to a lowered oxygen demand on the heart at a given work load. In more long-term training programs (greater than two years), more subtle beneficial factors—yet unverified—seem to be involved. Stroke volume tends to increase and myocardial work decrease at a given work load in relation to the pretraining state of the subject. In a subject who has shown some ischemic changes in the exercise ECG, a training program may produce better than a 50–50 chance of eliminating those ECG patterns.

The reduction in coronary blood flow, resulting in myocardial ischemia, may be stimulated by nervous anxiety. Thus it has been proposed that the beneficial effects of a conditioning program on the heart are simply the alleviation of anxiety. Future research in this area should show us exactly which conditions should be treated with how much and what kind of training.

It should be realized that there are also some contraindications of physical activity related to the cardiovascular system. Individuals with evidence of potential or recent infarction, congestive heart failure, myocarditis, aneurysm, aortic stenosis, arrhythmias, unexplained cardiac enlargement, hypertension, recent pulmonary embolism, or hepatic or renal insufficiency are suffering from conditions that could be complicated by increased physical activity.

Summary

Physical training has been shown to alter those factors associated with coronary heart disease by improving the circulation within the heart, improving the power output of the heart, decreasing the catecholamine content of the heart, decreasing the sympathetic tone of the heart, improving the peripheral circulation, and improving the oxygen transport system.

Drug Therapy

Drug therapy is as popular as exercise therapy—or more so—for the prevention, treatment, or rehabilitation of cardiovascular diseases. The drugs used most often are those related to decreasing hypertension and hyperlipidemia. The rationale for the administration of drug therapy is obviously based on knowledge of those high-risk factors associated with coronary heart disease. The hypothesis is that, if we can control blood pressure and lipid concentration in the blood, we should be able to reduce the risk of myocardial infarction. Indeed there is a long list of drugs used to control hypertension and hyperlipidemia.

Drugs such as *digitalis* (used to stimulate the strength of the contraction) and *nitroglycerine* (used to dilate coronary blood vessels) are commonly used in the treatment of *acute* cardiac overload.

Hypertension

Hypertension is found in nearly 20 million Americans. The term *hypertensive epidemic* is now common. Identification of general hypertension can be made from a blood pressure reading; however, the diagnosis of the etiology can be very complicated. Coronary heart disease is greatly complicated by hypertension, which acts to hasten the development of the disease. At an advanced stage of the disease, it is hard to separate the vascular pathology and clinical manifestations due to the elevation of the blood pressure from those due to the basic atherosclerotic process.

Blood pressure is essentially dependent upon the constriction or relaxation of the peripheral blood vessels in the body. The relaxation and constriction of these vessels have been found to be dependent upon: (1) emotional impulses from the brain, (2) heart regulation, (3) fluid regulation in the body, (4) hormonal balances, (5) sympathetic impulses, and (6) neurotransmitting substances from nerves to the muscles of the blood vessels.

One class of drugs used to control hypertension is the psychopharmacologic drugs (sedatives and tranquilizers). These are antihypertensive since they diminish the central nervous system vasopressor influences induced by emotional and environmental stress.

Other mechanisms to release hypertensive symptoms include: diuretic drugs (thiazides and chlorothiazide), ganglionic blocking agents (mecamylamine-HCL and pentolinium bitartrate), adrenergic blocking drugs (guaithidine-sulfate and phenoxy benzamine) and vasodilator drugs (hydralazine, diozoxide, and prostaglandins).

Hyperlipidemia

There is little doubt that a distinct correlation exists between a high level of circulating lipid levels and a high risk factor for coronary heart disease. The serum lipids circulate in association with proteins and are commonly classified as alpha- or beta-lipoproteins. The alpha-lipoproteins are heavier than the beta-lipoproteins and contain about 30% of the serum cholesterol. The beta-lipoproteins carry 70% of the serum cholesterol and about 85% of the triglycerides. In a discussion of degrees of hyperlipidemia, it would be necessary to consider total distribution of blood lipids and the relationships between alpha- and beta-lipoproteins.

There are at least 70 drugs on the market that a physician may prescribe for lowering the lipid content of the blood. Clofibrate is effective in reducing blood lipid levels; the use of this drug has been associated with a lower number of nonfatal heart attacks. It is thought that clofibrate inhibits the rate of cholesterol production in the liver. The expected reduction of cholesterol in the blood is 7–35%, and for triglycerides the reduction is 21–61%. Sodium Destrothyroxine, a derivative of thyroid hormone, has been widely used in reducing cholesterol levels. Its mode of action is thought to be to stimulate the liver to increase the oxidative catabolism and excretion of cholesterol and its degradation products via the biliary route into the feces. The expected reduction in cholesterol levels is roughly 20%.

Other drugs such as cholestyramine, estrogens, Triparanol, and nicotinic acid have been used in combatting hyperlipidemia; however, the side effects are contraindicative to many individuals. The use of estrogens was stimulated on the basis of a known lower incidence of atherosclerotic disease in women.

Summary

It is possible to grade ourselves on the cardiovascular risk factors. All of us will have some risk of coronary heart disease just because we are alive. However, the risk greatly increases when combinations of cigarette smoking, uncontrolled diet, and lack of physical activity are characteristic of our daily lives. The combination of these three risk factors will increase the risk far above that of any one of the three factors by itself.

Physical training has been shown to alter those factors associated with coronary heart disease by improving the circulation within the heart, improving the power output of the heart, decreasing the catecholamine content of the heart, decreasing the sympathetic tone of the heart, improving the peripheral circulation, and improving the oxygen transport system.

Within the medical profession it is unfortunate that drug therapy seems to be more popular than the use of increased physical activity as prevention against coronary heart disease. Drugs are used primarily as a method of reducing hypertension and hyperlipidemia.

Study Questions

1. How would you describe the feeling of being physically fit?
2. What physical activities are most likely to aid in the protection against cardiovascular disease?
3. What are some general guidelines for developing physical fitness?
4. How can cardiovascular fitness be measured?
5. What is the etiology of cardiovascular diseases?
6. How can exercise influence the risk factors associated with cardiovascular disease?
7. How can physical activity be used as a preventive measure against coronary heart disease (CHD)?

8. How can physical activity be used to detect CHD?
9. How can physical activity be used for the rehabilitation from CHD?
10. How does drug therapy control CHD?
11. How is hypertension related to CHD?
12. How is hyperlipidemia related to CHD?

Review References

Cooper, Kenneth H. *The New Aerobics.* New York: Bantam Books, 1970.

Enselberg, Charles D. Physical activity and coronary heart disease. *Amer. Heart J.* 80:137–141, 1970.

Larsen, O.A. and R.O. Malmborg, eds. *Coronary Heart Disease and Physical Fitness.* Baltimore: University Park Press, 1971.

Leon, Arthur S. Comparative cardiovascular adaptation to exercise in animals and man and its relevance to coronary heart disease. In *Comparative Pathophysiology of Circulatory Disturbances*, edited by Colin M. Bloor. New York: Plenum, 1972.

Naughton, J., H.K. Hellerstein, and I.C. Mohler, eds. *Exercise Testing and Exercise Training in Coronary Heart Disease.* New York: Academic Press, 1973.

Paffenbarger, Ralph S. Prevention of heart disease. *Postgraduate Medicine* 51:74–78, 1972.

Physician's Handbook for Evaluation of Cardiovascular and Physical Fitness. Prepared by the Tennessee Heart Association Physical Exercise Committee. Nashville, Tennessee, 1972.

Pollock, M. The quantification of endurance training programs. *Exercise and Sport Science Reviews*, vol. 1, edited by J.H. Wilmore. New York: Academic Press, 1973.

Schaper, W. The Collateral Circulation of the Heart. Amsterdam: North-Holland Publishing Company, 1971.

Sharkey, Brian J. *Physiological Fitness and Weight Control.* Missoula, Montana: Mountain Press, 1974.

Key Concepts

• Biological aging is the gradual loss of information processing in the cell.
• Known physiological responses to aging are influenced by exercise and training.
• Training adaptation becomes increasingly limited with advanced age.
• Can physical training retard the aging process?

Introduction

Aging, a process that all of us must face, can be defined as the gradual loss of the organism's ability to respond to the environment—the loss accompanied by an increase in an incidence of disease and in probability of death.

Throughout history, life expectancy has increased from the early Roman's life expectancy of 22 years, to the average American's 47 years in 1900, to the present average life expectancy of 75 (Figure 20–1). It should be noted that life expectancy, during earlier times, was reduced because of diseases, especially childhood diseases. Studies of the process of aging predict, through research and medical technology, an increase of 50 years in life expectancy by the year 2020—with ultimate longevity of up to 200 years. The goal of age-related research is not to extend the senescent years but to extent the active productive years (Figure 20–2).

The increase in human longevity presents a formidable problem for society. For physical education programs, the problems presented by an increased life expectancy are of very great concern. Some estimates indicate that the number of people over age 45 will be 45–50% of the population by 1980–2000. Of these people, at least 30% will be over 65. Should our present trend continue, those over the age of 65 will be expected to retire from active careers and jobs—only to be faced with a physically inactive retirement. Physical education programs advocate continued physical activity to maintain physical fitness for life.

CHAPTER TWENTY

Aging and physical activity

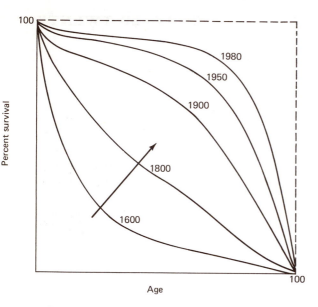

Figure 20–1. As medical care and technology overcome diseases, the survival curve more closely approaches the broken rectangular line.

Three areas of concern have been recognized in aging considerations: biological, clinical, and sociopsychological. Biological considerations include the physiological aging process and the methods to alter the basic process. Clinical considerations are concerned with the diseases of the aged—and their prevention or cure. The sociopsychological processes of the aged deal with the general welfare and outlook of these people. Appropriate applications of physical activity can make significant contributions to each of the above three areas in the following ways: (1) studies with animals have demonstrated that physically active animals live longer than their sedentary counterparts; (2) physical activity can contribute to the prevention of and rehabilitation from hypokinetic diseases that affect middle-aged and elderly persons; and (3) participation in physical activities can contribute to the well-being of elderly people and improve their sociopsychological adjustment.

A lifetime can be grossly divided into a growth and maturation period and an aging period. Since we study growth and development to implement physical education courses for school-age people, it is reasonable that we be introduced to the aging theories as a prelude to implementing programs of physical activity for mature and aging individuals.

Aging

Obviously aging is not an abrupt change in physiological functioning; it is a more gradual change occurring over several decades. Presently we do not know the mechanisms of aging, nor do we know how to reverse aging.

The remarkable increase in the mean life expectancy observed over the past 100 years has not been achieved through a retardation of the aging process but by an elimination or control of the diseases that cause premature death. It is estimated that the elimination of cardiovascular disease and cancer will increase the life expectancy in the United States to over 85 years. These breakthroughs, although increasing the number of people that live to be over 100 years old, will not add to the maximum lifespan of the human. Only by understanding the basic aging mechanisms and the application of retarding agents and preventive actions will the maximum age be extended.

Since there are several theories that attempt to explain the aging phenomena, this text will discuss only the most prominent theories involving the basic cellular mechanisms of aging.

Aging is most critical in nonreproducing cells like muscle and nerve cells. In reproducing cells or miotic cells, new cells are constantly being produced to replace the dying cells. Therefore we will confine our discussion to muscle and nerve cells.

Prominent Theories of the Aging Process

As a cell ages, it is continuously dependent upon the genetic information contained within the cell to provide the direction for protein

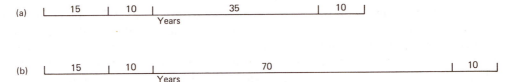

Figure 20–2. (a) General year distribution of a normal lifetime: 15 years as an adolescent, 10 years maturing, 35 years as an adult, and 10 years of "old age." (b) The goal of aging research is not only to extend the total lifetime but to increase the active-productive years as an adult.

synthesis. Any function or agent that interferes with this information processing will interfere with the operation of the cell. Alterations in genetic information can come about through radiation damage, mutations, misreading of the codes, or cross-linking of chemical bonds, as well as through other methods. These "errors" in genetic transcription and translation lead to inactive, nonfunctional proteins, which, in turn, lead to inactive, nonfunctional cells. All of the theories of aging attempt to account for this misuse of information.

Cross-linking is the formation of chemical bonds between chains of amino acids, nucleic acids, or other polymer-type molecular structures. A given amount of cross-linking is necessary for the correct biological function of proteins and nucleic acids. However, when the cross-linking becomes extensive, the behavior of these macro-molecules is altered. Especially affected in this process are proteins, which are dependent upon their three-dimensional shape for biological activity. An agent that alters the ability of the protein molecule to perform its function will alter the metabolism of the cell.

One of the primary causes of cross-linking has been related to the amount of *free radicals* formed within the cell. These highly reactive radicals are quite capable of initiating the formation of cross-linkages to form inactive molecules. Free radicals arise in the oxidative reactions within the mitochondria and are also related to the amount of unsaturated fats in the cell. The combination of inactive protein with unsaturated lipid forms a material labeled as *lipofusin*. This "age pigment" is found in increasing quantities in aged cardiac and skeletal muscle cells.

The role of exercise in the above "aging theory" is very unclear. We know that exercise increases the amount of protein within the muscle cell, which would be beneficial. On the other hand, exercise would increase the rate of genetic transcription and translation, increasing the chance for information-processing errors.

Physiological Correlates with Aging

When we examine the various living animal species, we observe a correlation between the maximum lifespan of the species and total body weight, total brain weight, and basal metabolic rate. The correlation of body weight and brain weight with maximum lifespan is a positive value: the larger the animal the longer the life expectancy. Conversely, there is a negative correlation between lifespan and metabolic rate: animals with high metabolic rates have shorter lifespans than those animals with lower metabolic rates. This relationship has been extended and shown to be valid within a given species.

In relation to body weight, there is evidence that indicates that, within certain limits, restricted caloric intake will increase the lifespan of laboratory animals. It is tempting to suggest that controlled body weights in humans will increase life expectancy. In regard to retarding the aging process, this statement cannot be made with any degree of certainty. However, there is increasing evidence that excessive body weight contributes to an increased susceptibility to cardiovascular diseases, which, if uncontrolled can lead to early death.

Observable Changes During the Aging Process

Although we know of no definite threshold age for deterioration of performance, several performance criteria are altered (reduced in most cases) with aging. In most instances, performance variables peak between 20 and 30 years of age and gradually decline approximately 40% over the next 40 years. This information should be encouraging to anyone contemplating the physical activity limitations of the aged person. An example of this decrease can be shown with the muscular strength measurements taken from various muscle groups (Figure 20–3). We should keep in mind that muscle strength is closely related to muscle cross-sectional area. (The increase in cross-sectional area accounts for most of the gain in strength during growth and development.) The decrease in strength during aging can be partly accounted for by decreased muscle size, which is probably due to loss of protein; this sequence agrees with our aging theory of age-related alterations in protein synthesis.

Endurance is altered during aging, but it is questionable how much of this decrease in endurance can be accounted for within the muscle. An equally possible explanation is a decrease in the capacity of the cardiovascular system to supply the muscle. During aging, vascular walls become increasingly rigid, with corresponding increases in blood pressure. The decreased efficiency of the vascular system would contribute to the decreased blood supply to working muscles.

That the functional capacity of the respiratory system is impaired during aging is reflected in decreased maximal breathing capacity and vital capacity.

In 1957 Simonson summarized the effects of age on performance-related physiological functions. These performance measurements show a decrease in cardiovascular functions and respiratory functions. The effects on the muscular system are primarily endurance characteristics, with little alteration of the mechanical generator other than that related to muscle strength and size (see Table 20–1). Simonson also concluded that the relationship of age to

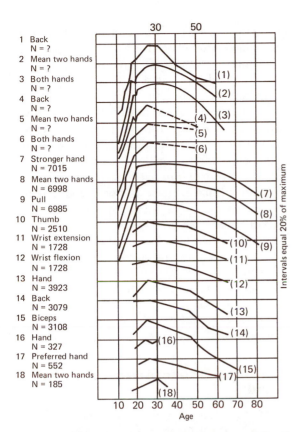

Figure 20-3. Effect of age on strength of various human muscles.

From K. Rodahl and B. Issekutz. Physical performance capacity of the older individual. In *Muscle as a Tissue* (K. Rodahl and S.M. Horvath, eds.) Copyright © 1962 by The McGraw-Hill Book Company, Inc. Used with permission of McGraw-Hill Book Company.

Table 20-1. Effects of Age on Performance and Related Physiological Functions

Performance/Physiological Function	Age
Motor coordination	
Small muscles	Unchanged
Larger muscles	Well maintained
Speed, repetitive movements, small muscle	Slight decrease
Muscle strength	Decreased
Endurance, static work	Unchanged
Endurance, moderately heavy work	Decreased
Pulse rate for recovery	Delayed
Oxidative recovery	Delayed
Respiratory efficiency	Decreased
Mechanical efficiency	Unchanged
Speed of initial increase of oxygen consumption in work	Delayed
Cardiac stroke volume	Decreased
Maximum oxygen intake	Decreased
Flicker fusion frequency	Decreased

Reproduced with permission from "Changes of physical fitness and cardiovascular functions with age" by Ernst Simonson, M.D. *Geriatrics* 12(1):28–39. Copyright 1957, The New York Times Media Company, Inc.

abnormal electrocardiograms is to a large extent due to latent coronary artery disease.

We known that, in animals, aging is manifested in the soleus muscle by a decreased number of muscle fibers. This decrease is not mediated by decreased motor units or nerve fibers. The morphological features of this decrease are unknown, but it may be due to increased invasion by connective tissue. In old animals, the muscle contraction time is increased (the muscle becomes slower); and myosin ATPase activity is decreased. All of these changes can be related to our aging theory of altered protein synthesis.

Exercising Conditions for the Aged

Now that we have a feeling for the physiological changes accompanying the aging process, we can examine the actual response of the aged person to exercising conditions. In all probability you can predict the exercise response from your knowledge of the preceding sections.

It is apparent, from research studies as well as observation, that low work rates will not show up the differences between people of differing ages. It is only when the work rate or the work time is increased that the age-related differences in people become apparent. During near maximal exercise, maximum oxygen uptake of the aged (approximately 70 years of age) is reduced to 60% that of the 25-year-old; maximum heart rate is reduced to 80% of the 25-year-old; maximum cardiac output is reduced; pulmonary ventilation is reduced; maximum respiratory frequency is reduced; and other energy support systems are impaired. Conditions in muscle and other peripheral tissues may be impaired to a great extent and may be the cause of the limitation of working capacity. It may be that the ability of the muscular system to extract oxygen is very low; therefore the exercising capability is limited.

In contrast to the above, during submaximal or low-level work, there appears to be no difference in ability, between the aged and the younger person, to take up and use oxygen. However, a decreased efficiency of work has been demonstrated: for the same amount of work, the aged person requires additional oxygen.

The age-related increase in the rigidity of the vascular system accounts for the increased blood pressure in aged persons at rest as well as during exercise. The aging arteries become incapable of the elasticity required to perform as a buffering system. Therefore, with the increased blood pressure during exercise, the arteries fail to absorb the total stroke volume from the heart. Thus, not only is blood pressure increased, but also peripheral resistance is increased, adding to the work of the heart.

It seems fair to conclude that the decrease in the ability of the aged person to perform in submaximal exercise is not accounted for by the cardiovascular or respiratory system. Maximal exercise presents a different case: it is possible that inability of the heart rate to rise above 160 in the aged may represent a limiting factor. However, it seems more reasonable to assume that the main exercising limitations in the aged, for submaximal and maximal exercise, are the muscular and nervous systems.

Training Expectations for the Aged

Regular exercise or training for young persons results in training adaptations; likewise there are training adaptations in the aged. In general the training adaptations in the aged are limited to those adaptations of an endurance nature; strength adaptations are less evident. Training adaptations can enable a trained man, of age 60, for example, to run longer than an untrained man of age 40. Maximal oxygen uptake can be expected to increase by 10–20%. Endurance exercise has been shown to increase the capillarity in animal skeletal muscle and heart. These data have not been confirmed in humans, but the increased number of capillaries could, of course, be a benefit in oxygen delivery to the active cells.

The most famous example of training in the aged is the case of the marathon runner Clarence DeMar, who continued training until his late sixties and at the age of 49 had a maximal oxygen consumption rate of 58 ml/kg/min.

deVries has found that a regiment of calisthenics, jogging, and stretching exercises for one hour three times per week results in improvements in several physiological parameters. The participants were limited to exertions with a heart rate of 145/min or less. He found that vital capacity was increased, minute volume increased, arm strength improved, and physical work capacity increased. Due to poor health, another group participated in a walking, calisthenics, and stretching program with maximum heart rate limit of 120/min. This latter group showed proportional improvements comparable to the first group.

Flexibility in joint movement and mobility have been shown to be altered by a physical training program. The ability of training to improve mobility in the aged should provide a significant incentive to increase training programs for the aged.

deVries has shown experimentally that men in their sixties and seventies, of average physical fitness, will derive benefits from a physical activity program if they raise their heart rates to near 100/min (threshold rate for training adaptations). Obviously, if a person is well trained at this age, the threshold heart rate would be higher; deVries lists a heart rate of 120 as the upper limit for a threshold rate.

Retardation of the Aging Process

In the previous section we discussed the benefits an aged person could expect from participating in an exercise program. We listed some of the training adaptations, but we did not claim these adaptations retarded the aging process. In essence training improves the functional capabilities of an aged person; we do not, however, have the evidence to claim an increased life expectancy due to physical training. Although there have been several animal studies indicating that participation in a lifetime physical activity program will increase the life expectancy of laboratory animals, these studies have not been verified in humans.

The process of retarding the aging process has captured the imagination of civilizations for centuries. For example the heart from the bodies of young virgins was believed to possess special curative and rejuvenating powers. In the *First Book of Kings* of the *Bible*, old and stricken King David is treated in this manner, apparently unsuccessfully. However, this significant failure does not seem to have diminished the popularity of similar types of therapy. Although these therapeutic regimens may not be amenable to proper scientific control and evaluation, their positive placebo effects should not be underestimated. The anticipation alone of impending treatment may well provide the driving force required of the senescent individual to survive from one day to the next.

Several investigators have attempted the task of examining the longevity and mortality of college athletes and nonathletes. The results indicate no difference between the two populations. These results are not very surprising in view of the necessary assumption in this research that college athletes are usually in "better physical condition" as a group than their nonathlete classmates. At first this assumption may seem reasonable, but when we consider the range of physical conditioning of college athletes, it soon becomes obvious that there would be only slight differences, if any.

In the animal studies we mentioned earlier, exercise programs were carried out throughout the lifetime of the animal. From the evidence available, if there is an exercising advantage in terms of longevity, the exercise program must be carried out on a regular basis. Residual or long-lasting effects of exercise have been very hard to identify.

The stress-stimulated response of the adrenal gland has been implicated as having a beneficial effect on longevity. Since we know that exercise stimulates the adrenal cortex and medulla, this could be an indication of a cellular mechanism responsible for an increased lifespan resulting from physical activity.

Summary

Although we do not propose that physical training retards the aging process, we know that physical training does increase the vitality of aged persons. The goal of physical training in the aged is to extend the duration of the active years. It has been demonstrated that the physiological changes brought about by training in the aged are similar to those changes occurring in young adults.

The goal of research on the aging phenomena is to increase the productive years. Theories on the cause of aging relate to the control of protein synthesis. To be able to control aging, we need to be able to alter protein metabolism.

The observable changes during aging show a decrement in most functions, especially after age 60. Maximal exercise is most effective, while the response to submaximal exercise in the aged population is relatively unchanged. Aged individuals can expect to achieve the same type of training responses as younger persons, although their percentage of improvement will be less.

Study Questions

1. What is aging?
2. What role can exercise play in aging?
3. Summarize your own theory of aging.
4. What physiological parameters correlate with aging?
5. How does muscular strength change with aging?
6. How does muscular endurance change with aging?

7. What considerations should be taken into account in exercises for the aged?
8. What are the training expectations for the aged?
9. Can physical activity alter the aging process?

Review References

Astrand, P.O. Physical performance as a function of age. *J. Amer. Med. Assoc.* 205:729–740, 1968.

Bender, A.D.; C.G. Kormendy; and R. Powell. Pharmacological control of aging. *Experimental Gerontology* 5:97–129, 1970.

Bullough, W.S. Aging of mammals. *Nature* 229:608–610, 1971.

deVries, H.A. Prescription of exercise for older men from telemetered exercise heart rate data. *Geriatrics* 26:102–111, 1971.

Edington, D.W.; A.C. Cosmas; and W.B. McCafferty. Exercise and longevity: evidence for a threshold age. *J. Gerontology* 27:341–343, 1972.

Gore, I. Physical activity and aging—a survey of soviet literature. *Gerontologia Clinica* 14:65–69, 1972.

Kohn, R.R. *Principles of Mammalian Aging.* Englewood Cliffs, N.J.: Prentice Hall, 1971.

Leaf, A. and J. Launois. Search for the oldest people. *National Geographic* 143:93–120, 1973.

Shock, N.W. Physiologic aspects of aging. *J. Amer. Dietetic Assoc.* 56:491–496, 1970.

Simonson, E. Changes of physical fitness and cardiovascular function with age. *Geriatrics* 12:28–39, 1957.

Art Credits

Cover photos courtesy of (left) Dr. William Considine and (right) Dr. Andrew Jackson.

Title page, from J.P. Schadé and D.H. Ford. *Basic Neurology*, 2d ed. Amsterdam: Elsevier, 1973.

Unit I, courtesy of Dr. Andrew Jackson

Unit II, photo by D.W. Edington

Unit III, photo by V.R. Edgerton

Unit IV, courtesy of Dr. Nathaniel F. Rodman

Unit V, photo by John T. Urban

Biological illustrations by Norman Archambault

Photomicrographs by D.W. Edington: Figures 2–3 (center and right), 2–4, 2–5, 2–7(b), 2–13, 2–15, 2–16

Photomicrographs by V.R. Edgerton: Figures 4–3 (a–e), 4–13, 4–15, 6–4

Figure 4–9 From J.R. Bendall. *Muscles, Molecules and Movement.* New York: American Elsevier, 1969. After (a) X. Aubert. Le couplage énergétique de la contraction musculaire . . . Thése d'agrégation, Université, Catholique de Louvain. Bruxelle, Editions Arscia, 1956. (b) K.A.P. Edman. The relation between sarcomere length and active tension in isolated semitendinosis fibres of the frog. *J. Physiol.* 183:407, 1966. (c) A.M. Gordon, A.F. Huxley, and F.J. Julian. The variation in isometric tension with sarcomere length in vertebrate muscle fibres. *J. Physiol.* 184:170, 1966.

Figure 4–10 From S. Bouisset, EMG and muscle force in normal motor activity. In *New Developments in Electromyography and Clinical Neurophysiology,* vol. 1 (J.E. Desmedt, ed.). Karger, Basel, 1973, pp. 547–583. Redrawn from T.C. Ruch, H.D. Patton, J.W. Woodbury, and A.L. Towe. *Neurophysiology.* Philadelphia: W. B. Saunders, 1966.

Figure 4–17 From S. Bouisset, EMG and muscle force in normal motor activity. In *New Developments in Electromyography and Clinical Neurophysiology,* vol. 1 (J.E. Desmedt, ed.). Karger, Basel, 1973, pp. 547–583. Redrawn from Herbert A. deVries. Efficiency of "electrical activity" as a physiological measure of the functional state of muscle tissue. *Am. J. Phys. Med.* 47:10–22, 1968. © 1968, The Williams & Wilkins Co., Baltimore.

Figures 13–2 through 13–5 From David C. Sinclair. *Human Growth After Birth.* London: Oxford University Press, 1969. Reprinted by permission of the publisher. Redrawn from J.M. Tanner, R.H. Whitehouse, and M. Takaishi. Standards from birth to maturity for height, weight, height velocity, and weight velocity: British children, 1965, I. *Arch. Dis. Childh.* 41:454–471, 1966. By permission of J.M. Tanner and Creaseys of Hertford Limited.

acetylcholine A neurotransmitter that is released at synapses and neuromuscular junctions. This chemical causes a depolarization of the postsynaptic membrane.

acetylcholinesterase An enzyme that degrades acetylcholine, hydrolyzing it into acetic acid and choline.

actin A protein located on the thin filament of muscle; combines with the myosin cross bridge to produce shortening tension.

action potential A rapid change in the electrical charge across an excitable membrane as a result in an alteration of the membrane permeability to selected ions.

active transport Process of an energy-dependent movement of a substance across a membrane.

actomyosin The name given to the protein-protein interaction when actin and myosin combine during contraction.

adenosine diphosphate (ADP) Formed through the enzymatic utilization of ATP. ADP is the substrate for phosphorylation and ADP availability is closely coupled to oxygen utilization.

adenosine triphosphate (ATP) The chemical molecule in the body most available for energy release. The last known chemical prior to the transfer of chemical energy to mechanical work. The splitting of ATP to ADP + Pi results in the liberation of 10,000 calories.

adenyl cyclic enzyme Enzyme responsible for the formation of cyclic AMP. Cyclic AMP appears to be the intracellular messenger of several hormones.

adipose tissue A group of cells characterized by large fat storage.

adrenalin *See* Epinephrine

adrenal medulla Central core of the adrenal gland. This neural tissue responds to sympathetic stimulation to release epinephrine (80%) and norepinephrine (20%).

Glossary

adrenocorticotrophic hormone (ACTH) A protein or polypeptide hormone released by the anterior pituitary and transported in the circulatory system to the adrenal cortex, where ACTH stimulates the release of cortisol.

aerobic Referring to utilization of oxygen.

afferent fibers Nerve fibers that conduct sensory information to the central nervous system.

alanine Amino acid closely related to pyruvate. May play an important role in gluconeogenesis.

aldosterone Steroid hormone released from the adrenal cortex in response to renin-angiotensin. Mineral corticoid that causes mineral retention by the kidney.

all-or-none principle A term used to imply that a motor unit "fires" or does not fire. There is no graded response within any one motor unit.

alpha-glycerol phosphate A chemical branching off from the mainstream of glycolysis. May serve as an energy reserve.

alpha-motoneuron A large motorneuron that innervates an extrafusal motor unit.

alveolar sac Small air sacs in the lung where gas exchange takes place between the alveolar space and the capillary.

alveolar ventilation The movement of air into and out of the alveoli.

amino acids Nitrogen-containing biochemicals that are structured in sequence to form proteins.

amphetamines A group of synthetic drugs related to epinephrine (adrenalin).

amylase An enzyme responsible for carbohydrate degradation during digestion.

anabolic steroids A group of synthetic drugs that have an anabolic effect in the body. Usually referred to as those drugs that result in increased muscle contractile tissue.

anabolism The building up of a cell.

anaerobic Without the use of oxygen.

androgenic hormones Hormones that stimulate male sex characteristics.

androgens That class of steroid hormones related to affecting the male sex characteristics.

anemia A decrease in the oxygen-carrying capacity of the cell resulting from a decreased RBC number or a decreased hemoglobin concentration.

anoxia A decreased oxygen supply.

antagonist A muscle or group of muscles that produces the opposite effect.

antidiuretic hormone (ADH) A protein or polypeptide hormone from the posterior pituitary that stimulates the kidney to conserve water.

arterial pressure The pressure within the arteries pushing outward; usually referred to as the systolic pressure. Systolic pressure is the pressure when the maximum amount of blood is flowing through the arteries. Dystolic pressure is the lowest pressure observed between peak systolic pressures.

asynchronous firing The firing of motor units in a random pattern rather than all at once.

atelectosis A partial collapse of the lungs.

ATPase Any enzyme that hydrolyzes ATP. The most common ATPase is the head of the myosin molecule when it splits ATP during muscle contraction.

atrophy A decrease, from the normal size, of an organelle, cell tissue, or organ.

autonomic nervous system (ANS) That part of the nervous system that controls the vital life functions such as visceral, cardiovascular, temperature regulation, ventilation, and so on.

autoregulation The ability of a tissue or organ to self-regulate its blood flow.

axo-axonal junction A functional synapse between an axon of one neuron and an axon of another neuron.

axons Extension or projection from the cell body of a neuron to some end-organ that it innervates. Impulses are conducted away from the cell body.

axon-somatic junction A functional synapse between an axon of one neuron and a cell body, or soma, of another neuron.

bag fibers Specialized fibers within the muscle spindle.

barbiturates A classification of drugs that produce a general depression.

basal metabolic rate (BMR) The least amount of oxygen needed by an individual. The BMR is measured during a resting state following at least eight hours of fasting.

beta-actinin One of several muscle proteins found in the Z-line.

betz cell Large cell bodies found in the motor cortex of the cerebrum. They play a role in the control of voluntary movement.

blood-brain barrier A selective barrier to the transport of molecules from the capillaries to the central nervous system.

bradycardia A decreased heart rate.

calcitonin A hormone secreted by the thyroid gland which alters calcium and potassium levels in the blood.

calorie A unit of heat. Foods and exercise are often described by their potential to produce or utilize calories.

capillary-fiber junction Where the capillary comes into close contact with a muscle fiber.

carbohydrates A group of organic compounds made up of a 6-carbon chain with oxygen and hydrogen atoms in a 2:1 ratio.

cardiac output The amount of blood pumped by the heart per minute. The product of the heart rate times the stroke volume.

carotid sinus A localized area of the carotid artery bifurcation that is sensitive to blood pressure. Neural impulses are transmitted to the CNS to record blood pressure. A decreased blood pressure results in an increased heart rate.

cartilage An elastic substance attached to bones.

catabolism The breakdown of biochemicals in metabolism. Often associated with the release of energy.

catecholamines A class of chemicals that include epinephrine and norepinephrine.

central nervous system (CNS) The nervous tissue in the brain and spinal cord.

cerebellum That part of the brain involved in coordination, posture, and equilibrium. It is located near the base of the skull.

cerebral cortex The gray matter of the brain; the outer layer of the cerebrum.

cerebrum Largest part of the brain; left and right hemispheres.

chain fibers Specialized fibers of the muscle spindle.

chemoreceptors Receptors (e.g., carotid bodies) that are sensitive to chemicals carried in the blood.

chloride shift Refers to the displacement of chloride ions from the RBC during CO_2 transport.

cholesterol A fatlike chemical made up of 26 carbons. Found in all animal tissues associated with membranes. Forms the base of all the steriod hormones.

cholinesterase *See* acetyl cholinesterase

citric acid cycle (Krebs cycle) A metabolic pathway; the first metabolite that is citrate. Acetyl groups enter the cycle and are converted to CO_2.

coactivation The process of simultaneously activating extra and intrafusal muscle fibers.

collagen A protein that is characterized by cross linkages which give it the strength necessary for tendons, ligaments, cartilage and bone.

concentric contractions When the muscle contracts and shortens at the same time.

core temperature Body temperature that is insulated from environmental temperature. The best measure is the rectal temperature. Core temperature is monitored by cells in the hypothalamus.

Cori cycle The sequence of events during which lactate is produced in the muscle, transported by the blood to the liver, converted in the liver to glucose, and transported back to the muscle for reutilization.

corticospinal tract Spinal tract through which nerve impulses are carried from the cerebral cortex to the spinal cord.

cortisol A steroid hormone released by the adrenal cortex.

cristae The folding of the inner membrane of the mitochondria.

cyclic adenosine-mono-phosphate (cyclic AMP) The intracellular transmitter of several hormonal actions.

cyclic nucleotides A special form of a nucleotide where the number 3 and 5 carbons of the ribose base are both bound to an oxygen molecule (e.g., cAMP).

cytochrome The proteins of the electron transport chain which become reduced and oxidized as an electron passes down the chain. These proteins contain iron as the electron acceptor.

cytoplasm The area of the cell excluding the nucleus, mitochondria, and other specialized structures. The myofibrils are located in the cytoplasm.

dendrites Projections of the cell body. Usually sense an impulse from some source and transmit the impulse toward the cell body.

depolarization A reduction in the voltage potential across the cell membrane.

diapedesis The passage of RBCs through the walls of the capillaries.

diffusion distance The distance from a capillary to the center of a cell.

disaccharide The combination of two simple sugars.

diving bradycardia The slowing of the heart during immersion of the face into water.

dopamine Biochemical intermediate in the synthesis of norepinephrine. Postulated to play a role in CNS neurotransmission.

duodenum The first part of the small intestine.

dynamic contractions Muscle contractions during which shortening or lengthening occur.

eccentric contractions Muscle contractions during which the resisting force is greater than the contraction force, resulting in muscle lengthening.

echocardiography The study of the heart through the use of sound waves.

electrocardiography (ECG) The recording of the electrical activity of the heart.

electrolytes A type of chemical that ionizes in solution (for example, NaCl) and can conduct a current.

electromyography (EMG) The electrical recording of the muscle fiber membrane potential. Usually recorded as the potential changes during contraction.

electron transport chain The respiratory chain primarily composed of cytochromes. Hydrogen transport molecules (NADH and FADH) are the substrates and oxygen is the final acceptor. The energy produced in this oxidative process is tied to the phosphorylation process of producing ATP.

end diastolic volume The amount of blood in the heart prior to a contraction.

endocardium The inner cell layers of the wall of the heart.

endoplasmic reticulum (ER) Intracellular network of membrane channels. In muscle this is referred to as the sarcoplasmic reticulum.

endothelium A single layer of epithelial cells that line the inside of the heart and blood vessels.

endurance The ability to continue physical activity.

enzymes Proteins upon which chemical reactions take place. Each enzyme is specific for a given biochemical reaction.

epicardium The outer layer of the heart wall; immediately outside the myocardium.

epinephrine A hormone from the adrenal medulla. Secreted in times of sympathetic nervous system stimulation.

epiphisitis Inflammation of the cartilage that joins the epiphysis to the ossified part of the bone.

epiphysis The growing part of the bone.

EPSP Excitatory postsynaptic potential. A subthreshold depolarization of a postsynaptic membrane resulting from the release of the neural transmitter by the presynaptic terminal. The magnitude of the depolarization may reach threshold as the number of synapse activations increase.

erythrocytes Red blood cells.

erythropoiesis The synthesis of new red blood cells.

erythropoietin A protein hormone secreted by the kidney which stimulates erythropoiesis.

estradiol A steroid hormone primarily secreted by the adrenal cortex and responsible for the development of the female sex characteristics.

estrogen A steriod hormone released from the adrenal cortex and corpus luteum. It is primarily a female hormone.

exhaustion The inability to continue, usually considered to be more final than fatigue.

expiratory reserve volume (ERV) That volume of air that can be forceably expelled after a normal expiration.

extrafusal fibers Those muscle fibers not located within the muscle spindle.

extrapancreatic insulin Either insulin or a substance similar to insulin secreted by cells located away from the pancreas.

extrapyramidal tracts Those spinal tracts descending from the brain other than the pyramidal tracts.

fast-twitch fibers Fibers that have a contraction time that is 2–3 times faster than slow-twitch fibers. Also called fast-twitch motor units.

fast-twitch glycolytic fibers Those fast-twitch fibers that have a low oxidative capacity and a high glycolytic capacity. Easy to fatigue.

fast-twitch oxidative-glycolytic fibers Those fast-twitch fibers that have a low oxidative capacity and a high glycolytic capacity. They are harder to fatigue than fast-twitch glycolytic fibers.

fatigue Loss of ability to continue a task. Fatigue is task-specific. Extreme fatigue is defined as exhaustion.

fat-soluble vitamins Those vitamins that are soluble in solution only when attached to fatty acids.

fatty acid oxidation The biochemical degradation of fatty acids to acetyl CoA.

fibrinolysis The process of breaking up a fibrin clot.

follicle-stimulating hormone (FSH) A hormone released by the anterior pituitary which stimulates follicle development in the ovary and testosterone release from the testes.

Frank-Starling mechanism The mechanism whereby the strength of the heart beat is responsive to the volume of the blood in the heart. That is, if venous return is such that a greater amount of blood returns to the heart resulting in increased filling and thus an increased stretch of the cardiac fibers, the heart contraction will be stronger.

free fatty acids (FFA) Fatty acids that are not esterfied to glycerol. FFA are carried in the blood stream in combination with proteins in the albumin fraction.

fusion frequency That stimulus frequency which results in a continuous smooth contraction.

gamma-aminobutyric acid (GABA) A biochemical that is an inhibitory neurotransmitter.

gamma-motoneuron A small motorneuron that activates the intrafusal muscle fibers of the muscle spindle. Intrafusal muscle fiber contractions are necessary to maintain tension on the sensory components of the spindle.

gamma system The motor efferents to the intrafusal muscle fibers.

genetic traits Those traits attributive to an individual resulting from parent inheritance.

genome That portion of the DNA strand that codes for a certain gene.

glial cells Cells of the nervous system that serve a metabolically supportive function, such as neuron insulators and probably other as yet undefined functions.

glucagon A protein hormone secreted by the pancreas in response to low blood glucose.

glucocorticoids Those steroid hormones involved with glucose metabolism, for example, cortisol.

glucogenic hormones Hormones affecting glucose metabolism.

gluconeogenesis The process of synthesizing glucose from lactate or amino acids. The new formation of glucose.

glucose The most common form of carbohydrate substrate used by the body. Most carbohydrates are derivatives of glucose or combinations of the various sugars.

glycogen A polymer of glucose, formed by linking glucose units together. Most common storage form of carbohydrate in muscle and liver.

glycogenolysis The process of breaking down glycogen into the individual glucose units. In the process of glycogenolysis inorganic phosphate is added so that the final product is glucose-1-phosphate.

glycolysis The enzymatic process of breaking down glucose-6-phosphate to pyruvate or lactate.

glycolytic capacity The capacity to utilize glucose-6-phosphate. Usually measured by the relative concentration of the enzymes of glycolysis.

glycolytic flux The rate of glycolytic operation.

Golgi apparatus A specialized complex of intracellular membrane folds that is associated with protein storage and/or cellular secretions.

growth hormone (GH) A small protein or polypeptide hormone secreted by the anterior pituitary. Primary functions are to promote amino acid uptake by tissues and to increase the concentration of fatty acids in the blood.

hematocrit (Hct) The packed cell volume from a sample of blood. The volume percent of red blood cells.

hemoglobin The iron-containing protein of the red blood cell that combines with oxygen.

hemoglobinuria The appearance of hemoglobin in the urine.

hexose Those 6-carbon, ring-structured chemicals such as glucose, fructose, and so forth.

histones Proteins that cover DNA. These proteins may be responsible for inhibiting transcription by binding tightly around the DNA.

homeostasis A state of physiological stability.

hypercholesteremia High levels of cholesterol in the blood.

hyperemia Increased blood flow.

hyperlipidemia High levels of lipids in the blood.

hyperplasia Increased number of cells or increased components of a cell.

hyperpnea Increased rate of ventilation.

hyperpolarization An increase in the difference between the membrane potential and the threshold level. Results in a less active cell.

hypertrophy An increased size of an organelle, cell, tissue, or organ.

hyperventilation An increased rate of ventilation. Usually referred to as a voluntarily increased rate of ventilation during the resting state.

hypoglycemia Low glucose levels in the blood.

hypopolarization A decrease in the difference between the membrane potential and the threshold level. Results in a more active cell.

hypothalamus That area of the brain below the thalamus but above the pituitary. Its function is to monitor several physiological parameters: temperature regulation, blood osmolarity, appetite, emotions, blood hormonal levels, and so on.

hypoxia A decreased oxygen supply, not as severe as anoxia.

impulse propagation The transmission of the nerve impulse.

innervate To form a neural connection.

inotropic Referring to the strength of the heart beat.

inspiratory reserve volume (IRV) That amount of air that can be inspired following a normal inspiration.

insulin A protein or small polypeptide hormone secreted by the pancreas and responsible for aiding the transport of glucose across cell membranes.

integrated EMG An averaging technique that "smooths" out the EMG signal.

intercalated disc The junction of two cardiac muscle cells. Its function may be to transmit the depolarizing wave.

intercostal muscle Those muscles located between the ribs.

interneurons Neurons that receive and transmit signals usually within one or among several spinal segments. The interneurons may be excitatory or inhibitory to motor neurons.

intrafusal muscle fibers Those muscle fibers within the muscle spindle. Activated by the gamma motor system to maintain tension or bias on the sensory organ of the spindle.

IPSP Inhibitory postsynaptic potential. A situation in which the postsynaptic membrane becomes hyperpolarized or stabilized, thus increasing the threshold level.

iron-deficiency anemia A classification of anemia directly attributable to a decreased level of available iron.

ischemia   A deficit in oxygen availability.

isohydric shift   A shift in ions into and out of the red blood cells during CO_2 loading and unloading.

isometric contractions   Activation of muscle fibers without allowing significant shortening of the muscle.

isotonic contractions   Activation of muscle fibers under conditions that permit the muscles to shorten or lengthen.

ketone bodies   Short fatty acid molecules of three or four carbons in length. Aceto-acetate and β-hydroxybutyrate are the most common; acetone is often included as a ketone body.

kilo-pond-meter (KPM)   A unit of measure on the bicycle ergometer; the amount of force required to accelerate a mass of 1 kilogram 1 meter per second.

kinase   A classification of enzymes that use ATP as a substrate.

kinesthetic sense   A term applied to the ability to sense body position through the sensory organs located in the muscles, tendons, and joints.

Krebs cycle   *See* Citric acid cycle

lactate   The end product of glycolysis. When glycolysis is operating maximally the amount of lactate produced is significantly increased. Can be used as an indicator of glycolytic flux.

lactate dehydrogenase (LDH)   The enzyme upon which pyruvate is changed to lactate and NADH to NAD^+.

leukocytes   White blood cell; a phagocytic cell transported in the blood stream.

ligament   A substance that binds bones together.

lipid   Fat, fatty acids, adipose tissue, and lipid are often used interchangeably. They have the general formula $CO\text{-}CH_2\text{-}(CH_2)_n$. Unsaturated fat has two or more hydrogens missing.

lipofusion   An age-related pigment in cells associated with the combination of inactive proteins and fatty acids.

lipolysis   The chemical breakdown of lipid molecules.

lipoprotein   Those proteins that have fatty acid chains attached.

lysosomes   Discrete organelles within the cell which contain a high concentration of hydrolytic enzymes having the potential of digesting the cell or cell particles.

macrophages   Large irregular-shaped cells that make up part of the body's disease defense by engulfing and digesting potentially harmful agents.

maximal aerobic capacity   The maximum amount of oxygen that an individual can utilize per unit time. Expressed as milliliters of oxygen per minute.

membrane permeability   The ability of substances to pass through the membrane. The membrane exerts selective permeability; that is, some substances pass through easily while others may be entirely impermeable.

metabolites   Usually referred to as the products of metabolism.

miniature endplate potential (MEPP)   The continuous subthreshold alteration in postsynaptic membrane potential as a result of random release of acetylcholine from presynaptic terminals of the NMJ.

mitochondrion   The organelle within the tissue that utilizes oxygen and produces the ATP.

mitosis   The normal division of cells. Involves a complete replication of the genetic material.

motoneurons   Neurons from ventral horn cells which innervate skeletal muscle fibers. The muscle fibers may be intrafusal or extrafusal.

motor cortex   That part of the cerebrum associated with voluntary movement.

motor endplate (MEP) That part of the sarcolemma with which the motor nerve ending makes a functional contact with the muscle fiber. Postsynaptic part of the NMJ.

motor unit A motorneuron and all the muscle fibers innervated by that motorneuron.

muscle spindle The sense organ located in the muscle responsible for recording the length of the muscle.

myelinated nerves Those nerves wrapped by Schwann cells (myelin sheath). Myelinated nerves conduct faster than nonmyelinated nerves.

myelin sheath An electrically and metabolically insulating sheath on most nerve fibers. It is developed by the Schwann cell being wrapped many times around the nerve fiber.

myoblasts Immature cells from which muscle fibers are formed.

myocardial infarction Necrosis of myocardial tissue due to lack of oxygen.

myocardium Heart muscle.

myofibril A subdivision of the muscle fiber composed of a bundle of thick and thin filaments arranged in hexagonal orientation. The myofibrils run in a parallel axis to the long axis of the muscle. This is the active subunit of muscle contraction.

myofilaments The actin and myosin chains.

myoglobin The iron-containing protein in muscle responsible for oxygen transport and storage.

myoglobinuria The appearance of myoglobin in the urine.

myosin A contractile protein molecule that forms the thick filament of the myofibril. The active head of the myosin molecule forms the ATPase activity.

neuromuscular junction (NMJ) *See* Motor endplate

neurons Cells of the nervous system that are electrically excitable and can transmit an impulse.

neurotransmitter A chemical that is released from a nerve ending and induces an alteration in the postsynaptic membrane potential.

neurotrophic The effect that a neuron has on its end organ besides that which is attributed to nerve impulses.

nicotinadenine dinucleotide (NAD) Oxidized or reduced; a chemical derivative of the vitamin niacin responsible for transporting hydrogen molecules to the electron transport chain.

node of Ranvier Junction between two Schwann cells on the nerve.

norepinephrine A hormone secreted at adrenergic nerve endings (sympathetic nervous system) on blood vessels and in the heart. Twenty percent of the hormone released from the adrenal medulla is norepinephrine, the rest being epinephrine.

nucleolus A dense area in the nucleus containing RNA.

nucleus The location of the DNA of the cell. Controls protein synthesis, thus, the metabolism of the cell.

obesity Excess body fat.

organelles Subcellular structures enclosed within a membrane.

osmolarity A method to report the number of particles in a solution.

ossification The hardening of bone.

osteoblast Those cells capable of forming bone.

overload The process of loading a cell or tissue with more than the usual stress.

oxidation The process of adding a proton or removing an electron from a chemical.

When one chemical is oxidized another one is reduced.

oxidative phosphorylation The process of coupling the oxidative process (utilization of oxygen) to the phosphorylation of ADP to form ATP.

oxygen uptake The amount of oxygen taken up by the body. Usually expressed as the amount of oxygen taken up per minute. Tissue oxygen uptake can be calculated by knowing the amount of oxygen taken up across a capillary bed.

pacing The ability to distribute one's effort throughout a performance.

Pacinian corpuscles (PC) Those sense organs responsible for recording pressure on the skin.

parasympathetic nervous system That part of the autonomic nervous system that tends to maintain homeostatic conditions.

parathyroid A hormone secreted by the parathyroid gland to regulate calcium and potassium concentrations in the blood.

P_{CO_2} The partial pressure due to CO_2 concentration.

peaking The act of preparing oneself for a peak performance on a given day.

peak-twitch tension Maximal tension produced by a muscle when given a single stimulus of a duration less than one millisecond.

peripheral nervous system (PNS) That part of the nervous system outside the brain and the spinal cord.

phagocytes Cells capable of digesting other cells or parts of cells.

phasic PTNs Pyramidal tract neurons of the motor cortex which are associated with rapid movements.

phosphofructokinase The enzyme responsible for the conversion of fructose-6-phosphate and ATP to fructose-1,6-phosphate and ADP. It is generally thought that the activity of this enzyme controls the glycolytic flux.

phosphorylase The enzyme responsible for glycogenolysis. It is activated by AMP and by the epinephrine-cyclic AMP effect.

pinocytosis Process of ingesting fluids or molecular particles in small vesicles through the cell membrane.

plasma The fluid part of the blood.

plasma membrane The cell membrane.

P_{O_2} The partial pressure due to oxygen concentration.

polarization Act of increasing the membrane potential.

polypeptide A chain of amino acids linked together identical to a protein chain but of a shorter length.

postsynaptic membrane Cell membrane that is receiving a synaptic input from a neuron.

power The amount of force applied per unit of time.

precapillary sphincter A physiological part of the microvascular anatomy that regulates blood flow through localized regions within tissue beds.

presynaptic membrane Membrane that surrounds the terminal which contains the neurotransmitter.

presynaptic terminal The nerve ending that contains the neurotransmitter.

proprioceptors Sensory receptors in muscle, tendons, and joints.

prostaglandins A classification of hormones to describe a group of cyclic 20-carbon unsaturated fatty acids that have a wide variety of physiological effects.

proteins Chain of amino acids that have a specific function.

protein synthesis The process of making new protein through the DNA-RNA-protein process.

proteinuria The appearance of protein in the urine.

proteolytic enzymes Those enzymes capable of degrading proteins.

prothrombin A precursor to thrombin.

Purkinje fibers The excitable fibers of the heart that conduct the impulse along the surface of the heart from the SA node to the ventricles.

pyramidal cells Large neurons of the motor cortex associated with voluntary movement. *See* Betz cell

pyramidal tract neurons (PTNs) These neurons originate in the motor cortex of the brain (pyramidal cells) and travel side-by-side throughout the spinal cord; they intervate motor units.

pyruvate A chemical located at the end of glycolysis. A branch point in metabolism. It can be converted to alanine, lactate, or acetyl CoA depending upon the metabolic conditions.

red blood cells (RBC) Erythrocytes; mature nonnucleated cells of the blood containing the enzymes of glycolysis and the hemoglobin protein.

renin A protein hormone secreted by the juxtaglomerular cells of the kidney. Stimulates the conversion of inactive angiotensin to active.

repolarization The process of reforming the polarization, or charged state, across a membrane after depolarization has taken place.

respiratory exchange index (R) Calculated as the ratio between the amount of CO_2 expired from the lungs and the amount of O_2 taken up.

respiratory quotient (RQ) Calculated as the ratio between the amount of CO_2 produced and the amount of O_2 utilized for the oxidation of glucose (RQ = 1), fatty acids (RQ = 0.7) and protein (RQ = 0.8). RQ is often used to denote R.

reticulospinal tract Spinal tract or bundle that contains neurons conducting impulses from the reticular formation to the spinal level.

ribosomes RNA structures, upon which the mRNA is translated to protein formation.

rubrospinal tract Spinal tract or bundle that contains neurons conducting impulses from the red nucleus to the spinal level.

saltatory Refers to the "jumping" action of nerve impulse conduction in myelinated neurons.

sarcolemma Muscle cell membrane.

sarcomeres The most basic functional contractile unit that extends from Z-line to Z-line.

sarcoplasma Cytoplasm of a muscle cell.

sarcoplasmic reticulum (SR) The endoplasmic reticulum in the muscle cell. The SR is associated with calcium release and uptake, protein synthesis, and glycogen metabolism.

satellite cells Small cells consisting almost entirely of nuclear material. They may be potential myoblasts. Normally these cells are situated between the sarcolemma and the plasma membranes of the muscle cell.

Schwann cell Cell that forms the myelin sheath of a nerve cell. The Schwann cell wraps around the nerve fiber to form many layers.

sensory fibers Nerve fibers that conduct sensory information to the spinal cord.

serum The fluid part of the blood after coagulation.

sino-atrial node (SA node) A group of cells on the surface of the right atrium from which originates the impulse to initiate the heart beat.

slow-twitch fibers Fibers that contract at a rate one-fifth that of a fast-twitch fiber. Also called slow-twitch motor units.

somatotype The classification scheme for body types according to muscularity, fatness, and thinness.

splanchnic Referring to the viscera.

static contractions Activation of a muscle without significant muscle shortening or lengthening; isometric contraction.

stimulants A classification of drugs that cause a general stimulation to the body.

strength The ability to move a resistance or resist a force.

stretch receptors A term used to describe the muscle spindle and tendon organ.

stroke volume The amount of blood expelled by the left ventricle during one heart beat.

ST segment The time of repolarization of the ventricles after contraction.

subcellular compartments Intracellular morphological or functional compartments that act relatively independent of other compartments. There is a relative barrier to the free movement of chemicals between compartments. This concept is of extreme importance in understanding metabolic control.

submaximal exercise Exercise at a rate less than maximal. An exercise can be less than maximal in terms of contraction strength, speed of movement, or duration of the exercise. Often submaximal exercise is expressed as a percentage of maximal rate of oxygen-uptake.

subneural folds Folding of the postsynaptic membrane at the motor endplate.

substrates Chemicals that are the initial reactants in an enzymatic reaction. Substrates are converted to products.

subthreshold An electrical potential less than the level of depolarization necessary (threshold) to initiate an action potential.

surfactant A protein produced by the epithelial cells of the lungs to decrease surface tension. Without it the resistance to air flow would be so great that air would not be able to move into the alveoli.

synapse Junctional contact between two neurons.

synergistic A muscle or group of muscles that produce the same or very similar effect.

tendon An extension of muscle which attaches to bone.

tendon organs (TO) Those sensory organs of the tendon and muscle that are sensitive to muscle tension.

tendon stretch reflex A simple reflex where a tendon stretch results in a muscle contraction.

tenotomy The cutting of a tendon.

terminal axon The end of an axon that contains a neurotransmitter.

testosterone A steriod hormone secreted from the adrenal cortex and testicles. The primary metabolic actions are androgenic (maintains male secondary sex characteristics) and anabolic (results in increased amino acid uptake and protein synthesis in muscle).

tetany A state of constant fused muscle contraction.

thermoreceptors Nerve endings responsive to temperature changes.

threshold Level of depolarization necessary to initiate an action potential.

thrombin A protein involved in the blood-clotting mechanism.

thyroid hormone An amino acid-derived hormone, containing iodine, which results in increased metabolic rate.

thyroxine The primary hormone secreted by the thyroid gland. Its main function is to increase metabolic rate.

tidal volume   The amount of air moved by the lungs during normal ventilation.

time-to-peak tension   The period of time necessary for a muscle to reach peak tension.

trainability   The potential for training.

tranquilizers   A class of drugs that result in a general depression or tranquil state.

treppe   The increase in muscle tension as the stimulation continues.

triglycerides   Three fatty acids on a glycerol molecule.

trophic   Refers to some sort of "nutritional" effect of a cell on another.

tropomyosin   A protein in muscle associated with the thin filament that has an inhibiting effect upon actin-myosin interaction.

troponin   A protein in muscle associated with the thin filament that is sensitive to calcium concentration. Troponin-calcium interactions result in an effect upon tropomyosin that releases the inhibition on actin-myosin interaction.

T-tubules   Tubular invaginations of the sarcolemma that form a transverse network throughout the muscle fiber.

tunica intima   The inner layers of large blood vessels which are associated with antherosclerotic plaques.

twitch   A contraction of a muscle or muscle fiber induced by a single stimulus.

unmyelinated nerves   Nerves that are surrounded by Schwann cells but where the Schwann cell did not form multiple wrapping; thus the nerves are not myelinated.

urea   The final product of nitrogen excretion from the body. It is eliminated from the body in the urine.

vagus nerve   Tenth cranial nerve which innervates several of the autonomic organs, e.g., the heart and the lungs.

valsalva maneuver   The forced expiration against a closed glottis.

$\dot{V}_{O_2}$ The amount of oxygen extracted per unit of time and percent of body weight (ml/min/kg).

vasoconstriction   The narrowing of the lumen of a blood vessel. Usually in response to sympathetic nervous system stimulation to the muscles surrounding the vessel.

vasodilation   An enlargement of the lumen of a blood vessel, usually in response to metabolic products but also caused by hormones, neural stimulation or lack of stimulation, temperature, and other factors.

warming-up   The process of preparing for an immediate performance. Used to increase the muscle or body temperature, mobilize metabolic pathways, stretch muscles, ligaments, or tendons or any other bodily function that will aid performance.

water-soluble vitamins   Those vitamins soluble in water.

Index